I0755234

E. C. Russell
Columbia College
1873.

LECTURES

ON

MINERALOGY.

DELIVERED AT THE

SCHOOL OF MINES,

COLUMBIA COLLEGE.

DESCRIPTIVE MINERALOGY.

By T. EGLESTON,
Professor of Mineralogy and Metallurgy,
School of Mines, Columbia College

WITH 34 LITHOGRAPHIC PLATES.

NEW YORK:
D. VAN NOSTRAND, PUBLISHER,
23 MURRAY AND 27 WARREN STREET.
1872.

ERRATA.

Page

xviii and xxvi. The note † does not apply to the plates of this edition.

xxiv. If the brachy axis is placed in front, the formulæ for the macro and brachy domes are right. If the macro axis is placed in front, the formulæ should read $m'\bar{P}'\infty$, $m_{,}\bar{P}_{,}\infty$ and $m_{,}\breve{P}'\infty$, $m'\breve{P}_{,}\infty$.

11. 14th line from bottom, Florentine should read Florentine (yellow.)

21. 11th line from bottom, for Borax read SPh.

27. 2d line from bottom, for Borax and SPh. read Fluor and $\dot{K}\ \dddot{S}^2$.

49. 15th line from top, for $\dot{N}a\ \dot{C}a$ read $\dot{C}a\ \dot{N}a$.

54. 14th line from bottom, for Ryacolite read Rhyacolite.

66. 14th line from bottom, for or read and.

83. 8th line from top, for red read reducing.

87. 21st line from bottom, for red read reducing.

92. 18th line from top, for base read hemi-pyramid.

92. 21st line from top, for P read —P.

92. 22d line from top, for —P read P.

92. 23d line from top, for obtuse read acute.

102. 16th line from bottom, for R read —½R.

103. 21st line from bottom, for with nitrate of Cobalt read by fusion with Soda.

121. 4th line from bottom, for flame of a candle read R. F. on coal.

127. 20th line from top, for globule read powder.

134. 6th line from bottom, for Chromium read Manganese.

139. 13th line from top, for orange colored read white.

139. 14th line from top, for AsS read $\ddot{A}s$.

144. 25th line from top, omit borax. For O. F. read R. F.

161. 25th line from top, for copper read baryta.

PREFACE.

These lectures are what their title indicates, the lectures on Mineralogy delivered at the School of Mines of Columbia College. They have been printed for the students, in order that more time might be given to the various methods of examining and determining minerals. The second part only has been printed. The first part, comprising crystallography and physical mineralogy, will be printed at some future time. Only those species have been described with which it is necessary for an Engineer of Mines to be familiar, or which present some peculiarity of interest. The list of synonyms is not complete; only the most familiar names used in Germany, France and the United States have been given. The lecture notes were written out somewhat hastily, amid the press of other duties, but it is hoped, that as they occupy a ground not hitherto entirely covered by other elementary works, they may prove acceptable to students wishing to commence the study of Mineralogy. As it has already proved serviceable at the School of Mines, it is hoped that it may be found useful elsewhere.

As it is quite essential that the notation of both Naumann and Dana should be understood, the first has been adopted in the text and the second used on the faces of the crystals. A table giving the comparison of the principal notations in use, has been given in the introduction.

In the preparation of the work I have been greatly indebted to my assistant, Mr. J. H. Caswell, who autographed the drawings and prepared the notations of the faces of the crystals.

T. EGLESTON.

School of Mines,
1871.

CONTENTS.

CLASSIFICATION OF THE SPECIES

ACCORDING TO

CRYSTALLINE FORM.

ISOMETRIC.

Diamond,
Garnet,
Lapis Lazuli,
Hauynite,
Leucite,
Analcite,
Halite,
Sal Ammoniac,
Fluorite,
Boracite,
Spinel,
Kalinite,
Iron,
Magnetite,
Franklinite,
Pyrite,
Pharmacosiderite,
Chromite,
Alabandite,
Linnæite,
Smaltite,
Cobaltite,
Ullmannite,
Sphalerite,
Lead,
Galenite,
Clausthalite,
Arsenolite,
Senarmontite,
Uraninite,
Copper,
Cuprite,
Bornite,
Tennantite,
Tetrahedrite,
Mercury,
Silver,
Amalgam,
Argentite,
Cerargyrite,
Bromyrite,
Embolite,
Gold,
Platinum.

TETRAGONAL.

Zircon,
Vesuvianite,
Wernerite,
Apophyllite,
Scheelite,
Braunite,
Hausmannite,
Cassiterite,
Stannite,
Rutile,
Octahedrite,
Stolzite,
Wulfenite,
Torbernite,
Chalcopyrite,
Calomel.

ORTHORHOMBIC.

Sulphur,
Chrysolite,
Iolite,
Muscovite,
Lepidolite,
Chondrodite,
Andalusite,
Topaz,
Staurolite,
Calamine,
Prehnite,
Natrolite,
Stilbite,
Talc,
Serpentine,
Nitre,
Aphthitalite,
Thenardite,
Mascagnite,
Barite,
Witherite,
Celestite,
Strontianite,
Anhydrite,
Aragonite,
Epsomite,
Diaspore,
Wavellite,
Chrysoberyl,
Goethite.
Marcasite,
Leucopyrite,

Arsenopyrite,
Scorodite,
Columbite,
Wolframite,
Pyrolusite,
Manganite,
Triplite,
Goslarite,
Brookite,
Bournonite,
Anglesite,
Cerussite,
Bismuthinite,
Aikenite,
Orpiment,
Valentinite,
Stibnite,
Autunite,
Molybdite,
Chalcosite,
Brochantite,
Atacamite,
Libethenite,
Olivenite,
Stephanite,
Polybasite.

MONOCLINIC.

Wollastonite,
Pyroxene,
Spodumene,
Petalite,
Amphibole,
Epidote,
Orthoclase,
Fibrolite,
Euclase,
Datolite,
Titanite,
Pectolite,
Laumontite,
Harmotome,
Heulandite,
Glauberite,
Mirabilite,
Borax,
Natron,
Barytocalcite,
Gypsum,
Pharmacolite,
Alunogen,
Melanterite,
Vivianite,
Bieberite,
Erythrite,
Annabergite,
Crocoite,
Realgar,
Kermesite,
Liroconite.
Malachite,
Azurite.

TRICLINIC.

Sassolite,
Rhodonite,
Anorthite,
Labradorite,
Oligoclase,
Albite,
Cyanite,
Cryolite,
Chalcanthite.

HEXAGONAL.

Water,
Tellurium,
Graphite,
Quartz,
Beryl,
Willemite,
Phenacite,
Biotite,
Nephelite,
Tourmaline,
Dioptase,
Chabazite,
Prochlorite,
Soda Nitre,
Apatite,
Calcite,
Dolomite,
Brucite,
Magnesite,
Corundum,
Alunite,
Hematite,
Pyrrhotite,
Copiapite,
Siderite,
Menaccanite,
Rhodochrosite,
Millerite,
Niccolite,
Zincite,
Smithsonite,
Pyromorphite,
Mimetite,
Bismuth,
Tetradymite,
Arsenic,
Antimony,
Molybdenite,
Cinnabar,
Proustite,
Pyrargyrite,
Iodyrite,
Iridosmine.

AMORPHOUS.

Carbonic Acid,
Opal,
Chrysocolla,
Chlorastrolite,
Sepiolite,
Aluminite,
Turquois,
Limonite,
Arseniosiderite,
Psilomelane,
Wad,
Remingtonite,
Zaratite,
Hydrozincite,
Minium.

CLASSIFICATION OF THE SPECIES

ACCORDING TO

HARDNESS.

H. < 1

Carbonic Acid,
Mercury,
Molybdite,
Sassolite,
Water.

H.=0·5—6

Wad.

H.=1—1·5

Cerargyrite,
Embolite,
Iodyrite,
Kermesite,
Molybdenite,
Natron,
Talc.

H.=1—2

Aluminite,
Arseniosiderite,
Bromyrite,
Calomel,
Graphite.

H.=1—2·5

Sulphur.

H.=1·5

Arsenolite,
Brucite,
Copiapite,
Ice,
Lead.

H.=1·5=2

Alunogen,
Gypsum,
Mirabilite,
Orpiment,
Realgar,
Sal-Ammoniac,
Soda-Nitre,
Tetradymite,
Vivianite.

H.=1·5—2·5

Erythrite.

H.=2

Melanterite,
Nitre,
Stibnite.

H.=2—2·25

Epsomite.

H.=2—2·5

Aikinite,
Argentite,
Autunite,
Bismuth,
Bismuthinite,
Borax,
Cinnabar,
Goslarite,
Hydrozincite,
Liroconite,
Mascagnite,
Pharmacolite,
Prochlorite,
Proustite,
Pyrargyrite,
Pyrolusite,
Senarmontite,
Stephanite,
Tellurium,
Thenardite,
Torbernite.

H.=2—3

Chrysocolla,
Kalinite,
Lepidolite,
Minium,
Polybasite.

H.=2·5

Chalcanthite.
Cryolite,
Halite,
Muscovite,
Pharmacosiderite,
Sepiolite.

H.=2·5—2·7

Galenite.

H.=2·5—3

Annabergite,
Bournonite,
Chalcocite,
Clausthalite,
Copper,
Crocoite,
Glauberite,
Gold,
Silver,
Valentinite.

H.=2·5—3·5

Barite,
Calcite.

H.=2·7—2·9

Biotite.

H.=2·7—3

Anglesite,
Wulfenite.

H.=2·75—3

Stolzite.

H.=3

Bornite,
Olivenite,
Serpentine.

H.=3—3·2

Zaratite.

H.=3—3·5

Amalgam,
Anhydrite,
Antimony,
Aphthitalite,
Atacamite,
Celestite,
Cerussite,
Millerite.

H.=3—4·5

Tetrahedrite.

H.=3·2—4

Wavellite.

H.=3·5

Arsenic,
Laumontite,
Mimetite.

H.=3·5—3·7

Witherite.

H.=3·5—4

Alabandite,
Alunite,
Aragonite,
Brochantite,
Chalcopyrite,
Cuprite,
Dolomite,
Heulandite,
Malachite,
Pyromorphite,
Scorodite,
Sphalerite,
Stilbite,
Strontianite,
Tennantite.

H.=3·5=4·2

Azurite.

H.=3·5—4·5

Magnesite,
Pyrrhotite,
Rhodochrosite,
Siderite.

H.=4

Barytocalcite,
Fluorite,
Libethenite,
Manganite,
Stannite.

H.=4—4·5

Chabazite,
Platinum,
Zincite.

H.=4—5

Pectolite.

H.=4·5

Iron,
Harmotome.

H.=4·5—5

Apatite,
Apophyllite,
Scheelite,
Wollastonite.

H.=5

Calamine,
Chromite,
Dioptase,
Franklinite,
Smithsonite.

H.=5—5·5

Goethite,
Hausmannite,
Leucopyrite,
Limonite,
Natrolite,
Niccolite,
Titanite,
Triplite,
Ullmannite,
Wolframite.

H.=5—6

Amphibole,
Cyanite,
Menaccanite,
Psilomelane,
Pyroxene,
Wernerite.

H.=5—6·5

Hematite.

H.=5·5

Analcite,
Chromite,
Cobaltite,
Datolite,
Lapis Lazuli,
Linnæite,
Uraninite,
Willemite.

H.=5·5—6

Arsenopyrite,
Brookite,
Chlorastrolite,
Haüynite,
Leucite,
Magnetite,
Nephelite,
Octahedrite,
Smaltite.

H.=5·5—6·5

Opal,
Rhodonite.

H.=6

Anorthite,
Columbite,
Labradorite,
Oligoclase,
Orthoclase,
Turquois.

H.=6—6·5

Albite,
Braunite,
Chondrodite,
Marcasite,
Petalite,
Pyrite,
Rutile.

H.=6—7

Cassiterite,
Fibrolite,
Iridosmine,
Prehnite.

H.=6·5

Epidote,
Vesuvianite.

H.=6·5—7

Chrysolite,
Diaspore,
Garnet,
Spodumene.

H.=7

Boracite,
Quartz.

H.=7—7·5

Iolite,
Staurolite,
Tourmaline.

H.=7·5

Andalusite,
Euclase,
Zircon.

H.=7·5—8

Beryl,
Phenacite.

H.=8

Spinel,
Topaz.

H.=8·5

Chrysoberyl.

H.=9

Corundum.

H.=10

Diamond.

H. undetermined.

Bieberite,
Remingtonite.

CLASSIFICATION OF THE SPECIES

ACCORDING TO

SPECIFIC GRAVITY.

G.=0·918—1
Water.

G.=1·2—1·6
Sepiolite.

G.=1·4
Mirabilite,
Natron.

G.=1·48
Sassolite.

G.=1·5
Carbonic Acid,
Sal-Ammoniac.

G.=1·5—1·8
Kalinite.

G.=1·6
Aluminite.

G.=1·6—1·8
Alunogen.

G.=1·7
Aphthitalite,
Borax,
Epsomite,
Mascagnite.

G.=1·8
Melanterite.

G.=1·9
Nitre.

G.=1·9—2·1
Goslarite.

G.=1·9—2·3
Opal.

G.=2
Graphite.

G.=2—2·1
Chabazite.

G.=2—2·2
Chrysocolla,
Soda-Nitre,
Stilbite.

G.=2—2·5
Biotite.

G.=2·072
Sulphur.

G.=2·1—2·2
Heulandite,
Natrolite.

G.=2·1—2·5
Halite.

G.=2·2
Analcite,
Chalcanthite.

G.=2·28—2·41
Laumontite.

G.=2·3
Brucite,
Gypsum,
Wavellite.

G.=2·3—2·4
Lapis Lazuli.

G.=2·35—2·39
Apophyllite.

G.=2·4
Harmotome.

G.=2·4—2·5
Leucite.

G.=2·4—2·8
Haüynite.

G.=2·42—2·45
Petalite.

G.=2.47—2·6
Serpentine.

G.=2·5—2·6
Iolite,
Nephelite,
Zaratite.

G.=2·5—2·7
Alunite,
Calcite,
Quartz.

G.=2·53—2·59
Orthoclase.

G.=2·54—2·64
Albite.

G.=2·6
Vivianite.

G.=2·6—2·7
Anorthite,
Beryl,
Labradorite,
Oligoclase,
Pharmacolite,
Wernerite.

G.=2·6—2·8
Glauberite,
Prochlorite,
Talc,
Turquois.

G.=2·6—3
Phenacite.

G.=2·7
Thenardite.

G.=2·7—2·8
Pectolite.

G.=2·7—2·9
Wollastonite.

G.=2·8—2·9
Anhydrite,
Dolomite,
Liroconite,
Prehnite.

G.=2·8—3
Datolite,
Lepidolite,
Magnesite.

G.=2·8—3·1
Muscovite.

G.=2·9
Aragonite,
Boracite,
Erythrite.

G.=2·9—3
Cryolite,
Pharmacosiderite.

G.=2·9—3·4
Amphibole.

G.=3—3·1
Annabergite,
Autunite,
Euclase.

G.=3—3·2
Tourmaline.

G.=3·1
Chondrodite,
Fluorite.

G.=3·1—3.2
Andalusite,
Spodumene.

G.=3·1—3·3
Scorodite.

G.=3·15—4
Garnet.

G.=3·18
Chlorastrolite.

G.=3·2
Fibrolite.

G.=3·2—3·3
Dioptase.

G.=3·2—3·5
Pyroxene.

G.=3·25
Apatite.

G.=3·3—3.4
Chrysolite,
Vesuvianite.

G.=3·3—3·5
Diaspore.

G.=3·3—3·7
Titanite.

G.=3·3—4
Epidote.

G.=3·35—3·5
Calamine.

G.=3·4
Orpiment.

G.=3·4—3·6
Realgar,
Rhodonite,
Rhodochrosite.

G.=3·4—3·8
Staurolite,
Triplite.

G.=3·5
Arseniosiderite,
Topaz.

G.=3·5—3·6
Cyanite,
Hydrozincite,
Torbernite.

G.=3·5—3·8
Azurite,
Chrysoberyl.

G.=3·5—4·9
Spinel.

G.=3·55
Diamond.

G.=3·6
Arsenolite,
Barytocalcite,

G.=3·6—3·7
Strontianite.

G.=3·6—3·8
Libethenite.

G.=3·6—4
Limonite.

G.=3·7—3·9
Siderite.

G.=3·7—4
Malachite.

G.=3·7—4·3
Psilomelane.

G.=3·8—3·9
Brochantite,
Octahedrite.

G.=3·8—4·1
Willemite.

G.=3·9
Celestite.

G.=3·9—4
Alabandite,
Corundum,
Sphalerite.

G.=4—4·3
Atacamite.

G.=4—4·4
Goethite,
Smithsonite.

G.=4—4·5
Bornite.

G.=4—4·7
Zircon.

G.=4—4·8
Barite.

G.=4·1
Brookite.

G.=4·1—4·2
Rutile.

G.=4·1—4·3
Chalcopyrite.

G.=4·1—4·4
Olivenite.

G.=4·2—4·3
Witherite.

G.=4·2—4·4
Manganite.

G.=4·3—4·4
Tennantite.

G.=4·3—4·5
Chromite,
Stannite.

G.=4·4—4·7
Pyrrhotite.

G.=4·4—4·8
Molybdenite.

G.=4·4—5
Bornite.

G.=4·5—4·6
Kermesite,
Stibnite.

G.=4·5—5
Menaccanite.

G.=4·5—5·1
Tetrahedrite.

G.=4·5—5·3
Hematite.

G.=4·6
Minium.

G.=4·6—4·8
Marcasite.

G.=4·7
Hausmannite.

G.=4·7—4·8
Braunite.

G.=4·8—4·9
Pyrolusite.

G.=4·8—5
Linnæite,
Pyrite.

G.=4·9—5·1
Magnetite.

G.=5·2—5·3
Senarmontite.

G.=5·2—5·6
Millerite.

G.=5·3—5·8

Embolite.

G.=5·4—5·5

Proustite.

G.=5·4—5·8

Zincite.

G.=5·4—6·4

Columbite.

G.=5·5

Cerargyrite,
Iodyrite,
Valentinite.

G.=5·5—5·8

Chalcocite.

G.=5·5—6·5

Franklinite.

G.=5·7—5·9

Bournonite,
Pyrargyrite.

G.=5·8—6

Bromyrite,
Cuprite.

G.=5·9

Arsenic.

G.=5·9—6·1

Crocoite.

G.=6

Scheelite.

G.=6—6·3

Cobaltite.

G.=6—6·4

Arsenopyrite.

G.=6·1—6·3

Tellurium.

G.=6·1—6·8

Aikinite.

G.=6·2

Anglesite,
Polyasite,
Stephanite.

G.=6·2—6·5

Ullmannite.

G.=6·3—6·9

Wulfenite.

G.=6·3—7·1

Cassiterite.

G.=6·4

Calomel,
Cerussite,
Uraninite.

G.=6·4—6·5

Bismuthinite.

G.=6·4—7

Smaltite.

G.=6·5—7

Pyromorphite.

G.=6·6—6·7

Antimony.

G.=7—8·7

Leucopyrite.

G.=7—8·8

Clausthalite.

G.=7·1—7·2

Mimetite.

G.=7·1—7·3

Argentite.

G.=7·1—7·5

Wolframite.

G·=7·2—7·7

Galenite.

G.=7·2—8·4

Tetradymite.

G.=7·3—7·6

Niccolite.

G.=7·3—7·8

Iron.

G.=7·8—8·1

Stolzite.

G.=8·9

Cinnabar,
Copper.

G.=9·7

Bismuth.

G·=10—14

Amalgam.

G.=1·01—11·1

Silver.

G.=11·4

Lead.

G.=13·5

Mercury.

G.=15—19

Gold.

G.=16—19

Platinum.

G.=19—21

Iridosmine.

Undetermined.

Bieberite,
Copiapite,
Molybdite,
Remingtonite,
Wad.

SYSTEMS OF CRYSTALLIZATION.

NO.	SIMPLE FORMS.	AXES.
1	Octahedron, or cube.	3 axes rectangular and equal.
2	Tetragonal pyramid, or right prism with a square base.	3 axes rectangular, 2 equal.
3	Rhombic pyramid, or right prism with a rhombic base.	3 axes rectangular, and unequal.
4	Monoclinic pyramid, or inclined rhombic prism.	3 axes unequal, 2 rectangular.
5	Triclinic pyramid, or doubly inclined rhomboidal prism.	3 axes unequal, and unequally inclined.
6	Hexagonal pyramid, hexagonal prism, or rhombohedron.	4 axes, 3 equal and equally inclined, 1 at right angles to the other three.

NAMES USED BY DIFFERENT AUTHORS.

NO.	MOHS.	WEISS & ROSE.	PHILLIPS.	NAUMANN.	DELAFOSSE.	DANA. 1854.	DANA. 1869.
1	Tessular.	Regular.	Cubic.	Tesseral.	Cubic.	Monometric.	Isometric.
2	Pyramidal.	2 and 1 axial.	Pyramidal.	Tetragonal.	Tetragonal.	Dimetric.	Tetragonal.
3	Orthotype.	1 and 1 axial.	Prismatic.	Rhombic.	Orthorhombic.	Trimetric.	Orthorhombic.
4	Hemiorthotype.	2 & 1 membered.	Oblique.	Monoclinohedric.	Clinorhombic.	Monoclinic.	Monoclinic.
5	Anorthotype.	1 & 1 membered.	Anorthic.	Triclinohedric.	Clinohedric.	Triclinic.	Triclinic.
6	Rhombohedral.	3 and 1 axial.	Rhombohedral.	Hexagonal.	Hexagonal.	Hexagonal.	Hexagonal.

SCALE OF HARDNESS.

1.—Talc. Laminated light green variety. Easily scratched by the nail.

2.—Gypsum. Crystallized variety. Not easily scratched by the nail. Does not scratch a copper coin.

3.—Calcite. Transparent variety. Scratches and is scratched by a copper coin.

4.—Fluor. Crystalline variety. Not scratched by a copper coin. Does not scratch glass.

5.—Apatite. Transparent variety. Scratches glass with difficulty. Easily scratched by the knife.

5.5.–Scapolite. Crystalline variety.

6.—Orthoclase. White cleavable variety. Scratches glass easily. Not easily scratched by the knife.

7.—Quartz. Transparent variety. Not scratched by the knife. Yields with difficulty to the file.

8.—Topaz. Transparent variety. Harder than flint.

9.—Sapphire. Cleavable varieties. Harder than flint.

10.—Diamond. Harder than flint.

COMPARISON OF NOTATIONS USED TO

ISOMETRIC

REGULAR

HOLOHEDRAL

	LEVY.
Octahedron,	a^1
Cube or hexahedron,	p
Rhombic dodecahedron,	b^1
Tetrahexahedron,	b^m
Trigonal trisoctahedron,	$a^{\frac{1}{m}}$
Tetragonal trisoctahedron,	a^m
Hexoctahedron,	$b^1\ b^{\frac{1}{m}}\ b^{\frac{1}{n}}$

HEMIHEDRAL

Hemi-octahedron, or tetrahedron,	a^1
Hemi-tetrahexahedron, or pentagonal dodecahedron,	b^m
Hemi-trigonal trisoctahedron,	$a^{\frac{1}{m}}$
Hemi-tetragonal trisoctahedron,	a^m
Hemi-hexoctahedron inclined,	$b^1\ b^{\frac{1}{m}}\ b^{\frac{1}{n}}$
Hemi-hexoctahedron parallel, or diploid,	$b^1\ b^{\frac{1}{m}}\ b^{\frac{1}{n}}$
Hemi-hexoctahedron parallel, or gyroid,	$b^1\ b^{\frac{1}{m}}\ b^{\frac{1}{n}}$

TETARTOHEDRAL

Tetarto-hexoctahedron, or tetartoid.	$b^1\ b^{\frac{1}{m}}\ b^{\frac{1}{n}}$

REPRESENT THE FACES OF CRYSTALS.

SYSTEM.

OCTAHEDRON.

FORMS.

MILLER.	WEISS.	NAUMANN.	DANA.
111	a : a : a	O	1
100	a : ∞a : ∞a	∞O∞	*O*
110	a : a : ∞a	∞O	*i or I*
h k o	a : na : ∞a	∞On	*in*
h h k	a : a : ma	mO	*m*
h k k	a : ma : ma	mOm	*mm*
h k l	a : ma : na	mOn	*mn*

FORMS.

± ϰ (111)	½rl (a : a : a)	$\pm\frac{O}{2}$	$\pm\frac{1}{2}$
± π (h k o)	½rl (a : na : ∞a)	$\pm\frac{\infty On}{2}$	$\pm\frac{in}{2}$
± ϰ (h h k)	½rl (a : a : ma)	$\pm\frac{mO}{2}$	$\pm\frac{m}{2}$
± ϰ (h k k)	½rl (a : ma : ma)	$\pm\frac{mOm}{2}$	$\pm\frac{mm}{2}$
± ϰ (h k l)	½rl (a : ma : na)	$\pm\frac{mOn}{2}$	$\pm\frac{mn}{2}$
± π (h k l)	½rl (a : ma : na)	$\pm\left[\frac{mOn}{2}\right]$	$\pm\left[\frac{mn}{2}\right]$
± π ϰ (h k l)	½rl (a : ma : na)	$\pm\left(\frac{mOn}{2}\right)$	$\pm\left(\frac{mn}{2}\right)$

FORMS.

π ϰ (h k l)		$\pm$r *or* l $\frac{mOn}{4}$	$\pm$r *or* l $\frac{mn}{4}$

TETRAGONAL

TETRAGONAL

HOLOHEDRAL

	LEVY.
Tetragonal pyramid of the first order,†	$a^{\frac{1}{m}}$
Tetragonal pyramid of the second order,	$b^{\frac{1}{m}}$
Ditetragonal pyramid,	$b^{\frac{1}{m}}\ b^{\frac{1}{n}}\ h^1$
Tetragonal prism of the first order,†	h^1
Tetragonal prism of the second order,	m
Ditetragonal prism,	$h^{\frac{1}{m}}$
Basal pinacoid,	p

HEMIHEDRAL

Sphenoid of the first order,	$a^{\frac{1}{m}}$
Sphenoid of the second order,	$b^{\frac{1}{m}}$
Tetragonal scalenohedron,	$b^{\frac{1}{m}}\ b^{\frac{1}{n}}\ h^1$
Tetragonal trapezohedron,	$b^{\frac{1}{m}}\ b^{\frac{1}{n}}\ h^1$
Tetragonal pyramid of the third order,	$b^{\frac{1}{m}}\ b^{\frac{1}{n}}\ h^1$
Tetragonal prism of the third order,	$h^{\frac{1}{m}}$

TETARTOHEDRAL

Sphenoid of the third order,

* r stands for right and l for left, to distinguish it

† These prisms and pyramids or octahedra of the first order

SYSTEM.

PYRAMID.

FORMS.

MILLER.	WEISS.	NAUMANN.	DANA.
h h l	a : a : mc	mP	*m*
h o l	a : ∞a : mc	mP∞	*mi*
h k l	a : na : mc	mPn	*mn*
110	a : a : ∞c	∞P	*i or 1*
100	a : ∞a : ∞c	∞P∞	*ii*
h k o	a : na : ∞c	∞Pn	*in*
001	∞a : ∞a : c	0P	*O*

FORMS.

$\varkappa$ (h h l)	r.l.½ (a : a : mc)	$\pm\frac{mP}{2}$	$\pm\frac{m}{2}$
$\varkappa$ (h o l)	r.l.½ (a : ∞a : mc)	$\pm\frac{mP\infty}{2}$	$\pm\frac{mi}{2}$
$\varkappa$ *or* $\varkappa'$ (h k l)	r.l.½ (ma : na : c)	$\pm\frac{mPn}{2}$	$\pm\frac{mn}{2}$
$\varkappa''$ (h k l)	r.l.±½ (ma : na : c)	*r *or* l $\frac{mPn}{2}$	r *or* l $\frac{mn}{2}$
π (h k l)	$\frac{r}{l}.\frac{l}{r}.\frac{1}{2}$ (ma : na : c)	$\frac{r}{l}$ *or* $\frac{l}{r}$ $\frac{mPn}{2}$	$\frac{r}{l}$ *or* $\frac{l}{r}$ $\frac{mn}{2}$
(h k o)	r.l.½ (a : na : ∞c)	$\frac{r}{l}$ *or* $\frac{l}{r}$ $\frac{\infty Pn}{2}$	$\frac{r}{l}$ *or* $\frac{l}{r}$ $\frac{in}{2}$

FORMS.

$\pm\pi\varkappa$ (h k l)		$\pm\frac{r}{l}$ *or* $\frac{l}{r}$ $\frac{mPn}{4}$	$\pm\frac{r}{l}$ *or* $\frac{l}{r}$ $\frac{mn}{4}$

from + right, and — left, in the previous forms.

are those of the second in the diagrams.

ORTHORHOMBIC

RHOMBIC

HOLOHEDRAL

	LEVY.
Rhombic pyramid,	$b^{\frac{1}{m}}$
Macropyramid,	$b^{\frac{1}{m}}$ $b^{\frac{1}{n}}$ g^1
Brachypyramid,	$b^{\frac{1}{m}}$ $b^{\frac{1}{n}}$ g^1
Rhombic prism,	m
Macroprism,	$h^{\frac{1}{m}}$
Brachyprism,	$g^{\frac{1}{m}}$
Macrodome,	$a^{\frac{1}{m}}$
Brachydome,	$e^{\frac{1}{m}}$
Macropinacoid,	h^1
Brachypinacoid,	g^1
Basal pinacoid,	p

HEMIHEDRAL

Rhombic sphenoid.	

The other hemihedral forms are single

SYSTEM.

PYRAMID.

FORMS.

MILLER.	WEISS.	NAUMANN.	DANA.
h h l	a : b : mc	mP	*m*
* h k l	na : b : mc	$m\bar{P}n$	$m\bar{n}$
† h k l	a : nb : mc	$m\breve{P}n$	$m\breve{n}$
110	a : b : ∞c	∞P	*i or I*
*h k o	na : b : ∞c	$\infty\bar{P}n$	$i\bar{n}$
†h k o	a : nb : ∞c	$\infty\breve{P}n$	$i\breve{n}$
o k l	∞a : b : mc	$m\bar{P}\infty$	$m\bar{i}$
h o l	a : ∞b : mc	$m\breve{P}\infty$	$m\breve{i}$
010	∞a : b : ∞c	$\infty\bar{P}\infty$	$i\bar{i}$
100	a : ∞b : ∞c	$\infty\breve{P}\infty$	$i\breve{i}$
001	∞a : ∞b : c	0P	*O*

FORMS.

ϰ (h k l)	na : b : mc a : nb : mc	r *or* l $\frac{m\breve{\bar{P}}n}{2}$	r *or* l $\frac{m\breve{\bar{n}}}{2}$

planes, or pairs of parallel planes.

* h < k † h > k

MONOCLINIC

MONOCLINIC

HOLOHEDRAL

		LEVY.
Monoclinic pyramid,	Hemi-pyramid,	$d^{\frac{1}{m}}$
	Hemi-pyramid,	$b^{\frac{1}{m}}$
Orthopyramid,	Hemi-orthopyramid,	$b^{\frac{1}{m}}\ d^{\frac{1}{n}}\ h^1$
	Hemi-orthopyramid,	$b^{\frac{1}{m}}\ b^{\frac{1}{n}}\ h^1_,$
Clinopyramid,	Hemi-clinopyramid,	$d^{\frac{1}{m}}\ b^{\frac{1}{n}}\ g$
	Hemi-clinopyramid,	$b^{\frac{1}{m}}\ d^{\frac{1}{n}}\ g^1$
Inclined rhombic prism,		m
Orthoprism,		$h^{\frac{1}{m}}$
Clinoprism,		$g^{\frac{1}{m}}$
Orthodome,	Hemi-orthodome,	$o^{\frac{1}{m}}$
	Hemi-orthodome,	$a^{\frac{1}{m}}$
Clinodome,		$e^{\frac{1}{m}}$
Orthopinacoid,		h^1
Clinopinacoid,		g^1
Basal pinacoid,		p

The hemihedral forms are single planes or pairs of parallel planes.

SYSTEM.

PYRAMID.

FORMS.

MILLER.	WEISS.	NAUMANN.	DANA.
h k k $\bar{h}$ k k	a : b : mc a′ : b : mc	$\pm mP$	$\pm m$
* h k l * h k l	a : nb : mc a′ : nb : mc	$\pm m\bar{P}n$	$\pm m\bar{n}$
† h k l † $\bar{h}$ k l	na : b : mc na′ : b : mc	$\pm m\grave{P}n$	$\pm m\grave{n}$
110	a : b : ∞c	∞P	*i or I*
* h k o	a : nb : ∞c	$\infty\bar{P}n$	$i\bar{n}$
† h k o	na : b : ∞c	$\infty\grave{P}n$	$i\grave{n}$
h o l $\bar{h}$ o l	a : ∞b : mc a′ : ∞b : mc	$\pm m\bar{P}\infty$	$\pm m\bar{i}$
o k l	∞a : b : mc	$m\grave{P}\infty$	$m\grave{i}$
100	a : ∞b : ∞c	$\infty\bar{P}\infty$	$i\bar{i}$
010	∞a : b : ∞c	$\infty\grave{P}\infty$	$i\grave{i}$
001	∞a : ∞b : c	0P	*O*

* h > k. † h < k.

TRICLINIC

TRICLINIC

HOLOHEDRAL

		LEVY.
Triclinic pyramid,	Tetarto-pyramid,	$f^{\frac{1}{2}}$
	Tetarto-pyramid,	$d^{\frac{1}{2}}$
	Tetarto-pyramid,	$b^{\frac{1}{2}}$
	Tetarto-pyramid,	$c^{\frac{1}{2}}$
Macropyramid,	Tetarto-pyramid,	$f^{\frac{1}{m}}\ d^{\frac{1}{n}}\ h^1$
	Tetarto-pyramid,	$d^{\frac{1}{m}}\ f^{\frac{1}{n}}\ h^1$
	Tetarto-pyramid,	$b^{\frac{1}{m}}\ c^{\frac{1}{n}}\ h^1$
	Tetarto-pyramid,	$c^{\frac{1}{m}}\ h^{\frac{1}{m}}\ h^1$
Brachypyramid,	Tetarto-pyramid,	$f^{\frac{1}{m}}\ c^{\frac{1}{n}}\ g^1$
	Tetarto-pyramid,	$d^{\frac{1}{m}}\ b^{\frac{1}{m}}\ g^1$
	Tetarto-pyramid,	$b^{\frac{1}{m}}\ d^{\frac{1}{n}}\ g^1$
	Tetarto-pyramid,	$c^{\frac{1}{m}}\ f^{\frac{1}{n}}\ g^1$
Doubly inclined rhombic prism,	Hemi-prism,	t
	Hemi-prism,	m
Macroprism,	Hemi-prism,	$h^{\frac{1}{m}}$
	Hemi-prism.	${}^{\frac{1}{m}}h$
Brachyprism,	Hemi-prism,	$g^{\frac{1}{m}}$
	Hemi-prism,	${}^{\frac{1}{m}}g$
Macrodome,	Hemi-dome,	$o^{\frac{1}{m}}$
	Hemi-dome,	$a^{\frac{1}{m}}$
Brachydome,	Hemi-dome,	$i^{\frac{1}{m}}$
	Hemi-dome,	$e^{\frac{1}{m}}$
Macropinacoid,		h^1
Brachypinacoid,		g^1
Basal pinacoid,		p

* $h > k$ † $< k$

SYSTEM.

PYRAMID.

FORMS.

MILLER.	WEISS.	NAUMANN.		DANA.
111	a : b : mc	m,' P,'	mP'	*m'*
$1\bar{1}1$	a : b' : mc		m'P	*'m*
$\bar{1}\bar{1}1$	a' : b' : mc'		mP,	*m,*
$\bar{1}11$	a' : b : mc		m,P	*,m*
* hkl	a : nb : mc	m,' P̄,' n	mP̄' n	*m'n̄*
* $h\bar{k}l$	a : nb' : mc		m'P̄n	*'mn̄*
* $\bar{h}\bar{k}l$	a' : nb' : mc		mP̄,n	*m,n̄*
* $\bar{h}kl$	a' : nb : mc		m P̄n	*,mn̄*
† hkl	na : b : mc	m,' P̆, 'n	mP̆' n	*m'n̆*
† $h\bar{k}l$	na : b' : mc		m'P̆n	*'mn̆*
† $\bar{h}\bar{k}l$	na' : b' : mc		mP̆,n	*m,n̆*
† $\bar{h}kl$	na' : b : mc		m,P̆n	*,mn̆*
110	a : b : ∞c	∞' P'	∞ P'	*i' or I'*
$1\bar{1}0$	a : b' : ∞c		∞' P	*'i or 'I*
* hko	a : nb : ∞c	∞' P̄' n	∞P̄' n	*i' n̄*
* $\bar{h}ko$	a' : nb : ∞c		∞' P n	*,in̄*
† hko	na : b : ∞c	∞' P̆' n	∞ P̆' n	*i' n̆*
† $\bar{h}ko$	na' : b : ∞c		∞' P̆ n	*'in̆*
hol	a ∞b : mc	m,'P̄,' ∞	m, P̄' ∞	*m, 'ī*
$\bar{h}ol$	a' : ∞b : mc		m' P̄, ∞	*m' ,ī*
okl	∞a : b : mc	m,'P̆,' ∞	m' P̆' ∞	*m'' ĭ*
$o\bar{k}l$	a : b' : mc		m, P̆, ∞	*m,, ĭ*
100	a : ∞b : ∞c	∞P̄∞		*iī*
010	∞a : b : ∞c	∞P̆∞		*iĭ*
001	∞a : ∞b : c	0P		*O*

HEXAGONAL

HEXAGONAL

HOLOHEDRAL

	LEVY. Hexagonal.
Pyramid of the first order,†	$b^{\frac{1}{m}}$
Pyramid of the second order,	a^1
Dihexagonal pyramid,	b^1 $b^{\frac{1}{n}}$ $b^{\frac{1}{p}}$
Hexagonal prism of the first order,	m
Hexagonal prism of the second order,	h^1
Dihexagonal prism,	$h^{\frac{1}{m}}$
Basal pinacoid,	p

HEMIHEDRAL

Rhombohedron,	$b^{\frac{1}{m}}$
Rhombohedron of the second order,	
Hexagonal scalenohedron,	
Hexagonal trapezohedron,	
Pyramid of the third order,	
Ditrigonal pyramid,	
Trigonal pyramid of the first order,	
Trigonal pyramid of the second order,	
Hexagonal prism of the third order,	
Ditrigonal prism,	
Trigonal prism of the first order,	
Trigonal prism of the second order,	

TETARTOHEDRAL

Rhombohedron of the third order,	
Trigonal trapezohedron (plagihedron),	
Trigonal pyramid of the third order,	

† These pyramids and prisms are those of the second in the diagrams.

SYSTEM.

PYRAMID.

FORMS.

MILLER. Rhombohedral.	WEISS. Hexagonal.	NAUMANN. Hexagonal.	NAUMANN. Rhombohedral.	DANA.
100	∞a : a : a : mc	mP	±mR	*m*
h k l	2a : a : 2a : mc	mP2	mP2	*m*2
(h k l) (p q r)	a : na : pa : mc	*mPn	*mPn	**mn*
$2\bar{1}\bar{1}$	∞a : a : a : ∞c	∞P	∞R	*i or I*
$1\,0\,\bar{1}$	2a : a : 2a : ∞c	∞P2	mR∞	*i*2
h k l	a : na : pa : ∞c	* ∞Pn	* ∞Rn	**in*
001	∞a : a ∞ : ∞a : c	0P	0R	*O*

FORMS.

MILLER.	WEISS.	NAUMANN. Hexagonal.	NAUMANN. Rhombohedral.	DANA.
(h k k) (p q q)	$\frac{1}{2}$ (∞a : a . a : mc) $\frac{1}{2}$ (∞a : a′ : a′ : mc)	$\pm\frac{mP}{2}$	±mR	$\pm\frac{m}{2}$
π (h k l)	$\frac{1}{2}$(2a : a : 2a : mc) $\frac{1}{2}$(2a′ : a : 2a′ : mc)	$\pm\frac{mP2}{2}$		$\pm\frac{m2}{2}$
(h k l) (p q r)	$\frac{1}{2}$ (a : na : pa : mc) $\frac{1}{2}$ (a′ : na′ : pa : mc)	$\pm\frac{mPn}{2}$	± mRn	$\pm\frac{mn}{2}$
α (h k l) (p q r)		r *or* l $\frac{mPn}{2}$		r *or* l $\frac{mn}{2}$
± π (h k l) (p q r)		$\frac{r}{l}$ *or* $\frac{l}{r}$ $\frac{mPn}{2}$		$\frac{r}{l}$ *or* $\frac{l}{r}$ $\frac{mn}{2}$
		r *or* l $\left[\frac{mPn}{2}\right]$		r *or* l $\left[\frac{mn}{2}\right]$
± κ (h k k) ± κ (p q q)		r *or* l $\frac{mP}{2}$		r *or* l $\frac{m}{2}$
± κ (h k l)		r *or* l $\frac{mP2}{2}$		r *or* l $\frac{m2}{2}$
π (h k l) (p q r)		$\frac{r}{l}$ *or* $\frac{l}{r}$ $\frac{\infty Pn}{2}$		$\frac{r}{l}$ *or* $\frac{l}{r}$ $\frac{in}{2}$
α (h k l) (p q r)		r *or* l $\left[\frac{\infty Pn}{2}\right]$		r *or* l $\left[\frac{in}{2}\right]$
		$\pm\frac{\infty P}{2}$		$\pm\frac{i}{2}$
		$\pm\frac{\infty P2}{2}$		$\pm\frac{i2}{2}$

FORMS.

MILLER.	WEISS.	NAUMANN.	DANA.
± π κ (h k l) (p q r)		$\pm\frac{r}{l}$ *or* $\frac{l}{r}$ $\frac{mPn}{4}$	$\pm\frac{r}{l}$ *or* $\frac{l}{r}$ $\frac{mn}{4}$
α (h k l)		± r *or* l $\frac{mPn}{4}$	± r *or* l $\frac{mn}{4}$
α (p q r)		± r *or* l $\left[\frac{mPn}{4}\right]$	± r *or* l $\left[\frac{mn}{4}\right]$

$* \ n < 2$

DESCRIPTIVE

MINERALOGY.

LECTURES

ON

MINERALOGY.

HYDROGEN.

Water. Ḣ. HEXAGONAL.

SYN.—Snow, Ice, Wasser, Eis, Eau, Glace.

Water is found in three different states; solid when it takes the form of ice or snow, both of which may be crystallized or crystalline; liquid at ordinary temperature; or in the form of vapor. When water solidifies, it crystallizes, taking forms which can be referred to the hexagonal system. These crystals are oftenest seen in frost or snow, but it is somewhat doubtful whether the primitive form is the prism or the rhombohedron. A rhombohedron of about 120° has been observed, but generally the crystals are skeleton planes. The hexagonal prism is sometimes found in northern countries, but elsewhere it is rare. These crystals are sometimes very large, but they have rarely been found terminated. Generally they are more or less hollow and made up of concentric layers, which are are not in contact, but united by little planes, radiating from the center towards the angles. Smithson describes double hexagonal pyramids derived from a rhombohedron of 112° 21′. These modified crystals are rarer than the others. In lower latitudes ice does not show any traces of crystallization, except occasionally when breaking up in the spring it assumes hexagonal forms. In ordinary climates like our own, snow is less perfectly crystallized and appears under the shape of hexagonal stars, *Pl.* I. *Figs.* 1–10, which may sometimes be triangular in form from the carrying out of certain planes at the expense of others. The axes of these stars are at angles of 60°. On these axes others are often found inclined to them at 60°, which are arranged like the feathers on a quill. Sometimes the extremities of the axes are terminated by another set of hexagonal axes, forming another star. Crystals of snow suspended in the atmosphere produce the meteorological phenomena, called halos and mock suns. These phenomena are owing to the fact that polarized light does not penetrate all the angles of the crystals. It goes through the angles of 60° but not of 120°.

Water in a liquid state has exactly the same composition as ice. When it is perfectly pure its specific gravity is 1, being the unit to which all others are referred. But because the volume and the density of water vary with the temperature, it is necessary always to indicate a temperature to which all observations must be reduced, in order to get a series of comparative results. In physics the temperature is 4° C. In mineralogy it is generally about 15° C. which is about the mean temperature of the air. Bodies generally increase in density as they become colder, but water forms an exception to this rule. Below the temperature of 4° C. it dilates, so that the density, of ice is only 0.93, which is the reason why ice floats on water.

When pure, water is inodorous, colorless and tasteless; it is liquid above 0°C. and 0.762 of the barometer. A cub. decim. of water at the latitude of Paris and temperature of 15.5° C. weighs 1K. A cubic inch weighs 252.569 grains. Water is most dense at 3.982° C. or 39.176° F. Composition O 88.89, H 11.11.

Water, according to its temperature in springs, is known as cold or thermal. The temperature of cold springs is usually about 17°-18° C. Beyond this point they are called thermal. Water is rarely ever pure, but according to the nature of the elements dissolved is called drinkable (potable), or mineral water. Drinkable waters are those in which only a small proportion of mineral matter is dissolved, and which contain oxygen. They are good for domestic purposes. When they dissolve soap they are called soft waters. Such is usually the water of springs which do not come from a limestone formation. When they do not dissolve soap they are called hard waters.

Some analyses of various city waters are given on pp. 3, 4.

Mineral waters are those which contain so large a proportion of mineral matter as to render them unfit for ordinary domestic purposes; they are usually divided into Salt and Medicinal waters.

Salt waters may come from the sea or from salt springs. The water of the sea is the most abundant. They are generally those masses of water which have no outlet and can only diminish their volume by evaporation. Except at the outlets of rivers, sea water is of about the same general composition. The quantity of salts dissolved varies from 3–4%, in which chloride of sodium is the principal part, but it rarely exceeds 2.6% of the weight of the water or about two thirds of the whole of the solid material. The other substances are $\dot{N}a\dddot{S}$, MgCl, CaCl, and a very small quantity of NaBr, MgBr, NaI and MgI. The largest amount of solid substances in the Atlantic is at the equator, and is 3.66%. Inland seas contain much less than this. The bitter taste of sea water is owing to salts of magnesia. For analyses of sea water, v. p. 3.

For variations of the saltness of water, v. p. 4.

The salt lakes, which are sometimes found in the interior of continents, are generally owing to the infiltration of fresh water, through deposits which contain salt. The quantity of salt which such waters may contain is as high as 25%, as in the Dead Sea; of this amount from 7–10% is salt. The water of the Great Salt Lake, Utah, contains 20.196% of salt.

Analyses are given on p. 3.

Analyses of River Waters.

	ĊaC̈	ṀgC̈	S̈i	F̶e	M̶n	A̶l	ĊaS̈	ṀgS̈	ṄaS̈	NaCl	KCl	CaCl	MgCl	Organic Matter.	Total.
Rhine, near Basle.	12.79	1.35	0.21	*tr.*	...	*tr.*	1.54	0.39	0.18	0.15 (NaCl, KCl)		...	...	0.33	=16.94
Rhine, near Bonn.	9.46	0.65	0.89	0.28	...	...	2.38	1.81	0.16	1.45 (NaCl, KCl)		...	...	...	=17.08
Elbe, near Hamburg.	6.98	0.39	0.54	0.12 (F̶e, M̶n, A̶l)			0.72 (ĊaS̈, ṀgS̈)		..	3.94 (NaCl, KCl, CaCl, MgCl)				...	=12.42
Lake of Geneva.	7.2	0.7	0.1	...	...	...	2.6	3.1	...	...	...	...	0.9	0.6	=15.2
Rhone, near Lyons.	15.00	...	...	...	...	...	2.00	0.70	...	0.70 (NaCl, KCl, CaCl, MgCl)				*tr.*	=18.40
Thames, near London Bridge.	11.56	...	0.18	*tr.*	...	*tr.*	4.56		...	3.65	0.33	6.24	...	10.00	=36.52
Croton Water.	(ĊaḢC̈²) 4.53	(ṀgḢC̈²) 3.25	1.05	*tr.*	...	*tr.*	0.26	(K̇ S̈) 0.30	0.44	0.68	...	...	...	1.13	=11.67

The quantity of the several constituents is calculated for 100,000 parts of water.

Analyses of Sea Water.

In 100 parts of the solid matter.

	Solids in 100 pts. of Ḣ.	NaCl	MgCl	KCl	CaCl	NaBr	MgBr	ĊaS̈	ṀgS̈	ĊaC̈	ṄaS̈i	
German Ocean.	3.15	77.78	8.12	0.97	1.39	...	0.47	3.47	6.79	0.56	0.25	=100
Atlantic Ocean.	3.57	78.14	6.54	4.33	...	1.46	...	4.36	5.17	...	...	=100

Analyses of Water from the Dead Sea.

	NaCl	MgCl	CaCl	KCl	MnCl	AlCl	MgBr	NH⁴Cl	ĊaS̈	S̈i	Total	Ḣ
Gmelin.	7.078	11.773	3.214	1.674	0.212	0.090	0.439	0.008	0.053	....	24.541	75.459=100.000
Marchand.	6.578	10.543	2.894	1.398	....	0.018	0.251	....	0.088	0.003	21.773	78.227=100.000

PURITY OF CITY WATERS.

Impurities contained in one wine gallon of 231 *cubic inches expressed in grains.*

City.	Source.	Inorganic Matter.	Organic and Volatile Matter.	Total Solids.
New York.....	Croton, 1869...........	4.11	0.67	4.78
" "	Well, 8th Av...........	38.95	4.59	43.54
Brooklyn.......	Ridgewood, 1869........	3.37	0.59	3.92
Jersey City.....	Passaic River...........	4.58	2.86	7.44
Trenton........	Delaware River.........	2.93	0.55	3.48
Philadelphia....	Schuylkill River.........	2.30	1.20	3.50
Boston.........	Cochituate Lake........	2.40	0.71	3.11
Albany.........	Hydrant...............	8.47	2.31	10.78
Troy...........	Hydrant...............	6.09	1.34	7.43
Schenectady....	Well, State St..........	46.88	2.33	49.21
Utica..........	Hydrant...............	5.50	0.96	6.46
Syracuse.......	New Reservoir..........	12.13	1.80	13.93
Rochester......	Genesee River..........	12.02	1.23	13.25
Cleveland......	Lake Erie..............	4.74	1.53	6.27
Chicago.......	Lake Michigan..........	5.62	1.06	6.68
Dublin.........	Lough Vartry...........	1.77	1.34	3.11
London........	Thames River...........	15.55	0.83	16.38
"	Well, Leadenhall St.....	90.38	9.59	99.97
Paris..........	River Seine............	7.83	1.00	8.83
Amsterdam.....	River Vecht............	14.45	2.13	16.58
"	Well..................	64.55	4.38	68.93

Variation of the Saltness of the Water near New York.

I.—North River, on the South and West Sides of the Island.

Locality.	Depth.	Salt contained in one U. S. Gallon.
Between Castle Garden and Governor's Island, below the City.....	Surface............	807 grains.
Off 25th street, 300 ft. from pier head.	Bottom, depth, 26 feet	1,037 "
Off 25th street, 300 ft. from pier head.	Surface.............	328 "
Off 25th street, midchannel........	Bottom, depth 36 ft	1,140 "
Off 51st street, 400 ft. from pier head.	Bottom, depth 36 ft..	812 "
Off 51st street, 400 ft. from pier head.	Surface............	266 "

II.—East River.

Locality.	Depth.	Salt contained in one U. S. Gallon.
Off 35th street..................	Surface............	1.088 "

Where the substance dissolved is not salt, the water is generally called medicinal, if it is not fit for ordinary uses. Several kinds of medicinal waters are distinguished.

1. Waters, with solids in solution without gas.
2. Waters, with solids in solution with gases and not acids.
3. Acid waters.
4. Alkaline waters.
5. Ferruginous or chalybeate waters.
6. Sulphurous waters.
7. Waters with iodine.

In almost all of these waters a little silica and some nitrogenous substances are found.

Water in the state of vapor is always present in the lower part of the atmosphere. In this condition it is completely invisible, and its presence is only ascertained by hygrometric or meteorological phenomena which depend on the quantity of the vapor at a given time, and the relative temperatures of the earth and the different strata of air. Occasionally it becomes visible in the state of fogs and mists. It is frequently observed issuing from volcanoes as condensed steam, and also from fissures in the earth's surface. These vapors frequently carry off with them other substances either in solution or as a mechanical mixture. It is also found in the state of vapor, but condensed at a high pressure, mixed with lava in a state of fusion, from which it is given off in jets of steam as the lava cools.

SULPHUR.

Sulphur. S. Orthorhombic.

Syn.—Natürlicher Schwefel, Soufre.

Sulphur crystallizes as a right rhombic prism of 101° 46′. It has traces of cleavage parallel to the faces of the prism and rhombic octahedron It is not found with the primitive form, but generally as rhombic octahedra. Many of the crystals have the basal pinacoids and some trace of the prism. Combinations of two octahedra with the brachydome and macropinacoid are not unfrequent, and sometimes the brachypinacoid and macrodomes are found. When crystallized below 0°C. it sometimes takes the form of a pseudo-hexagonal prism, with a rhombic octahedron. Fracture, conchoidal. The lustre of Sulphur is variable and is often quite bright. It is always somewhat adamantine and somewhat resinous, but is not always the same on all the faces. Thus, those of the basal pinacoid and the macropinacoid are always less bright than the others. The lustre of the fracture is very bright. Transparent, translucent and opaque. When it is quite pure, it is of a yellow color, called sulphur-yellow, sometimes having a greenish tint. It is sometimes of a reddish color, which has been attributed to traces of selenium. Streak, sulphur-yellow, reddish or greenish. It possesses the property of double refraction in a very high degree, the phenomenon manifesting itself even between two parallel faces. It is very brittle, breaking with the heat of the hand. **H.**=1.5–2.5. **G.**=2.072, when pure, but it often contains other substances.

Pyr. &c. B. P. Heated in a closed tube it fuses and volatilizes, leaving no residue if it is pure, which however is quite rare in natural specimens; a residue, which is more or less large, is almost always left according to the degree of purity of the specimen. In an open tube it burns with a blue flame, and gives off S̈. If rubbed, it shows resinous electricity, which it preserves for a long time. Insoluble in water and acids.

Pure Sulphur is met with quite frequently in nature and is sometimes crystallized and sometimes amorphous. It is sometimes found as stalactites radiated about a central axis. It is also found as incrustations, which are sometimes fibrous and interlaced. The density of such varieties varies from 1.98–2. They are very easily broken as the fibers separate, and have a very feeble, resinous lustre. Their color is generally sulphur-yellow, tending to a grayish-white. They are hardly translucent on the edges. When it is amorphous, Sulphur has a variety of shapes; sometimes it seems to have been melted and appears as stalactites, in which state it is usually found in calcareous or volcanic rocks. It is opaque and is sometimes entirely brown and could not be immediately recognized. This coloration is sometimes owing to not more than 0.1% of foreign substances. It is sometimes found as cavernous and drusy masses, in which state it is generally produced by thermal springs. In this state it is in the shape of a whitish powder that has little or no consistency. It is also found in whitish, compact masses, with occasional yellow spots. This same variety is also found pulverulent. Such varieties sometimes occur in the nodules of Silex found in the Jurassic formation.

Sulphur was the first substance in which dimorphism was discovered. Mitscherlich was the first to discover this fact. In nature Sulphur crystallizes in the right rhombic prism, it may however be crystallized artificially by fusion and it then assumes a form derived from the monoclinic system, the angle of the prism being 90° 32′. The inclination of the axes is 95° 46′. These crystals are of a dark orange color. They are soluble in CS^2 as natural Sulphur is. If they are touched, they become opaque and fall into octahedra of the orthorhombic system. These crystals may be produced directly by evaporating a solution of Sulphur in CS^2. This allotropic condition is very unstable at the ordinary temperature. Deville obtained permanent oblique crystals by dissolving Sulphur in chloroform and evaporating; this produces a mixture of oblique octahedra and prisms of an orange color. If the solution of CS^2 is quite exempt from Sulphur in excess, these prisms may be obtained. If not, the crystals become transformed giving out a considerable amount of heat. These two allotropic conditions of Sulphur may become mixed by crystallization in different proportions. When the oblique form is in only a very small proportion in the mass, the transformation of the whole is very slow. The mass is viscous and of a dark color and of lower density. It sometimes takes several years for it to resume its natural density. There are still other allotropic conditions of Sulphur. Some are insoluble in CS^2 and others soluble, but do not crystallize. One of them is the elastic Sulphur which is obtained in laboratories.

Sulphur appears to be formed under several distinct conditions. Sometimes it is evidently a volcanic product, at others it results from the decomposition of HS, which is contained in the soil of certain countries, which becoming exhaled, undergoes a partial combustion, and deposits

Sulphur in little crystals. These varieties of Sulphur are easily soluble in CS^2, while those of volcanic origin are much less so. It is also produced by the decomposition of sulphates, which in contact with organic matter, undergo a partial decomposition, and give rise to sulphides which are often again decomposed by the $\dot{O}$ in the air, so that the Sulphur is deposited in a pulverulent form. This phenomenon is reproduced every day in the bleacheries, and also in the pipes which conduct water charged with sulphates. A remarkable instance of this kind is seen at Engleisen, the lake of which is supplied by artesian wells. At a certain depth the well pierces a bed of peat. All the water coming from below this bed, is charged with sulphates, and all the water above is charged with sulphides. The flints of the Jurassic formation, which have been mentioned, are generally the fossil remains of some of the inferior orders of animals, whose viscose bodies have been replaced by silica, in waters which were charged with sulphates; in presence of the organic matter, these sulphates have been decomposed, and have given up their Sulphur. This phenomenon takes place frequently on the sea shore, where the water contains sulphates in solution. All the iron nails which fall from vessels, become in this way after a time, transformed into sulphide of iron.

Sulphur also exists in nature as HS, and is often discharged with more or less vapor of water. Many of the miasmas of Italy are of this kind. $\ddot{S}$ is also given off in nature, but may come from the HS, for it may undergo different kinds of combustion by which S, $\ddot{S}$ and even $\dddot{S}$ are produced. In the presence of certain porous rocks and certain organic matter, $\dddot{S}$ is formed. Thus, at Aix-les-Bains, the bathing rooms are frequently rotted by the $\dddot{S}$ produced by the decomposition of the HS. Springs charged with $\dddot{S}$ have been discovered in this country. Boussingault says that the Rio Madeira of Brazil is charged at certain points with considerable quantities of $\dddot{S}$.

FORMULÆ OF THE CRYSTALS.

Pl. I.

Fig. 11. P; with the base, occurs quite frequently. *Fig.* 12. P. $\infty P \infty$. *Fig.* 13. P. 0P; often tabular. *Fig.* 14. P. $\frac{1}{8}$P. *Fig.* 15. P. ∞P. *Fig.* 16. P. $\breve{P}\infty$. *Fig* 17. P. $\breve{P}\infty$. $\frac{1}{8}$P. 0P.

TELLURIUM.

Tellurium. Te. Hexagonal.

Syn.—Gediegen Tellur, Gediegen Sylvan, Tellure-auro-natif ferrifère.

It crystallizes as hexagonal plates, the basal edges of which are modified, or as a rhombohedron with an angle very near 90°. Cleavage perfect, parallel to the prism, imperfect, parallel to the base. Lustre, metallic, but becomes dull on exposure. Color and streak, tin-white, sometimes yellowish, rosy or indistinct, and not always uniform. Quite brittle. **H.**=2–2.5. **G.**=6.1–6.3. Composition, Te 92–97, Fe 7, Au 0.25–2.785.

Pyr. &c. B. P. Fuses easily when heated. It will even melt in the flame of a candle. In an open tube it burns with a blue flame. Sometimes it gives a little sulphur, and generally the smell of selenium. The result of the combustion is a fusible oxide, which is white if pure, and red if mixed with selenium, or yellow if soiled with bismuth. On Ch., fuses, coloring the flame green, and gives a white coating, and almost always leaves a little gold. The non-oxidizing acids are without action. $\dddot{\dot{N}}$ dissolves it, and the solution becomes turbid on the addition of water, on account of the bismuth present.

It was found in Transylvania some years ago, in large quantities, where it was smelted to extract the gold. It is now quite rare. The blowpipe distinguishes it from Antimony, Bismuth and analogous substances. It is usually found in crystalline, granular masses.

CARBON.

Diamond C. ISOMETRIC.

SYN.—Demant, Diamant.

Pure carbon is found as two minerals, Graphite and Diamond. The Diamond is carbon almost chemically pure. It is found in nature in different states, sometimes crystallized, sometimes as a concretion, and sometimes with an intermediate structure. The Diamond crystallizes in the isometric system, like most of the simple bodies. Its usual forms are the octahedron and the hexoctahedron. These two forms are frequently combined, *Pl.* I. *Fig.* 21. The forms of the Diamond are various. The cube, by itself, is quite rare and is almost always opaque. It is however found combined with the hexoctahedron. Frequently the faces of the hexoctahedron become rounded and the lines corresponding to the diagonals of the rhombic dodecahedron are obliterated; the figure then appears to be a rounded dodecahedron, *Fig.* 19. When the faces of the octahedron are shown on the hexoctahedron, they are always plane and brilliant, though small. This allows of their being easily distinguished from the others, which are generally rough and convex. The octahedron with plane and curved faces is sometimes found. The cube with faces of the tetrahexahedron also occurs, *Fig.* 23. Rounded hemihedral forms are quite common; these, compressed so as to obliterate some of the faces, form quite complicated crystals, *Fig.* 25. Hemitrope crystals of Diamond are also found; the most usual one is that of the octahedron, *Figs.* 26 and 27. The hemitrope of the hexoctahedron in which a part of the faces are wanting is a characteristic form, *Fig.* 28. Macles of two tetrahedra are common, *Figs.* 29, 30, 31, and also those of the tetrahedron and hemi-hexoctahedron inclined, *Fig.* 32; two cubes interpenetrated are also found, *Fig.* 33. Fracture, conchoidal. Lustre adamantine. Cleavage perfect, parallel to the octahedron. Transparent, translucent, or opaque. Color, white or colorless; sometimes it shows yellow, red, orange, green, blue, brown, or black tints, which in certain specimens may become quite dark. Index of refraction, 2.439. **H.**=10. **G.**=3.5295–3.55. Composition, pure carbon.

Pyr. &c. B. P. May be entirely burned in the oxyhydrogen flame, forming $\ddot{C}$. Insoluble in acids.

The various forms of the Diamond present the remarkable peculiarity of rounded faces. This is probably owing to the circumstances in which it crystallizes. If a number of artificial salts are crystallized from a viscous liquid, some of the crystals will have plane faces, but others curved faces. If now when the crystals are formed, the mother liquor is rendered less viscous, some, but not all, of the faces will become plane. Those which become plane at the same time, are generally those which are produced by the same law, and reciprocally. While these phenomena are taking place, the cleavage plane of the crystal remains plane, whatsoever may be the convexity of the crystal.

Even the best crystallized Diamonds have a rough look and a peculiar lustre, which is characteristic. The variation in density is owing to the fact that it is sometimes mixed with foreign substances. It is the hardest of all known bodies, but is however very fragile, owing to the ease with which it cleaves parallel to the face of the octahedron. It is most valued when it is colorless. Rarely it has quite bright colors, which, instead of diminishing its lustre and value, increase them. It has a very peculiar lustre, and in order to distinguish it from all others, this lustre is called adamantine. Its intensity is very variable. It may be very strong or weak, according to the state of crystallization, but its peculiar character is always preserved. This lustre is brighter in the fracture than in the cleavage, which is lamellar, while the fracture is conchoidal. The Diamond possesses simple refraction in the highest degree. Its dispersive power is very great, which is the cause of its peculiar brilliancy called the fire.

With the exception of its form, the concretionary Diamond is in every respect similar to the crystallized Diamond. Its form is frequently spherical, with surfaces which are sometimes drusy or pointed with little crystals. It does not show any distinct cleavage. Its fracture is almost fibrous and frequently radiating from a center. This variety is frequently colored. The amorphous Diamond has no very definite structure, and is remarkable only for its extreme hardness. It is found in irregular masses, which, when examined with a glass, are found to be made up of minute cells separated by thin partitions, which have a kind of transparency, that gives to the mass the appearance of coke. These masses are often brown, or almost black, with a very feeble adamantine lustre. There appears to be at times, a passage from the amorphous to the crystalline variety. Its density varies from 3–3.4, being very much diminished by its interior cavities. It is lighter and less fragile than the two preceding varieties. The powder is grayish, almost the same color as that of the crystallized variety. It is sometimes called the Black Diamond. It is easily recognized by its hardness and lustre. M. Mène heated Anthracite from Creusot, having 2 % of ash, in a crucible for a long time, when it assumed a metallic lustre and cut glass. The Diamond may be considered as pure Carbon. This fact was discovered in 1694, when it was burned for the first time in the academy at Florence, by a powerful burning glass. The crystalline colorless varieties, gave only 0.01 % of ash. In the colored varieties the proportion is larger, the Black Diamond giving 2–3 %. The appearance of the ash varies; that of the crystallized varieties is crystalline, and that of the Black Diamond resembles the ashes of vegetable fuel.

For many years the Diamond was supposed to be the product of organic action, but it is now everywhere admitted that it is a mineral. What proves it, is that some Diamonds contain other mineral substances such as Gold, &c.; besides minerals, it contains germs of Fungi, and fibers of vegetable structure of a higher organization, from which it does not seem to have been formed at a very high temperature. Some Diamonds contain red, white and black spots, and if they are heated to redness, and protected from the air, these spots disappear. This property, which is well known to and used by jewelers, seems to prove that the temperature at which the Diamond was formed, was below red heat.

Another experiment made by Jacquelin, in which he transforms the Diamond into Graphite, by exposing it to an electric current, seems to prove that Diamond and Graphite are only allotropic conditions of carbon, differing in the temperature at which they were formed. The Diamond has probably been formed like coal, by a slow decomposition of substances containing carbon, whether vegetable or mineral. It has been formed under the same conditions as regards metamorphism, that have produced the argillaceous shales and gold-bearing rocks which accompany it. The schists which were altered, may have been shales impregnated with hydocarbons. The formation of Diamond from a hydrocarbon gas is quite analogous to the formation of sulphur from HS in the wet way, where, the hydrogen becoming oxidized, only a part of the sulphur becomes $\ddot{S}$, the rest remaining as Sulphur; so in the humid oxidation of the hydrocarbons the hydrogen is oxidized, part of the carbon becomes $\ddot{C}$, while the rest of the carbon may crystallize as Diamond.

Formerly the Diamond was cut in flat stones, having large parallel faces; this method however gave only a little fire, and now the methods which are generally adopted are the forms called Rose and Brilliants. The Rose has usually two plane parallel faces, the one below being larger than the one above, but the upper table is sometimes surmounted with a pyramid, *Fig.* 35. This method of cutting, however, diminishes the amount of fire and it becomes necessary to set the stone with an opaque back, which is generally of silver, thus throwing back the rays of light and increasing the fire. This method of cutting is only used for flat stones which do not have a good form. The method of cutting in Brilliants, *Fig.* 34, is much better adapted to stones of value. It consists of an upper plane, which is quite large and generally square, with the angles cut off. This is surmounted by a border made up of triangular or lozenge-shaped faces. This border generally occupies one-third of the whole height of the stone. There is sometimes another border below called the pavilion, which occupies the other two-thirds of the height, the faces of which have the same general shape, but are longer. This is terminated by a lower plane similar to the upper one, but much smaller. This method of cutting is the best for the fire, since the rays of light no longer traverse the stone but are thrown out of it by the upper face through which they enter. For this reason the stones are always set open. There are special methods of cutting for those stones, which would lose too much of their weight by being cut in the ordinary way. It is expected that a rough Diamond will lose $\frac{9}{20}$ of its weight in cutting. When the stone is of a remarkable size a mould is generally taken of it in plaster. This model is then cut, so as to allow the largest possible brilliant. In this way the method of cutting which will make the stone lose the least possible weight is determined. When the stone is very large and would lose too much of its weight by cleavage, it is simply worn down and then polished, but it would take three or four years to cut a large-sized stone by friction only. A cut Diamond sometimes seems to have defects; this may be owing to the fact that some of the natural faces have been left in order not to reduce the size of the stone too much. Such faces are always very small. This is the case with the stone known as the "Star of the South."

The ancients knew nothing about cutting the Diamond, but wore as ornaments only the natural stones. They chose always the most regular ones, especially those with sharp edges. It was not until 1456 that Louis

Berquen, of Bruges, in Belgium, discovered the method of cutting the Diamond so as to increase its lustre. The first step in cutting is to take off the outer coating of the stone by means of the natural cleavages; if in this operation it becomes apparent that the texture of the stone is uneven, or that it is covered with knots, it is thrown into the refuse to be powdered. If the stone passes this test, it is ground according to the form that it is to receive, by rubbing two stones together under water. When the approximative form has been given in this way it is fastened to a stem of copper, terminated by an instrument destined to hold the stone which is held in position by casting fusible metal around it. The face is now finished by bringing it against the surface of a wheel made of soft steel and covered with diamond powder and oil. This wheel is horizontal and the stone is inclined upon it at the angle which is intended for the face. In this way the stone is polished, each face being successively produced. The Diamonds which are not fit to cut and which are not needed for powder are used for the ends of tools for drilling or turning hard rocks, such as Granite and Porphyry. The small stones which have very sharp edges are used for cutting glass; those which are best adapted for this purpose are those which have the terminal angles of the octahedron, because in this case there will be but a single point of the Diamond in contact with the glass which cuts the latter, without crushing it. Wollaston has shown that every stone which scratches glass, as Corundum, Quartz, etc., cut in the same way, will also cut glass giving a clear cut. The clear stones of Diamond have for a long time been used as jewels for watches. The Black Diamond has also been used for a long time for turning, and lately in this country for drilling the harder rocks.

A Diamond of 5–6 carats is a very large stone, those of 12–20 are very rare and very few are known that weigh more than 100 carats.

WEIGHT OF HISTORIC DIAMONDS.

	Carats.		Carats.
Rajah	367	Piggott	$82\frac{1}{4}$
Great Mogul*	$279\frac{9}{16}$	Nassac‖	$78\frac{5}{8}$
Orloff	$194\frac{1}{4}$	Dresden	$76\frac{1}{2}$
Koh-i-noor†	186	Sancy	$53\frac{1}{2}$
Portuguese	148	Eugenie	51
Florentine	$139\frac{1}{2}$	Pasha	49
Regent‡	$136\frac{3}{4}$	Dresden (green)	$48\frac{1}{2}$
Star of the South§	$125\frac{1}{4}$	Hope (blue)	$44\frac{1}{2}$
Koh-i-noor (recut)	$106\frac{1}{16}$	Polar Star	40
Shah	95	Cumberland	82
Sultan of Turkey	84	Russian (red)	10

Jewelers acquire by practice the habit of distinguishing between the Diamond and all other stones which are used to imitate it. It requires long experience and practice to acquire this power and it is better to use some scientific means than to trust to the judgment alone. These means are given in the following table.

* Uncut it weighed 900 carats. † Uncut it weighed 793 carats; it it supposed that another stone weighing 130 carats was cut off the original. ‡ Uncut it weighed 410 carats. § Uncut it weighed $251\frac{1}{2}$ carats. ‖ Uncut it weighed $89\frac{3}{4}$ carats.

TABLE FOR DISTINGUISHING PRECIOUS STONES.

	Density.	*Refraction.*	*Index of Refraction.*	*Electricity*
Diamond.	3.52–3.55	Simple.	2.455	Positive, not durable.
Ruby, Sapphire and Oriental Amethyst.	3.9–4.3	Double. 1 axis.	1.765	Lasts several hours.
Chrysoberyl.	3.5–3.8	Double.	1.760	Lasts several hours.
White Topaz.	3.4–3.6	Double. 2 axes.	1.635	More than 24 hours.
Chrysolite.	3.3–3.5	Double.	1.660	Positive.
Emerald.	2.6–2.8	Double. 1 axis.	1.585	Positive.
Spinel.	3.4–3.8	Simple.	1.755	Not tried.
Zircon.	4.4–4.6	Double. 1 axis.	1.990	Positive, not durable.
Quartz.	2.6–2.8	Double. 1 axis.	1.549	Positive, not durable.
Strass.	var. 3.5	Simple.		Not durable, variable.

The Diamond has never yet been found in place, but only in alluvial formations. In Brazil, in the province of Minas Geraes, it is found in a sandstone, which is probably a metamorphic rock, called Itacolumyte or flexible sandstone. It is often pierced by minerals which may have penetrated it after its formation. It is made up of crystals of Quartz that interlock with each other so as to allow of a little movement. It is possible that the Diamond may also have been formed and penetrated this rock like the other minerals it contains. Fragments of Itacolumyte containing Diamonds have been offered for sale at Rio Janeiro.

Many attempts have been made to reproduce Diamond artificially but only very small crystals, if any, have ever been made. Any attempt to reproduce it is made in the dark, because we know so little about its natural production, and this is probably the secret of the constant failure to make it artificially. The mines of Brazil were first opened in 1727. It is estimated that since then, they have produced two tons of Diamonds. The unit of weight adopted to determine the value of the Diamond is the carat. The name is taken from a bean of the East Indies, for which the weight of 2½ centigr. has been substituted. To determine the value of the stone, the price of the first carat is determined by the purity of the stone and the probable ease with which it can be cut. In England, a Diamond weighing one carat and of the purest water is worth, when cut and polished, £12. From this as a starting point the price increases with the square of the weight multiplied by 12, but in larger stones the value is generally arbitrarily fixed. The rough Diamonds are divided into several categories. Those which contain no defects are called crystallized, or of the first water; those which are only slightly colored and have only a few defects are called of the second or third water. Beyond this point, they are placed among the refuse and are reduced to powder to polish those of the other classes. The dealers in rough stones acquire the power of distinguishing the water of a rough stone by simply breathing upon it. Not only the water, but the color may be ascertained in this way after practice. If the stone breathed upon is light green, the Diamond is colorless and of the first water. If it is pale yellow, it is fawn color and of the second water. A milky white will give a bluish tinge called celestial-blue, of the second water; gold-yellow gives a reddish tinge of the second water. Stones which have deep colors are called fancy stones. They may be green, red, blue, or black and have a high value on account of their rarity. In an invoice of rough Diamonds there is generally one-third of the first water, one-third of the second, and one-third of a quality for reducing to powder for polishing. This proportion will of course vary, with the place from which the stones are sent. The stones which came from the mines of India, now no longer worked, contained more of the first water than those from the Urals and Brazil.

The Diamond is found in this country in Franklin, Rutherford and Hall counties, North Carolina, in Manchester, Va., and in several places in California, and it is reported from Idaho.

FORMULÆ OF THE CRYSTALS.

Pl. I.

Fig. 18. O. *Fig.* 19. ∞O, with curved faces. *Fig.* 20. $3O\frac{3}{2}$, with curved faces; characteristic form. *Fig.* 21. $3O\frac{3}{2}$. O. *Fig.* 22. $\infty O\frac{4}{3}$. *Fig.* 23. $\infty O\frac{4}{3}$. $\infty O \infty$. *Fig* 24. $\frac{3O\frac{3}{2}}{2}$. *Fig.* 25. Distorted form of *Fig.* 20. *Fig.* 26. Hemitrope; composition-face O. *Fig.* 27. The same. *Fig.* 28. Hemitrope of *Fig.* 21; the middle portion of the crystal, between the opposite sets of six planes, is wanting. *Fig.* 29. Interpenetrating tetrahedra; composition-face O. *Fig.* 30. The preceding, with the octahedron. *Fig.* 31. Interpenetrating tetrahedra. *Fig.* 32. Twin crystal formed by interpenetrating crystals, with the combination $\frac{3O\frac{3}{2}}{2}$. $\frac{O}{2}$. *Fig.* 33. Interpenetrating cubes. *Fig.* 34. Form of the Brilliant. *Fig.* 25. Form of the Rose Diamond.

Graphite. C. HEXAGONAL.

SYN—Plumbago, Black Lead, Reissblei, Fer carburé.

It crystallizes in hexagonal tables, with traces of a rhombohedron of 85° 29′. Other forms are sometimes found, but it is generally not very well crystallized. Cleavage very easy, parallel to the base. Lustre, metallic. Opaque. Color, black. Streak, black and shining. Flexible in thin lamellæ. Sectile; soils the fingers. **H.**=1–2. **G.**=2.0891. Composition, pure carbon.

Pyr. &c. B. P. Infusible. At a high temperature burns, leaving generally a little red ash. Not acted on by acids.

It is rarely ever pure, and when used must undergo a careful preparation to separate it from foreign substances. By many it is considered as the ultimate form of the coals of vegetable origin. It appears to be produced by the decomposition of some organic matter. It is frequently found associated with Calcite and the granular variety of Amphibole, called Pargasite. It can only be confounded with Molybdenite, which is of a bluish color, and which gives a greenish streak on glazed porcelain and the molybdenum reactions before the blowpipe. It is largely used in the arts for the manufacture of lead pencils and crucibles, and also as a lubricator. It has been found in the U. S. in Massachusetts, Connecticut, Vermont, New York, North Carolina, and elsewhere.

Carbonic Acid. $\ddot{C}$.

SYN.—Kohlensäure, Acide carbonique.

Carbonic acid is, given off from many mineral waters, and near some volcanoes. **G.**=1.5245. Taste, slightly acid and pungent. Destroys life. Composition C 27.27, O 72.73.

The Saratoga and Ballston waters owe their sparkling to this gas. Sometimes its disengagement is distinctly traceable to the decomposition of carbonates, but generally this phenomenon is more or less intimately connected with volcanic action. In some countries, as in France, the soil is impregnated with Carbonic acid, which is such a serious obstacle as to entirely prevent the working of certain mines. A pit which was full of Carbonic acid has been known to empty itself after standing for years, the gas following the track of a road and suffocating horses in a stable more than 100 meters from the orifice of the pit. In the island of Java, there is a valley filled with Carbonic acid which is called the "Valley of Death," the sides of which are strewn with the bones of men and animals. It is probably the crater of some extinct volcano, or is connected by fissures with some permanent source of Carbonic Acid.

BORON.

Sassolite. $\dot{H}^3 \dddot{B}$. TRICLINIC.

SYN.—Sassoline, Boric acid, Borsäure, Acide Boracique.

It crystallizes as a doubly inclined rhombic prism of 118° 30′. Crystals are however quite rare. It has a very perfect basal cleavage. Lustre, pearly. Translucent. Color, white, or yellowish, sometimes gray when associated with Sulphur. Taste, slightly acid, saline and bitter. Feels smooth and unctuous. **H.**=1. **G.**=1.48. Composition, $\dddot{B}$ 56.4, $\dot{H}$ 43,6.

Pyr. &c. B. P. In a closed tube gives off water. Fuses to a clear glass, coloring the flame green. Soluble at 100°C. in 2.97 parts of water.

It is quite rare in nature. The eruptions of some volcanoes have produced deposits of little white crystals of Sassolite sometimes associated with Sulphur, which have a peculiar and waxy lustre. The crystallization is usually very imperfect, and they are generally nothing more than silvery-white scales. It occurs also in considerable quantities in the lagoons of Tuscany. It is easily recognized by its lustre and the green flame before the blowpipe.

SILICON.

Oxygen Compounds.

Quartz. $\dddot{S}i$. HEXAGONAL.

SYN.—Bergkrystall, Quarz, Kiesel.

Silicic acid alone constitutes the mineral Quartz. This mineral is susceptible of a large number of very remarkable allotropic conditions, which are distinct by their mineralogical, physical and chemical properties. They might constitute a number of distinct mineral species. It would however, be difficult to separate them, on account of the gradual and imperceptible passage of one into the other. They are therefore united under the name Quartz and regarded simply as varieties of the same species.

We shall thus consider:

Crystallized Quartz,

Concretionary Quartz Agate or Chalcedony,

Jasper,

Silex or Flint, which is more easily attacked by alkalies than the other varieties It is never pure.

Earthy Quartz, sometimes in the shape of flour and, in every way analogous to the silicic acid produced in the laboratories. It is often formed of the skeletons of infusoria.

Quartzites and Sand.

Quartz crystallizes in rhombohedra of 94° 15′. It has a very difficult cleavage, parallel to the faces of the primitive rhombohedron. In order to produce it the crystal must be heated and thrown into mercury; if thrown into water it would split, without cleaving. The primitive rhombohedron of Quartz is very rarely found, and is always in very small crystals. A form which is sometimes found is the rhombohedron, with the hexagonal prism, *Pl.* II. *Fig.* 3. The faces of the prism are then always striated parallel to the horizontal diagonal of the rhombohedron, i. e. across the faces of the prisms. The most general form however is the combination of the two rhombohedra, R and -R, by which the prism is apparently terminated by an hexagonal pyramid, *Fig.* 2. These faces are frequently very unequally developed as in *Figs.* 9 and 33, and generally show a difference in lustre. In some varieties, especially the opaque ones, the two rhombohedra are equally developed, which gives an hexagonal pyramid with equal faces, *Fig.* 2. When the hexagonal prism disappears, we have a complete hexagonal pyramid, *Fig.* 1. It sometimes happens that two and even three rhombohedra at different angles are placed together, giving rise to pointed forms, *Figs.* 10, 11, 12, 13, and 32. The faces of different rhombohedra usually have different lustres.

Crystals of Quartz are capable of a peculiar kind of hemihedry. Sometimes the faces of the tetarto-dihexagonal pyramid (plagihedron) are observed, *Figs.* 5, 9, 11. The right and left forms, when they occur together, make a scalenohedron, *Fig.* 14. It may happen that a series of two and even three plagihedral faces occur at the same time. These two series of plagihedral faces have very remarkable and different optical properties. They show the plane of polarization turning in opposite directions. This hemihedry is frequently the cause of peculiar phenomena, and sometimes changes the lustre of the faces. Some varieties, especially those which are violet or dark yellow, frequently have a velvety lustre in their transverse fracture, instead of being vitreous as they generally are. If we examine this fracture with the microscope, we shall find that it is produced by little faces fitting the one in the other. These faces are alternately plagihedral right and left, so that the fracture has undulations which give it the velvety appearance, the different faces reflecting the light sometimes in one direction and sometimes in another. Faces of the trigonal pyamid are often observed, *Figs.* 5–7. Hemimorphic crystals, however, are very rare, *Fig.* 15. Twin crystals of Quartz occur very frequently, the most usual ones being those formed by interpenetration, *Figs.* 16, 17, 28 and 30, or by juxtaposition parallel to the vertical axis, *Figs.* 18 and 19. More generally, however, they are very irregular, so that one rhombohedron is distributed unevenly through the faces of the other, as in *Fig.* 31, in which the shaded parts of the pyramidal faces are −R and the unshaded parts R. The same formation can be very often observed upon the faces of the prism, *Fig.* 29. Geniculated twins also occur, *Fig.* 20. Crystals of Quartz are often found which have been compressed, and the real crystalline form is very often concealed by this deformity. This compression may take place parallel to two of the faces of the hexagonal prism, as in *Figs.* 21 and 27; when at the same time it is lengthened in the direction of the horizontal axis, forms similar to *Figs.* 25 and 26 result. Sometimes the flattening takes place parallel to the faces of the hexagonal pyramid, *Figs.* 22 and 23, and sometimes one face of the rhombohedron is developed at the expense of the others, *Fig.* 24. Besides these, crystals variously distorted and bent are very often found.

Its fracture is conchoidal or scaly, the pieces having very sharp edges. Its lustre is vitreous, but some colored varieties have a velvety lustre caused by the interpenetration of fibrous minerals. It may possess every possible degree of transparency. The colors of Quartz are very variable. It may be colorless, or different shades of yellow, brown, violet, red, green, blue or black. The colorless varieties are sometimes called Hyaline Quartz, and the yellow False Topaz. The brown and black are called Smoky Quartz, and the violet varieties Amethyst. The other permanent colors are white and rose. Almost all of these colors admit of being transparent. There are a few, such as red and yellow, which are always opaque. A variety much esteemed is known as Aventurine, which is Quartz filled with yellow mica; it is often called Gold Stone. This stone is very often imitated and is not to be confounded with the variety of Feldspar, that has the same name. Streak, white or lighter than color. It acquires vitreous electricity by friction, but loses it very quickly. **H.**=7. **G.**=2.5–2.8. Composition, $\dot{S}i$ 46.67, O 53.33.

Pyr. &c. B. P. It is infusible by the ordinary manipulations with the blowpipe, and with difficulty fusible in the oxyhydrogen flame. It is not acted upon by any acid except HFl.

When it crystallized the Quartz may have become associated with other substances. Thus the blood-red varieties, as Hyacinth Quartz, which are generally doubly terminated, are impregnated with anhydrous sesquioxide of iron, which sometimes only forms an interior or exterior film. The violet varieties, Amethyst, are often colored by organic matter, but generally by manganese. The brown, smoky varieties appear to be colored by organic matter. Wöhler announced some years ago that he had discovered free silicon in these varieties, which however existed only in traces. The brown color however does not seem to be owing to this cause, for it disappears or at least becomes lighter when the crystal is heated. The same is true of the rose and yellow varieties. This property is made use of by the jewelers, who heat them until they have assumed the exact color of real Topaz. The mineral may be colored artificially. To effect this it must first be heated and thrown into water, by which fissures are produced. It is then placed in oil or honey and there results a feeble coloration which is most apparent in the fissures. The colored varieties produced by plunging Quartz into metals in fusion are called *Rubasses.* The color is produced by the oxidation of the thin films of metal that penetrate the cracks. Ebelmen succeeded in making silicic acid enter into combination with certain organic substances; a substance was thus formed which gradually deposited its silica.

Quartz frequently envelops crystals of other substances such as Barite, Rutile, Pyrite, Asbestus, Chlorite, Epidote, or even substances which would be affected by heat, as Stibnite and some carbonates. Sometimes it contains cavities, which are filled with a liquid that makes itself apparent by little bubbles of gas. Davy after examining this liquid, and Brewster after studying its optical properties, came to the conclusion that it was a very volatile liquid, so volatile that when the crystals were cut, the sides of the cavity becoming thin, the crystal burst with a slight explosion and the liquid was immediately volatilized. Besides this we have seen that under the influence of heat the colorations of Quartz changed; these circumstances would seem to indicate that Quartz was at least sometimes, if not always, formed at a relatively low temperature. Within a few years very distinct crystals of Quartz have been produced artificially at a temperature of 150° or 160°C. Crystals of Quartz are also found in the Chalk where they must have been formed at a low temperature. The appearance of crystallized Quartz is often modified by the foreign substances which it contains and we sometimes see faces of increase in the interior which have been covered with other substances. These faces oftentimes do not adhere very closely and can be separated by a blow. This phenomenon produces the crystals known as *cap crystals*. Generally the material which thus covers the crystal is oxide of iron or chlorite.

FORMULÆ OF THE CRYSTALS.

Pl. II.

Fig. 1. R. –R. *Fig.* 2. ∞P. R. –R. *Fig.* 3. ∞P. R. 4R. *Fig.* 4. ∞P. R. –R. 2P2; the latter faces occur frequently and are usually very brilliant. *Fig.* 5. ∞P. R. –R. 2P2. 6P$\frac{5}{6}$; the latter faces form the plagihedron or trigonal trapezohedron. *Figs.* 6 and 7. ∞P. R. –R. 2P2; showing the difference between, and the position of the *left-handed* and *right-handed* trigonal pyramids. In *Fig.* 6 the face 2P2 is *left* of the face R, and in *Fig.* 7 *right.* *Fig.* 8. ∞P. R. –R. 4R. 6P$\frac{5}{6}$. 2P2. *Fig.* 9. ∞P. ∞P2. R. –R. –7R. 6P$\frac{5}{6}$, showing the tetartohedral face of ∞P2; from Carrara. *Fig.* 10. ∞P. R. –R. –7R. 6P$\frac{5}{6}$. 2P2; from Dauphiny. *Fig.* 11. ∞P. R. –R. –11R. 6P$\frac{5}{6}$. *Fig.* 12. –11R. R. –R; with the rhombohedron –11R predominating, also from Dauphiny. *Fig.* 13. ∞P. R. –R. 3R. –$\frac{7}{2}$R. 6P$\frac{5}{6}$. 4P$\frac{4}{3}$; from Switzerland. *Fig.* 14. ∞P. R. –R. 6P$\frac{5}{6}$; remarkable because 6P$\frac{5}{6}$ is *hemihedral,* forming a scalenohedron, as both the right and left forms are present; from Brazil. *Fig.* 15. ∞P. 0P. R. –R; a very rare hemimorphic crystal. *Fig.* 16. Interpenetration twin. *Fig.* 17. Hemitrope; composition-face 0P. *Fig.* 18. Twin crystal by juxtaposition. *Fig.* 19. The same. *Fig.* 20. Geniculated twin; composition-face P2. *Figs.* 21–27. Distorted crystals. *Fig.* 28. Group of crystals, showing the formation of the striations upon the prism. *Fig.* 29. Interpenetration twin crystal. *Figs.* 30

and 31. Twin crystals, showing regular and irregular interpenetration. The unshaded parts are R, and the shaded parts -R. *Fig.* 32. Pointed crystal, with a series of different rhombohedra. *Fig.* 33. R. -R. ∞ P; with R predominating.

Concretionary Quartz or Agate is less pure than the crystallized Quartz. It contains 1% of metallic oxides and sometimes alkalies, which appear to remain as witnesses of the solutions which have deposited the Quartz. It has a variable hardness, but always much less than that of crystallized Quartz. Its fracture is generally scaly and rarely fibrous. When examined by polarized light small needles are sometimes seen, which would appear to indicate an incomplete crystallization. It is generally mamelonated and stratified. Sometimes it looks like a membrane, as if it had been a gelatinous substance resting on a few points and had afterwards contracted. It frequently moulds itself into cavities, left empty by the solution of crystals of Fluorite, Calcite and Gypsum, and takes the shapes of organisms, and especially of fresh water shells. It is also found lining large geodes in concentric layers, each layer being of a different color. This is the variety sometimes called Ribbon Agate. Some Agates remarkable for their colors are made use of in the arts, such as the blue variety called Sapphirine, which is the variety that usually fills the cavities left by Fluorite. The Sardine stone, which is a brownish-yellow, Carnelian, clear red, Chrysoprase, clear apple green, and Prase, dark green, are also varieties that are often made use of for ornamental purposes. These colors are sometimes caused by metallic oxides, but more frequently by organic matter. It may be colored by imbibation like crystallized Quartz. The varieties which contain crystalline substances ramified like vegetation, are called Moss Agate. This disposition is often produced by metallic oxides, such as those of iron and manganese, and sometimes Stibnite, which are crystallized in dendrites. Sometimes the same effects are produced by fractures which were formed while the Agate was being deposited. Agate may have every degree of transparency, and it often replaces the structure of organisms. This is done in a peculiar way and by a method of petrifaction which nature has not explained, and which is entirely distinct from the filling up of the cavities left by other crystals, of which we have already spoken and which often takes place in fossil shells. Concretionary Quartz takes the place of the organism almost molecule for molecule, the empty spaces remaining empty and the full parts only being replaced. This is often seen in wood and animal organisms, which are entirely converted into Quartz, but which retain their own proper structure. With the Madrepora and other species this replacement is so perfect, that even the name of the species can be determined. The Moss Agate may be artificially imitated, by causing nitrate of silver to penetrate ordinary Agate, when the former crystallizes and turns black under the influence of light. The zoned or Ribbon Agate is much used in the arts. When the zones or strata are in parallel layers, and the colors in great contrast, this variety is called Onyx. Ordinary Onyx is a combination of black and white, while the Sardonyx consists of a combination of the Sard with white. The variety which is penetrated by Asbestus, when cut in carbuncle form, is known as Cat's Eye. Agates are often used for cameos, in which case they must have parallel layers of different colors. These are often produced artificially.

Jasper is the name given to all the highly-colored opaque varieties of

Quartz. One of the most beautiful kinds is the variety called Heliotrope or Bloodstone, which is green, spotted with red. The variety found in parallel bands of different colors, called Ribbon Jasper, is not always composed entirely of Quartz. It contains fusible portions which are probably Feldspar. The red variety of Jasper, when polished, is much used for ornaments. Jasper is a state which may be affected by any one of the varieties, and is especially interesting on account of its use in the arts.

In the variety of Quartz called SILEX or FLINT there is no trace of crystallization to be distinguished, even under the microscope. The silicic acid is still less pure than in Agate. It would seem as if it was in a different condition from that of the two other varieties. It even differs a little from that of Agate in its chemical characters, for it is easily attacked by alkaline solutions, a property which is used in certain localities for preparing on a large scale the soluble silicates used in painting. It frequently shows a passage into Agate, and even into crystallized Quartz. In the Silex or Flints that occur in the Chalk, some parts are found which are quite analogous to Agate, sometimes even resembling little crystals. Silex is found in two distinct states. The first is called Pyromac, but is generally known as Flint. It is found in the Chalk formation, the layers of Flint being distributed in extensive beds in which the nodules are not contiguous. Their forms are very irregular, sometimes ramified and having rounded edges, and generally the outside is covered with Chalk. The fracture is quite perfect, with very sharp edges; somewhat translucent. The fracture and the colors are always dull, generally white, gray or black. The dark varieties are the most frequent. These colors appear to be owing to organic matter, for when it is heated it becomes white and opaque, giving off water. The black varieties treated with bichromate of lead give off carbonic acid. In most of the arenaceous formations it has undergone a very singular transformation, the sand being composed of very small crystals of Quartz and Silex which have become farinaceous, though still preserving their form. The second variety is known as Quartz-millstone. It has a peculiar look and is full of cavities, sometimes resembling a sponge. It frequently resembles the reticulated tissue of bone. This variety is found in regular beds, especially in the Tertiary formation, which contains the Fontainebleau Sandstone. They are used for millstones when the texture is fine, and for building purposes when it is coarser. When the texture becomes spongy, they are used for fortifications. They become crushed under the stroke of the projectile and thus the ball loses a large part of its living force. It replaces organisms, especially the Echinoderms. It is however done by filling up the empty spaces and preserves no trace of the organization.

The variety called EARTHY QUARTZ is an allotropic condition which is entirely distinct from the preceding. This earthy silica is sometimes called Flowers of Silica and is almost entirely soluble in alkalies. It is found in a number of varieties. The Flints of the Chalk formation are covered over with a white crust which appears at first sight to be chalk. It does not however effervesce, and it is in fact earthy silica, which is almost entirely soluble in alkalies. These Flints often rattle when they are shaken. This is caused by the interior being incompletely filled with this earthy silica, which is entirely analogous to that which forms the exterior crust. This decomposition is often carried so far as to make a light porous mass, which floats upon water and is called Float-stone. This same phenomenon is sometimes observed in some calcareous silicates, from which the lime has been dissolved. The Geysers of Iceland deposit a variety of silica, very analogous to this form, called Geyserite. The waters of these springs are charged with alkaline silicates, which decompose and deposit silica. When this rock is examined with the microscope, it is found to be made up of little concretionary spheres quite analogous to geodes of Agate. All these varieties of Earthy Quartz are rough to the touch, but there is another variety, made up of the skeletons of infusoria which is much smoother. The skeletons of these animals consist of silica, instead of phosphate of lime. This variety is known under the name of Tripoli, and is very much used for polishing. Under the earthy aspects, silica has only negative characters. It is known by its infusibility and not dissolving in acids, and it is not plastic when moistened, as clays are.

Sand is the name given to Quartz when in small fragments. Sands may be of different kinds; sometimes each grain is a complete crystal; sometimes it is rounded or concretionary; and sometimes it appears to have no form, but to be made up of fragments of crystals. These grains of Sand are often united together by a cement. The cement may be metallic as sesquioxide of iron, or it may be lime or silicic acid. This produces a Sandstone. When the fragments are large and round, the rock is called Pudding-stone, if angular, Breccia. When the cement is silicic acid it forms a rock which is called Quartzite. In this case a metamorphic action has often taken place which changes the condition of the Quartz, so that the mixture becomes entirely uniform and the fracture gives sharp edges. At Fontainebleau the Sands contain sufficient lime to cause them to crystallize with the form of Calcite, even when they contain as much as 80–85 % of silicic acid. Quartz is found in great abundance in all parts of U. S.

Opal. $\ddot{S}i$.

Syn.—Fire-opal, Feueropal, Menilite, Hyalite, Fiorite, Kieselsinter, Float Stone, Schwimmstein, Tripolite, Tripoli, Kieselguhr, Alumocalcite, Hydrophane.

Opal is sometimes called Quartz-resinite. It is however distinct from Quartz, both in its chemical reactions and in its composition. It is a dimorphous form of Quartz. Its fracture is smooth conchoidal like Silex, but it is brilliant, almost sparkling, with a peculiar lustre known as opaline. The lustre of Opal is vitreous, pearly or resinous. Transparent, translucent, opaque. Color, white, yellow, red, brown, green, often with a very bright play of colors. Streak, white. **H.**=5.5—6.5. **G.**=1.9-2.3. Composition, $\ddot{S}i$ as for Quartz; but it contains a variable proportion of water, from 3–21 %, which it may sometimes be made to give up and take again. In a vacuum it loses its water and becomes entirely opaque.

Pyr. &c. B. P. It is infusible before the blowpipe, but loses water and becomes opaque. In some varieties, the transparency may be made to reappear by plunging it into water. Sulphuric acid frequently turns it black, which shows that it contains organic matter. Some varieties turn red, owing to the presence of sesquioxide of iron. It is soluble in alkalies, but not so easily as the calcined silica of the laboratories.

The transparent varieties are called Hyalite. They have a mamelonated and stalactitic form. The variety known as precious Opal is generally found disseminated in trachytic or porphyritic rocks and has frequently very brilliant colors. Such Opals are highly prized as objects of ornament. The exact cause of the play of colors is not perfectly understood. It appears to be owing to the hydration of the silicic acid, for if an Opal is heated it loses the fire, but sometimes regains it in a less degree when it is put into water. Fire Opal is the name of a red variety, with a bright play of colors, from Mexico. Jewelers recognize two varieties of precious Opal; one of them, called *hard Opal*, retains its fire without any sensible diminution; the other, called *soft Opal*, loses its fire after a time altogether and changes color. It is always most brilliant after an exposure to a damp atmosphere for a certain time. The varieties which show no play of colors are called Semi-Opal. Some milk-white varieties which are sometimes a little yellowish, but never transparent, become transparent when put into water and give off little bubbles of air, and are called Hydrophane; when it is put into different liquids, as water and alcohol, their angle of refraction changes, but this refraction does not appear to have any relation with the liquid which has been absorbed. A very impure variety is found near Paris, which is called Menilite. It is very fragile and seems to have been formed in a pasty state. Opal has every variety of color; one of the most curious is the rose variety called Quincite. It occurs disseminated through limestone and contains considerable magnesia.

Opal may, like Agate, replace organisms and especially the woods of the Monocotyledons, such as the palms; such pseudomorphs are called Wood Opals. The impure varieties are found in many places in the U. S.

SILICATES.

ANHYDROUS SILICATES.

I. Bisilicates.

Amphibole Group.

Wollastonite. $\dot{C}a\ \ddot{S}i$. MONOCLINIC.

SYN.—Tabular Spar, Tafelspath.

It crystallizes as an inclined rhombic prism of 87° 28′. It is rarely found in crystals, *Pl.* II. *Fig.* 34, and these are not often simple. It is generally in crystalline masses, which are lamellar or bacillary. Cleavage easy, parallel to the base. Fracture, uneven. Lustre, vitreous or pearly on the cleavage faces. Translucent. Color, white, yellowish, brownish or reddish. Streak, white. **H.**=4.5–5. **G.**=2.78–2.9. Composition, $\dot{C}a$ 48.3, $\ddot{S}i$, 51.7.

Pyr. &c. B. P. Fuses easily on the edges. With borax, it dissolves and gives a skeleton of silica. When heated it becomes phosphorescent. With acids, it gelatinizes. Most varieties effervesce in acids, owing to the presence of a little Calcite.

It will be seen that this mineral has the same form and the same formula as Pyroxene. The angle of the primitive form is different, however, and its optical properties are so distinct, that it cannot be regarded as isomorphous with it. A compact variety, which is pinkish, has been found in Lake Superior. It is distinguished from Tremolite by gelatinizing with acids. From the Zeolitic minerals and the hydrous silicates, by the absence of water. From Asbestus and the fibrous varieties of Brucite, by the action of acids. It is found in several localities in New York, Pennsylvania and Michigan.

FORMULÆ OF THE CRYSTALS.

Pl. II

Fig. 34. ∞P. $\infty \bar{P} \frac{3}{2}$. $\infty \grave{P} \infty$. $\infty \grave{P} 2$. $0P$. $-5 \grave{P} \infty$. $-\grave{P} \infty$. $-\frac{1}{5} \grave{P} \infty$. $\frac{1}{3} \grave{P} \infty$. $\grave{P} \infty$. $3 \grave{P} \infty$. $2 \grave{P} \infty$. $-2P$. $2P$. $-2 \grave{P} 2$. $2 \grave{P} 2$; from Vesuvius.

Pyroxene.

Pyroxene embraces several varieties of minerals, all having the same general formula $\dot{R}\,\ddot{Si}$. Generally only bases having one atom of oxygen are included under this species. The relation of the oxygen of the base to the oxygen of the acid is as 1 : 2. The bases which make up $\dot{R}$ are $\dot{C}a$, $\dot{M}g$, $\dot{F}e$, and rarely $\dot{M}n$ and $\dot{Z}n$, which are never present in large proportions; $\dot{N}a$ and $\dot{K}$ have been found in some varieties, but they appear to be accidental, and not essential elements of its composition. $\dot{C}a$ is the only base that is always found in large proportions. Besides these substances, analysis has shown that the mineral sometimes contains $\dddot{Al}$. The presence of this substance cannot be explained according to the laws of multiple proportions, for it cannot exist here, except as replacing a certain quantity of the $\dot{R}$ or $\ddot{Si}$, or as a mechanical mixture. If we admit as some chemists do, that it is possible for $\dddot{Al}$ to replace $\ddot{Si}$, we should be obliged to change the equivalent of aluminium. There is no other reason for doing this, and it should not be done on such an unsubstantial foundation. The explanation of a certain quantity of $\dddot{Al}$ as a mixture in the $\ddot{Si}$ is a sufficient and a rational explanation. The species will therefore be divided into two general families. Those containing $\dddot{Al}$, and those which do not. We shall take no occount of the $\dddot{Al}$, which is rarely present and always in very small quantities, never exceeding 13 %, and shall consider it as correct to adopt for the formula of the family $\dot{R}\,\ddot{Si}$.

Pyroxene. $\dot{R}\,\ddot{Si}$. Monoclinic.

It crystallizes as an inclined rhombic prism of 87° 5′, with the inclination of the prism 100° 25′. This difference allows of distinguishing it from Amphibole, one of the angles of which is obtuse and the other very acute. It has four cleavages, two of which are parallel to the faces of the prism and are easy. The third, parallel to the orthopinacoid is less easy, and the fourth, parallel to the clinopinacoid is difficult. In a single variety an easy cleavage is found parallel to the base. Sometimes the fracture is conchoidal, but generally it is lamellar or fibrous, owing to the easy cleavage. Parallel to the vertical faces, it is fibrous. The lustre is always vitreous when it is not decomposed. Transparent, opaque. It has every variety of color, from white, through green to black. There are also colorless, clear transparent green, dark green, brown and black varieties. Streak, white, gray, or greenish. **H.**=5–6. **G.**=3.23-3.5.

Pyr. &c. B. P. Fusibility from 2.5–3.75. The varieties rich in iron give a magnetic scoria. The reactions are very different according to the composition of the varieties. Usually they fuse to a glass darker than the specimen. They are not generally affected by acids.

MALACOLITE.

Lime-Magnesia Pyroxene.

SYN.—Diopside, Alalite, Traversellite, Mussite, White Coccolite.

The colorless varieties are called Malacolite or Diopside. Their color is white, yellowish, grayish-white or green, but the color of the crystals is not always uniform. They are usually transparent, or at least translucent. All the images seen through the transparent varieties are lengthened. The part adhering to the rock is generally greener than the point, which is usually colorless. The crystalline form is usually a combination of the rectangular and rhombic prism, with the hemi-orthodome, basal pinacoid, hemi-pyramid and clinodome, *Pl.* III. *Figs.* 7, 8, 9. The crystals are generally long, having the form of a rectangular prism, which predominates over the rhombic prism. They are almost always striated and sometimes rounded. **G.**=3.2–3.38. Composition, for (Ċa, Ṁg) S̈i, Ċa 25.8, Ṁg 18.5, S̈i 55.7.

These slightly colored varieties frequently become bacillary, having a very decided cleavage parallel to the base, and the crystals are often very much flattened in the direction of the orthopinacoid. In masses the cleavages and structure produce striations in every direction. The cleavage parallel to the basal pinacoid, when the crystal is very much flattened, is shown in steps; for this reason these lamellar masses are easily recognized. This variety exists also in granular masses disseminated in limestone. In order to distinguish it from the Amphibole family, recourse must be had to the measurement of the angles of cleavage. From its form Malacolite might be confounded with some varieties of Epidote, but it can be distinguished because it is regular and symmetrical, while the crystals of Epidote though in the same system are not so much so; they are generally lengthened in the direction of the orthodiagonal.

FORMULÆ OF THE CRYSTALS.

Pl. III.

Fig. 7. ∞P∞. ∞P̀∞. ∞P. 2P. -P. P. 0P. *Fig.* 8. ∞P∞. ∞P̀∞. ∞P. ∞P3. P. 2P. *Fig.* 9. ∞P. ∞P∞. 2P. P. 0P. P̀∞. 2P̀∞.

SAHLITE.

Lime-Magnesia-Iron Pyroxene.

SYN.—Baicalite.

Some of the green varieties of Pyroxene are called Sahlite. They are intermediate in color between Malacolite and the darker varieties, being usually grayish-green, deep green or blackish-green. It is usually remarkable for its crystalline form, which is generally such a combination of rectangular and rhombic prisms with the basal pinacoid and hemi-orthodome, as to seem at first sight to belong to the orthorhombic system, *Pl.* III. *Figs.* 13, 14, 15. It sometimes has a perfect basal cleavage. These green varieties are often found in crystalline masses, among which some crystals can be seen. The fracture of these masses is generally lamellar, but sometimes granular. **G.**=3.25–3.4. Composition, for (Ċa, Ṁg, Ḟe) S̈i, Ċa 24.9, Ṁg 13.4, Ḟe 8, S̈i 53.7.

It is found in granular masses usually of a bottle-green color, which allows of distinguishing them from the Amphiboles, which are generally bluish-green. These masses are made up of grains formed of incomplete crystals. It is impossible at first sight, when there is neither crystal nor cleavage, to be able to distinguish them from Amphibole. If, however, a cleavage or an angle can be found, they can readily be distinguished.

FORMULÆ OF THE CRYSTALS.

Pl. III.

Fig. 13. $\infty P \infty$. $\infty \grave{P} \infty$. ∞P. $\grave{P} \infty$. $0P$. *Fig.* 14 $\infty P \infty$. $\infty \grave{P} \infty$. ∞P. $\grave{P} \infty$; Baicalite. *Fig.* 15. The preceding, with $\infty P 3$. $\infty \grave{P} 3$.

HEDENBERGITE.

Iron-Lime Pyroxene.

Syn.—Bolopherite.

Certain black varieties are called Hedenbergite; these crystals are usually radiated around a centre, or are rectangular and present particularly the faces of the rectangular prism, and have an easy cleavage parallel to the orthopinacoid. They generally grow larger from the centre, and are frequently terminated by two hemi-orthodomes. **G.**=3.5–3.58. Composition, for ($\frac{1}{2}\dot{C}a+\frac{1}{2}\dot{F}e$) $\ddot{S}i$, $\dot{F}e$ 27.01, $\dot{C}a$ 22.95, $\ddot{S}i$ 47 78.

AUGITE.

Lime-Magnesia-Alumina-Iron Pyroxene.

Syn.—Fassaite, Pyrgom.

The black varieties are called Augite. It is almost always a volcanic product. Its crystalline form differs a little from that of the preceding varieties. They are generally the figures given in *Pl.* III. *Figs.* 1–6. *Figs.* 4 and 5, which have a part of the basal pinacoid remaining, are less frequent. The cleavage parallel to the clinopinacoid is very indistinct. The two cleavages parallel to the faces of the rhombic prism are much easier than those parallel to the orthopinacoid, so that the form most readily produced by cleavage is the rhombic prism. The crystals frequently have other smaller faces. These crystals of Augite are sometimes very large. A rare variety called Fassaite is sometimes found with pyramids, *Figs.* 9, 10, 11, 12. It is found in crystalline masses, or as isolated crystals in lavas, basalts and other volcanic rocks. **G.**=3.25–3.5. Composition, for ($\dot{C}a$, $\dot{M}g$, $\dot{F}e$) ($\ddot{S}i$, $\ddot{\bar{A}}l\frac{2}{3}$), $\ddot{S}i$ 44.40–51.79, $\dot{C}a$ 14.0–24.0, $\dot{M}g$ 8.75–21.11, $\dot{F}e$ 4.24–13.02, $\ddot{\bar{A}}l$ 3.38–8.63.

Pyroxene has two kinds of hemitropy, sometimes it takes place in the direction of a normal to the surface of the orthopinacoid; the crystal then appears to be entirely dissymmetric, *Fig.* 16. The second kind of hemitropy does not change the general form of the crystals. It is then produced by a rotation around the normal to the basal pinacoid. In this case, the irregularity is made apparent by striations and irregularities on the vertical faces. It is probably to this hemitropy, that the cleavage parallel to the base, which has been mentioned, is owing. In this case it is not a cleavage properly speaking, but a disaggregation of the hemitrope faces.

FORMULÆ OF THE CRYSTALS.

Pl. III.

Fig. 1. ∞P. $\infty \bar{P} \infty$. $\infty \grave{P} \infty$. P; the most usual form of Augite. *Fig.* 2. The preceding, with $\bar{P} \infty$. *Fig.* 3. The preceding, with $-P$. *Fig.* 4. The preceding, with the base $0P$. *Fig.* 5. ∞P. $\infty \bar{P} \infty$. $\infty \grave{P} \infty$. (P. $\frac{1}{2} \bar{P} \infty$. *Fig.* 6. The combination *Fig.* 1, with $2 P$ and $2 \grave{P} \infty$. *Fig.* 9. ∞P. $\infty \bar{P} \infty$. $2 P$. P. $0P$. $\bar{P} \infty$. $2 \grave{P} \infty$; Fassaite. *Fig.* 10. ∞P. $\infty \bar{P} \infty$. $\pm 2 P$. $\pm P$; Fassaite and Pyrgom. *Fig.* 11. ∞P. $2 P$. $2 \grave{P} \infty$; Fassaite. *Fig.* 12. The preceding, with $\infty \bar{P} \infty$. $3 P$; Fassaite. *Fig.* 16. Twin crystal; composition-face $\infty \bar{P} \infty$.

LHERZOLYTE.

When Pyroxene exists in large masses, it becomes a rock and has received the name of Lherzolyte. It is found largely in the Pyrenees and elsewhere, resembling the compact masses of Amphibole, which are described under the name of *Corneine* or Aphanyte. It is a rock, with a composition very near to Pyroxene, and which has sometimes served as a magma, from which the crystals have been developed in the interior of the mass when the surrounding circumstances were favorable to its production. In their exterior aspect, these masses do not differ much from the Corneine; they are compact, green, black or dark brown, but they resist the hammer much less. From the looks of such a mass, it is generally impossible to say whether it is Lherzolyte or Corneine, but it is quite rare that some portion of the mass does not show a crystal, or is at least granular, in which case the cleavages may be distinguished.

DIALLAGE.

SYN.—Bronzite.

Diallage has until recently been considered as a variety of Pyroxene. It has two cleavages of the rectangular prism; it is lamellar and has a lustre peculiar to itself. Its color is a brownish-green somewhat like bronze, and it is sometimes called from this Bronzite. Some of the varieties are apple-green. It has a changeable lustre. It is frequently found associated with Feldspar, with which it forms a rock called Euphotide. It is almost always found in small lamellæ. A mineral associated with Serpentine has been called Diallage, but it belongs to the Serpentines. It is generally of a bronze color.

HYPERSTHENE.

SYN.—Paulite.

Hypersthene very much resembles Diallage and is found under like circumstances. It has the two cleavages of the rectangular prism and sometimes that of the rhombic prism, but this is more difficult. It has a metallic-glistening lustre on the cleavage faces, and often a play of colors inclining to copper-red. It is always found in large crystals and in larger masses than Diallage, but the crystals are not distinct. It is frequently associated with Labradorite.

Some varieties of Pyroxene have a remarkable peculiarity; they contain a certain amount of water which cannot be driven off except at a very high temperature. This fact has not yet been satisfactorily explained; it is probable that it is sometimes owing to the mode of formation and sometimes to a commencement of decomposition. If a silicate which contains no bases capable of a higher degree of oxidation commences to decompose through the influence of some interior solvent, it frequently happens that the soluble bases carry off with them a part of the silica in solution. A certain quantity of water may thus become fixed in the crystal, without any sensible alteration in the crystalline form. This is all the more likely to take place if there are a certain number of easy cleavages. It is probable that the minerals Diallage and Hypersthene were formed in this way, the changing lustre being probably due to the presence of water.

It sometimes happens that Pyroxene and Amphibole, which are two species very nearly allied, exist together in the same rock so that the characters of these massive specimens may sometimes be quite vague. Besides Amphibole, there are some other substances such as Epidote, Vesuvianite and Garnet, which are found in masses analogous to Pyroxene, but the tint of the color will generally furnish a guide for distinguishing them.

Pyroxene is bottle-green; Amphibole is bluish-green; Epidote is pistachio-green; Vesuvianite and Garnet are yellowish-green, but the blowpipe distinguishes them. Pyroxene and Amphibole melt easily to a colored glass, while Epidote melts with difficulty to a radiated scoria. Vesuvianite melts with intumescence, but Garnet is only slightly fusible in the dark varieties.

Rhodonite. $\dot{M}n\,\ddot{S}i$. TRICLINIC.

SYN.—Fowlerite, Bustamite, Mangankiesel, Kieselmangan, Paisbergite.

It has the same oxygen ratio as Pyroxene. It would seem to be a Pyroxene in which the $\dot{R}$ is $\dot{M}n$, but the crystalline system is different. The crystals are rarely ever perfect and have as yet only been found at Paisberg, Sweden, *Pl.* III. *Fig.* 17. It has an easy cleavage, parallel to the prism and a less easy one parallel to the base. The fracture of the mass is lamellar or fibrous. The fibers are diverging or bent. It is sometimes found in masses, having a saccharoidal, conchoidal or compact fracture, which pass into the other varieties. When it is compact and fibrous at the same time, it frequently breaks up into pyramidal shapes. Lustre, vitreous. Transparent, opaque. Color, light brownish-red, sometimes greenish or yellowish. The color is quite characteristic when the mineral is unaltered. When this color is not caused by the presence of manganese it is only produced by cobalt; but in the latter case there is always a violet tinge which distinguishes it from the red of manganese. It frequently shows on the outside stains of black, brown or pale gray, owing to the decomposition of the silicate, which becomes an oxide, generally Braunite. Streak, white. When massive, it is very tough. **H.**=5.5–6.5. **G.**=3.4–3.68. Composition, $\dot{M}n$ 54.1, $\ddot{S}i$ 45.9.

Pyr. &c. B. P. Blackens and fuses with a slight intumescence at 2.5. In the O. F., it becomes reddish-brown and forms $\ddot{\bar{M}}n$, and then gives the reactions for manganese with the different fluxes. Partially soluble in acids.

It can only be confounded with Rhodochrosite, but it is easily distinguished by the actions of acids, as Rhodonite is only slightly acted on. The hardness is also a sufficient characteristic, as it is much harder than the carbonate. It is found in considerable quantities in Maine, New Hampshire, Mass., and R. I. A variety containing 5% of zinc, called Fowlerite, is found at Stirling, N. J. Bustamite, from Mexico, is a Rhodonite, in which part of the Mn is replaced by lime. It contains 9–15% of $\dot{C}a$.

FORMULÆ OF THE CRYSTALS.

Pl. III.

Fig. 17. $\infty P.\ \ \infty P'.\ \ \infty \breve{P}\infty.\ \ \infty \bar{P}\infty.\ \ 2'P'.\ \ 2{,}P{,}.\ \ 0P.$

Spodumene. $(\dot{L}i^3+\dddot{A}l)\ \dddot{S}i^3$. MONOCLINIC.

It crystallizes as an inclined rhombic prism of 87°. It was considered for a long time a Lithia-Pyroxene, as its crystalline form very much resembles that of Pyroxene. Crystals are usually macled, and show a predominance of the orthopinacoid, with the prism and one or two clinoprisms, *Fig.* 18. Cleavage very easy, parallel to the orthopinacoid. Fracture, uneven. Lustre, pearly, vitreous on the fracture; dull when altered. Translucent, opaque. Color, grayish-green to greenish-white, slightly reddish. Streak, white. **H.**=6.5–7. **G.**=3.13–3.19. Composition, $\dot{L}i$, 6.4, $\dddot{A}l$ 29.4, $\dddot{S}i$ 64.2.

Pyr. &c. B. P. Becomes white and opaque. Intumesces and melts at 3.5 to a white glass, coloring the flame red from the Li. Not acted on by acids.

It is generally found in lamellar masses of a slightly greenish tint, which are translucent on the edges Although resembling Pyroxene, it contains the $\dddot{A}l$ which would ally it to the Feldspars. This identity of form with another mineral, is an example of pleisomorphism, which is isomorphism independent of the chemical composition. It resembles some minerals of the Feldspar and Scapolite families, but its lustre, density, cleavage and lithia reaction distinguish it. It has been found at Norwich and Sterling, Mass., in very large crystals.

FORMULÆ OF THE CRYSTALS.

Pl. III.

Fig. 18. $\infty \bar{P}\infty.\ \ \infty P.\ \ \infty \grave{P}3.\ \ \infty \grave{P}2.\ \ P.\ \ 2P.\ \ 2\grave{P}2.\ \ 0P.$ $2\grave{P}\infty.$

Petalite. $((\dot{L}i,\ \dot{N}a)^3+\dddot{A}l)\ \dddot{S}i^3+3\ \dddot{S}i$. MONOCLINIC.

SYN.—Castorite, Kastor.

It crystallizes as an inclined rhombic prism of 85° 20′, but crystals are exceedingly rare. The usual form is shown in *Pl.* III. *Fig.* 19. It has an easy cleavage parallel to the base and a less easy one parallel to an orthodome. Lustre, pearly on the base, elsewhere vitreous. Translucent in thin plates; opaque. Colorless, or white and gray, with a slight greenish or rosy tint These tints can almost always be seen in some point even of the gray varieties. Streak, colorless. Fracture, splintery or imperfectly conchoidal. **H.**=6–6.5. **G.**=2.39–2.5. Composition, $\dot{L}i$ 33, $\dot{N}a$ 1.2, $\dddot{A}l$ 17.8, $\dddot{S}i$ 77.7.

Pyr. &c. B. P. Heated gently, phosphoresces. Melts on the edges and becomes more transparent. It colors the flame red from lithia, With borax and S. Ph., the coloration is more intense, almost purple. It is not attacked by acids.

It is generally found in lamellar masses which have four cleavages, three of which are parallel to the same straight line and the fourth inclined to it; the last is the easiest of the three. From its composition, it has been called the Lithia-Feldspar. It is easily distinguished from the other Feldspars by its blowpipe characters. Its reaction for lithia allies it to Spodumene; its density is however much higher, its lustre is more vitreous and it is much more fusible. It is a very rare mineral, being found mostly at the island of Utö, in Sweden, and in Bolton, Mass. The variety called Castorite, comes from the island of Elba.

FORMULÆ OF THE CRYSTALS.

Pl. III.

Fig. 19. ∞P. $\infty\bar{P}\infty$. $\infty\grave{P}2$. $\infty\grave{P}\infty$. $-2\bar{P}\infty$. $-\frac{4}{3}\bar{P}\infty$. $0P$. $2\grave{P}\infty$.

Amphibole.

AMPHIBOLE is the name given to a number of minerals, which may be regarded as made up of at least three varieties very near one another. They are quite distinct in their composition and characters, but belong to the same species. They are generally composed of bases, which have only one atom of oxygen, the relation of the oxygen of the base to that of the acid being as 1 : 2. The formula is $\dot{R}\,\ddot{Si}$. Some authors have admitted the relation of 4 : 9. The bases represented by $\dot{R}$, are $\dot{Na}$, $\dot{K}$, $\dot{Ca}$, $\dot{Mg}$, $\dot{Fe}$ and $\dot{Mn}$, which may replace each other in any proportion within the limits of the formula. It is the more or less greater proportion of one or the other, that constitutes the different varieties. Most varieties contain alumina, but generally in small quantities, and replacing the $\ddot{Si}$; such varieties often contain a little Fl, and others show the presence of $\dot{H}$, which has been referred to under Pyroxene. $\dot{Ca}$ is not so universally present as in the Pyroxenes, and in some cases may be almost entirely wanting.

Amphibole. $\dot{R}\,\ddot{Si}$. MONOCLINIC.

SYN.—Hornblende, Tremolite, Grammatite, Calamite, Asbestus, Anthophyllite, Actinolite, Strahlstein, Pargasite, Smaragdite, Richterite, Uralite.

It crystallizes as an inclined rhombic prism of 124° 30′, with an inclination of 103° 12′. There is thus one obtuse and one very acute angle. It has two cleavages, which are parallel to the faces of the prism and which consequently have the same angles, one obtuse and the other acute. These two cleavages are at the same time fibrous and lamellar. In some varieties they are almost exclusively fibrous; in other varieties there is a cleavage parallel to the ortho-and clinopinacoids. Fracture, subconchoidal or uneven. Lustre, vitreous or pearly on the faces of the crystal and cleavage faces ; silky in the fibrous varieties. Transparent, translucent, opaque. The colors of Amphibole vary and present every gradation from colorless to black, passing through green. The black is the extreme limit of both the green and brown varieties, as may be seen by looking through a thin plate of the mineral. Some Amphiboles have a violet color, due to the presence of a little manganese. Streak, white. **H.**=5–6.5. **G.**=2.9–3.4. It is not always easily scratched with the knife. This property, however, is difficult of trial, as the mineral is found almost always in fibers which separate under pressure.

Pyr. &c. B. P. Fusible more or less easily, according to the varieties, giving a more or less colored glass which is usually darker than the specimen. With borax reacts according to the variety. Some varieties are acted on by acids.

TREMOLITE.

Lime-Magnesia Amphibole.

SYN.—Calamite, Grammatite, Nordenskiöldite.

The white varieties are called Tremolite; they contain $\dot{C}a$ and $\dot{M}g$. The $\dot{C}a$ appears to be a necessary constituent, for if the simple silicate of magnesia is crystallized with the proportion $\dot{M}g\ \ddot{S}i$, it takes a form which resembles Amphibole, but which is not the general form of Tremolite. The crystals of Tremolite are generally more or less distorted, and rarely ever terminated. The most usual forms are combinations of the prism, clinopinacoid, orthopinacoid and clinodome, *Pl.* III. *Figs.* 20–24. Sometimes the prism, ortho-, clino-and basal pinacoids and clinodome occur, *Fig.* 24. The faces of the rhombic prism are generally very bright, and striated in the direction of the vertical axis. The basal pinacoids and the domes are generally matt. Colors, white and gray. All the faces which are parallel to the vertical axis usually have a silky lustre. **H.**=5–6.5. **H.**=2.9–3.1. Composition, for ($\dot{C}a$, $\dot{M}g$) $\ddot{S}i$, $\dot{C}a$ 12–15, $\dot{M}g$ 24–26, $\ddot{S}i$ 57–59.

Tremolite generally occurs in crystalline masses, in which no terminations can be distinguished. The arrangement is frequently in diverging crystals. It is usually found imbedded in masses of Calcite or Dolomite. When the latter decomposes, isolated crystals are frequently found in the sand. In some localities the Tremolite is enveloped in a Dolomite which has stains, either black or brown, made by bituminous matter. This same material has been absorbed by, and surrounds the crystals of Tremolite, so that they are often yellow or even black.

The bacillary varieties are sometimes called Grammatite, when the crystals become mere fibers, which are generally divergent; they show the silky lustre and the fibrous structure, which distinguishes them. The variety colored with manganese belongs to this species. It is almost always found in bacillary fibers, with a silky structure. It is quite a rare variety, being only found in mines of manganese, or near manganese minerals. The tint may become slightly greenish owing to the presence of a small quantity of $\dot{F}e$, partially transformed into $\bar{F}e$; for the protosalts of iron, which are pure, are generally colorless.

FORMULÆ OF THE CRYSTALS.

Pl. III.

Fig. 20. ∞P. $\infty \bar{P} \infty$. $\infty \grave{P} \infty$. $0P$; usual form of Tremolite. *Fig.* 21. ∞P. $\grave{P} \infty$. *Fig.* 22. The preceding, with $\infty \grave{P} \infty$. *Fig* 23. The combination *Fig.* 21, with $\infty \bar{P} \infty$ and 0P. *Fig.* 24. The preceding, with $\infty \grave{P} \infty$.

ACTINOLITE.

Lime-Magnesia-Iron Amphibole.

SYN.—Strahlstein, Actinote.

The green varieties are called Actinolite or Actinote. They are rarely found in complete crystals; when they are the basal pinacoid without the clinodome is the usual form, *Pl.* III. *Fig.* 20. The faces of the rhombic prism are sharper and less fibrous, though the fracture parallel to them is fibrous. It is often found in talcose slates in long crystals, which are rarely terminated, but which have very brilliant prismatic faces. They are easily recognized by their very obtuse angle. Their lustre is much brighter than that of Tremolite and the crystals are much more perfect. **G.**=3-3.2. Composition, for ($\dot{C}a$, $\dot{M}g$, $\dot{F}e$) $\ddot{S}i$, $\ddot{S}i$ 55-59, $\dot{M}g$ 9-24, $\dot{C}a$ 9-21, $\dot{F}e$ 3-11.

They are also found in fibers, sometimes interlaced and sometimes placed parallel to each other. It is frequently found in capillary masses, but instead of being divergent as in Tremolite they are usually more or less parallel, and are sometimes interlaced. There is often an insensible passage from Tremolite into Actinolite.

HORNBLENDE.

Lime-Magnesia-Iron-Alumina Amphibole.

SYN.—Pargasite, Basaltische Hornblende.

This variety which is called Hornblende is generally found in volcanic formations. It is easily distinguished by its color and also by its crystalline form and terminations, which are quite different from any of the other varieties, *Pl.* IV. *Figs.* 1-6. One of these forms, *Fig.* 1, appears to be an hexagonal prism with a rhombohedral termination and might be mistaken for a crystal of Tourmaline; it is however made up of the prism, clinopinacoid, hemipyramid and base. A hemitropy occurs parallel to the orthopinacoid, *Fig.* 6. There is no cleavage parallel to the clinopinacoid in Hornblende. Parallel to the rhombic prism, the fracture is lamellar. **G.**=3.05-3.47. Composition is variable; three varieties are recognized according to the quantity of iron contained, $\dot{C}a$ 10-14, $\dot{M}g$ 5-23, $\dddot{A}l$ 5-15, $\dot{F}e$ 3-29, $\ddot{S}i$ 40-55.

It is very often found in lamellar masses, with a very brilliant cleavage parallel to the faces of the prism. It is less fibrous than the paler varieties. These masses are often composed of distinct crystals which are interlaced. The black varieties are also found in formations, which are not volcanic. Thus in Sweden and Norway, large, short crystals are found in the limestone showing faces of the hemipyramid and base. These crystals might be mistaken for Tourmaline, but are easily distinguished by the inequality of the angles, and the cleavage. In the same place green and colorless varieties are also found, with the same forms, but the crystals are rarely ever distinct, their faces being generally rounded. This variety, which is more granular than crystallized and which is scattered through limestone, has been called Pargasite from the name of the locality, Pargas, where it was first found.

FORMULÆ OF THE CRYSTALS.

Pl. IV.

Fig. 1. ∞P. $\infty \grave{P} \infty$. P. 0P; the usual and characteristic form of Hornblende. *Fig.* 2. The preceding, with $2 \grave{P} \infty$. *Fig.* 3. The preceding, with the hemi-clinopyramid $3 \grave{P} 3$. *Fig.* 4. The preceding, with -P. *Fig.* 5. The preceding, with $-3 \grave{P} 3$. *Fig.* 6. Twin crystal, composition-face $\infty \bar{P} \infty$; remarkable for the different terminations of the extremities, one being formed by the faces of the two hemipyramids, and the other by the bases.

CORNEINE.

SYN.—Aphanyte.

Certain compact varieties of Amphibole are called Corneine or Aphanyte. Their color is blackish, greenish, sometimes violet, but never white, with an irregular, compact and sometimes granular fracture, which is often almost saccharoidal. It resists the hammer like horn, and hence the name Corneine, from *cornu* a horn. This material has been referred to Amphibole, because its composition although variable, resembles that of Amphibole, and because Amphibole crystallizes out of certain parts of it. It is probably the matrix of the Amphibole, a sort of magma in which all the elements are united very nearly as they are in Amphibole, and from which the latter has been formed, when the circumstances were favorable for its production. Some of the varieties of a clear color, which are semi-translucent with a scaly fracture, are remarkable for their sonority.

ASBESTUS.

SYN—Amianth, Asbest, Byssolite, Bergleder, Bergkork.

Certain varieties called Asbestus or Amianthus, are the products of the decomposition of Amphibole and Pyroxene. They are generally more or less decomposed, capillary crystals of these two minerals. It is impossible at first sight to decide to which of these species Asbestus belongs; it can only be distinguished by analysis, and this, on account of the partial decomposition, is frequently but a very uncertain guide. Like Pyroxene and Amphibole, it is fusible.

It is found in more or less long fine fibers without elasticity, and sometimes closely joined together. These fibers have a very high silky lustre. Sometimes they are knitted together in every direction, and give to the mass the appearance of pasteboard or leather. It is then called Mountain Paper, Mountain Cork, Mountain Leather or Mountain Wood, according as it seems to resemble one or the other of these substances. When the fibers are very long and fine and do not hold together, they appear like silk, and are called Amianthus. These fibers may be spun into yarn and made into a cloth which is incombustible, and undergoes no change in an ordinary fire. It sometimes occurs as fine fibers covering other crystals, or penetrating them and lining cavities in the rock like a sort of cloth. The fibers are then usually short and fine, like silk. It is frequently found penetrating crystals of Quartz and Pyroxene. In the latter case the fibers are usually in the direction of the length of the crystals, which however, does not necessitate that the composition should be any more Pyroxene than Amphibole. Asbestus sometimes undergoes a partial decomposition, with fixation of a certain quantity of water as in Hypersthene and Diallage. Some of the so-called Asbestus is the variety of Serpentine called Chrysotil.

The two species, Amphibole and Pyroxene, are very much alike, and generally difficult to distinguish. If the material is massive and has no crystalline structure, there is no possible way of distinguishing them, for such masses have very rarely any definite chemical composition. Distinct crystals can easily be distinguished, since their forms are not the same. If the mass is granular, or only imperfectly crystallized, there is but one method of distinction possible, which is the measurement of the cleavage and angles. It is quite easy to distinguish the angles of cleavage of Amphibole, since one of them is acute and the other obtuse, one being 124° and the other 56°. In Pyroxene the angles of the cleavage parallel to the prism are 87° and 93°. The two angles between the prism and the ortho- and clinopinacoids, which are 133° and 47°, are easily confounded without measurement with the angles of the prism of Amphibole. The different appearance, however, of the faces will frequently allow of distinguishing them. As they are not of the same order, they have different lustres. In the white varieties Pyroxene may be distinguished by its cleavage parallel to the base, which does not exist in Amphibole. When the two varieties are found as Asbestus, it will be impossible to distinguish them, unless the analysis shows a difference in composition. Amphibole is found extensively in the U. S.

Beryl Group.

Beryl. $(\frac{1}{2}\,\dot{\bar{B}}e^3+\frac{1}{2}\,\dddot{\bar{A}}l)\,\ddot{S}i^3$. HEXAGONAL.

SYN.—Emerald, Aquamarine, Smaragd.

It crystallizes as an hexagonal prism, with a cleavage parallel to the base. Traces have been also said to have been found parallel to the faces of the prism. It is however almost impossible to determine them. It is always found in crystals or crystalline masses. Its usual form is the hexagonal prism and base, which combination may occur alone. Combinations of the prism and base with pyramids of the first and second orders, are found alone, as in *Pl.* IV. *Figs.* 8 and 9, and together, *Fig.* 10, or with several pyramids of the same order, *Figs.* 11 and 12. The di-hexagonal pyramid and the prism of the second order are sometimes found, *Figs.* 12 and 7. The bright, emerald-green varieties are called Emerald; those which are not highly colored are known as Beryl or Aquamarine. These crystals may be quite large. They are usually more or less distorted and are often cylindrical. In the cross section, however, the hexagonal form can be distinctly seen, outside of which there appears to be a kind of coating made up of diverging fibers, which are generally more opaque than the crystal. There is often a great difference in the color. Crystals sometimes occur in sections joined together with Quartz, *Fig.* 13. In Siberia it has been found as fibers fastened together, forming large hexagonal crystals. These varieties often have a variation in the color, sometimes parallel to the fibers, and sometimes in beds perpendicular to them. Frequently one of the extremities of the crystal is green, while the other is yellow. Fracture, conchoidal, uneven. Lustre, vitreous or resinous; the opaque varieties, however, have no lustre. Transparent, translucent, opaque. The colors of Beryl are very variable; they are green, blue, yellow or colorless. A red variety is sometimes found, which is probably owing to a mechanical mixture of iron. Streak, white. Brittle. **H.**=7.5-8. **G.**=2.63-2.76. Composition, $\dot{\bar{B}}e$ 14.1, $\dddot{\bar{A}}l$ 19.1, $\ddot{S}i$ 66.8. The relation of the $\dddot{\bar{R}}$ to $\ddot{S}i$ is as 1:2. It sometimes contains $\dddot{\bar{F}}e$; $\dddot{\bar{C}}r$ is found also, but very rarely. The color of the Emerald may sometimes be owing to chromium.

Pyr. &c. B. P. At a high temperature, the edges become rounded. Fuses at 5.5. The colored varieties become white when heated, and lose some thousandths of their weight, which would seem to indicate, that their color is due to organic matter. With borax, a slightly green bead is obtained. Not acted on by acids.

The earthy varieties are usually found in large columnar crystals, which may be hexagonal or cylindrical. The color of these masses is rarely uniform. Some parts are whitish or yellowish, while others are greenish and translucent. These varieties are usually striated parallel to the prism and frequently show indications of cleavage parallel to the base. It is sometimes found passing into porcelain clay. The cause of this decomposition is not known. On account of its hexagonal form, Beryl might be confounded with Apatite or Tourmaline. It is, however, much harder. It might resemble Topaz and Euclase, but is distinguished by its imperfect cleavage. In earthy or stony varieties, it cannot be confounded with anything, on account of its hexagonal form and cleavage. It might possibly at times be confounded with some of the earthy varieties of Topaz, but it is distinguished by its density; the density of Topaz is 3.5, while that of Beryl is 2.6. The rolled crystals of the colorless or slightly colored varieties might easily be mistaken for Quartz, but can be distinguished by the cleavage and fracture. On the island of Elba, crystals of Beryl are found, which are perfectly colorless, and which are sometimes cut to imitate Diamond. Such stones have much less fire than the Diamond and their lustre is not adamantine; it is simply vitreous. The green and limpid varieties are much sought after by jewelers. They do not have much, if any, fire, but their color gives them value. Most of the Emeralds come from Muso in New Granada. Very large Beryls have been found in the U. S. One from Grafton, N. H., weighed 2,900 lbs. and another weighed 1,076 lbs.

FORMULÆ OF THE CRYSTALS.

Pl. IV.

Fig. 7. ∞P. ∞P 2. 0 P; very frequent, the oscillatory combinations often causing vertical striations. *Fig.* 8. ∞P. 0 P. P. *Fig.* 9. ∞P. 2 P 2. 0 P. *Fig.* 10. The combination *Fig.* 8, with 2 P 2. *Fig.* 11. The preceding, with 2 P. *Fig.* 12. The preceding, with 3 P $\frac{3}{2}$. *Fig.* 13. Crystal, consisting of displaced sections joined together by Quartz.

II. Unisilicates.

Chrysolite Group.

Chrysolite. $(\dot{M}g, \dot{F}e)^2 \ddot{S}i$. ORTHORHOMBIC.

SYN.—Olivine, Peridot, Glinkite.

It crystallizes as a right rhombic prism of 94° 2′, and has a cleavage parallel to the brachypinacoid. The usual forms are combinations of the rhombic and rectangular prisms, the macro- and brachydomes and octahedron, with or without the base, *Pl.* IV. *Figs.* 14–18. Fracture, conchoidal. Lustre, vitreous and generally feeble, but much more pronounced on the fracture. Transparent, translucent, opaque. The colors are very different, but are generally various shades of green. There are some varieties, which are yellowish-green, almost colorless; they usually contain $\dot{M}g$. Others are green, with a slight tint of yellow; others again are almost black. Some crystals have undergone alteration, which may be greater or less. They are then red or iridescent. Streak, colorless or yellowish. **H.**=6–7. **G.**=3.33–3.5. Composition, $\dot{M}g$ 50.28, $\dot{F}e$ 9.36, $\ddot{S}i$ 40.75. The relation of the oxygen of the base to that of the silica is as 1 : 1. The formula is therefore $\dot{R}^2\ddot{S}i$.

Pyr. &c. B. P. The clear varieties whiten, but are infusible. The dark varieties rich in iron fuse with difficulty to a magnetic globule. Gelatinizes in HCl and $\dddot{S}$.

It is rarely found, like Pyroxene and Amphibole, containing $\dddot{\bar{A}}l$. The bases which it contains are $\dot{M}g$, $\dot{F}e$, $\dot{M}n$ and rarely $\dot{C}a$ and $\dot{N}i$. It is here to be remarked that $\dot{Z}n$, which is usually isomorphous with $\dot{M}g$, is distinctly separated from it in its silicate. There is a silicate having the formula $\dot{Z}n^2\ddot{S}i$, called Willemite, but it crystallizes in the hexagonal system. Its density varies; in the dark varieties, especially in those which crystallize in the iron scorias (Fayalite), the density may be as high as 4.2. Its hardness is also very variable, the clear varieties being the hardest. It is found usually as little crystals in the rocks of volcanic or igneous origin and in those thrown out of modern volcanoes. Quite large crystals are found in Ceylon. They are, however, generally rolled and it is quite difficult to detect their form. It is also found well crystallized in the cavities of Meteorites and in the scorias of iron forges. It is also found in granular masses, which contain cavities filled with crystals. These masses are sometimes whitish, but generally they are pale yellowish-green. The color may vary from one grain to another and the same grain may be in different parts, green, yellow and black. The lustre is quite bright. Such masses are generally made up of a collection of not very distinct crystals, which are sometimes large. They are usually found in pockets in volcanic rocks, such as Basalts and Trachytes. These grains might be confounded with the grains of Pyroxene, but they are usually of a clearer color, which is generally yellowish or yellowish-green. They may also be distinguished by their infusibility, and by gelatinizing with acids. It is especially in these granular masses, that the red color is found. This is produced by a commencement of decomposition, which is accompanied by a partial solution of the $\dot{M}g$ and $\ddot{S}i$, the fixation of $\dot{H}$ and the peroxidation of the $\dot{F}e$. This kind of Chrysolite forms a peculiar rock. Associated with Augite, it forms Hyalosiderite; in this rock the Augite is in distinct crystals, recognizable by their form and color, which are implanted in the red granular Chrysolite. Crystals of Chrysolite which have undergone such a decomposition as to be iridescent are frequently found in the midst of these crystals. It has been found associated with other minerals in Pennsylvania, Virginia and N. Carolina.

FORMULÆ OF THE CRYSTALS.

Pl. IV.

Fig. 14. $\infty\bar{P}\infty$. ∞P. $\infty\breve{P}\infty$. P. $\bar{P}\infty$. $0P$. *Fig.* 15. The preceding, with $2\breve{P}\infty$. *Fig.* 16. The combination *Fig.* 14, with $2\breve{P}\infty$ and $\bar{P}\infty$. *Fig.* 17. ∞P. $\infty\breve{P}2$. $\infty\breve{P}\infty$. $2\breve{P}\infty$. P. $\bar{P}\infty$. *Fig.* 18. ∞P, $\infty\bar{P}\infty$. $\infty\breve{P}\infty$. P. $\bar{P}\infty$. $2P2$. $2\breve{P}\infty$. $\breve{P}\infty$. $0P$.

Phenacite Group.

Willemite. $\dot{Z}n^2\ddot{S}i$. HEXAGONAL.

SYN.—Troostite.

It crystallizes as a rhombohedron of 116° 1′, with an easy cleavage parallel to the base, and another in the American varieties parallel to the prism. The crystals are usually small, very brilliant and distinct. They rarely show the hexagonal pyramid, although two rhombohedra sometimes occur, *Pl.* IV. *Fig.* 19. They are usually terminated by the rhombohedron. They are usually found in cavities in the compact varieties of Calamine and Smithsonite, sometimes in Hematite and Limonite; rarely they are associated with ores of lead. When found with iron they are usually colored red, yellow or brown. Fracture, conchoidal. Lustre, vitreous, resinous or dull. Transparent, translucent, opaque. Color, white, greenish-yellow, green, black, red or dark-brown when impure. Streak, colorless. Brittle. **H.**=5.5. **G.**=3.89–4.18. Composition, $\dot{Z}n$ 72.9, $\dddot{S}i$ 27.1.

Pyr. &c. B. P. Fuses with great difficulty to a white enamel. With soda in the R. F., it is reduced with difficulty and gives zinc, which is volatilized, burned, and deposits itself as a ring, yellow while hot, and white when cold. Gelatinizes with acids.

In N. J. a variety called Troostite has been found in large crystals, which have the same form as that described. They are rarely translucent, and sometimes have the appearance of having been melted. They occur associated with Franklinite. The outside of these crystals is brownish or reddish. In limestone they are usually of a flesh color. It is easily distinguished from the minerals which it resembles by its hardness, fusibility and action with acids. It is found in the U. S. at Franklin and Stirling, N. J.

FORMULÆ OF THE CRYSTALS.

Pl. IV.
Fig. 19. ∞P2. R. $-\frac{1}{2}$R; often without $-\frac{1}{2}$R.

Phenacite. $\dot{B}e^2 \dddot{S}i$. HEXAGONAL.

SYN.—Phenakit.

It crystallizes in the hexagonal system, having generally a rhombohedral termination, *Pl.* IV. *Fig.* 20. Its crystalline forms are very similar to those of Beryl. It has however no cleavage and its lustre is vitreous. The crystals are usually transparent, but sometimes translucent or opaque. Color, white, yellowish and brownish. Streak, colorless. **H.**=7.5–8. **G.**=2.96–3. Composition, $\dot{B}e$ 45.8, $\dddot{S}i$ 54.2.

Pyr. &c. B. P. Infusible. Gives a blue color with cobalt solution. Not attacked by acids.

It is difficult to distinguish it from Beryl, with which it was for a very long time confounded. The base rarely occurs in Phenacite, while it is very usual in Beryl. It might be mistaken for Quartz, but its terminations are much lower. It is found under the same circumstances as Quartz and Beryl.

FORMULÆ OF THE CRYSTALS.

Pl. IV.
Fig. 20. ∞P2. R. $\frac{4}{3}$R2; the prism is often entirely wanting.

Garnet Group.

Garnet. $(\dot{R}^3)^2 \dddot{S}i^3 + \ddot{R}^2 \dddot{S}i^3$. ISOMETRIC.

SYN.—Granat, Grenat.

Garnet is generally found crystallized. The usual forms are the rhombic dodecahedron and the tetragonal trisoctahedron, *Pl.* IV. *Figs.* 21 and 22. It is quite frequent to find the combination of these two forms, *Fig.* 24, but it is usually the rhombic dodecahedron which predominates. It has a cleavage, that is sometimes distinct, parallel to the rhombic dodecahedron. Crystals are very often distorted as shown in *Pl.* V. *Figs.* 2–5. Fracture, conchoidal or uneven. Transparent, translucent, opaque. Color, red, brown, yellow, white, green or black. Streak, white. **H.**=6.5–7.5. **G.**=3.15–4.31. Composition varies with the varieties. Its composition is exactly the same as that of Vesuvianite, a silicate which crystallizes in a different system. These two substances thus present a remarkable example of dimorphism. It it probable, that the latter is owing to the presence of water, which perhaps plays the part of $\dot{H}$ in Vesuvianite, while Garnet never contains any.

Pyr. &c. B. P. Most varieties fuse to a light brown or black glass and often become magnetic in the R. F., owing to the presence of iron. The dark red varieties are easily fusible to a magnetic scoria, as they contain more iron. The clear varieties are more or less easily attacked by acids. A green variety contains $\ddot{\bar{C}}r$; it is infusible, insoluble and very rare.

Its hardness is very variable. The pale varieties scratch Quartz with difficulty, while the slightly-colored varieties have almost an equal hardness, and the black varieties are scratched by it. Those having a color between clear and dark red are more easily attacked, when they have been melted.

PYROPE.

Magnesia-Alumina Garnet.

SYN.—Bohemian Garnet, Böhmischer Granat.

Pyrope, or Bohemian Garnet, is the one from which carbuncles are cut. Its color is red. Transparent or entirely opaque. Sometimes they have a violet tinge, owing to the presence of a little manganese. **G.**=3.7–3.76. Composition, for $(\frac{1}{2}(\dot{Mg}, \dot{Ca}, \dot{Fe}, \dot{Mn})^3+\frac{1}{2}\ddot{\bar{Al}})^2\dddot{Si}^3$, $\dot{Mg}$ 13.43, $\ddot{\bar{Al}}$ 22.47. $\dot{Ca}$ 6.53, $\dot{Fe}$ 9.29, $\dot{Mn}$ 6.27, $\dddot{Si}$ 42.45.

GROSSULARITE.

Lime-Alumina Garnet.

SYN.—Essonite, Cinnamon Stone, Grossular, Kaneelstein, Grossulaire.

Color, white, pale green, yellow, brown, but rarely green. **G.**=3.4–3.7. Composition, for $(\frac{1}{2}\dot{Ca}^3\ \frac{1}{2}\ddot{\bar{Al}})^2\dddot{Si}^3$, $\dot{Ca}$ 37.2, $\ddot{\bar{Al}}$ 22.7, $\dddot{Si}$ 40.1. $\dot{Fe}$ is sometimes present, replacing the $\dot{Ca}$. This variety is usually transparent.

ALMANDITE.

Iron-Alumina Garnet.

SYN.—Almandine, Edler Granat.

This variety is the oriental Garnet. When the color is a fine deep red, it is used as an ornament. It may also be brownish-red and translucent. Composition, for $(\frac{1}{2}\dot{Fe}^3+\frac{1}{2}\ddot{\bar{Al}})^2\dddot{Si}^3$, $\dot{Fe}$ 43.3, $\ddot{\bar{Al}}$ 20.5, $\dddot{Si}$ 36.1. It is often opaque, and is one of the varieties most frequently found.

SPESSARTITE.

Manganese-Alumina Garnet.

Syn.—Spessartine.

Color, dark hyacinth-red, violet, or brownish-red. **G.**=3.7–4.4. Composition, for $(\frac{1}{3}(\dot{M}n, \dot{F}e)^3+\frac{1}{3}\dddot{A}l)^2 \dddot{S}i^3$, $\dot{M}n$ 30.96, $\dot{F}e$ 14.93, $\dddot{A}l$ 18.06, $\dddot{S}i$ 35.83.

ANDRADITE.

Lime-Iron Garnet.

Syn.—Allochroite, Melanite, Aplome, Colophonite, Topazolite, Kalkgranat, Grenat résinite, Rothhoffite, Polyadelphite.

Color, topaz or greenish-yellow, apple-green, brownish-red, yellow, grayish, dark green, brown, black. This includes the varieties, Topazolite from its resemblance to Topaz in color, Colophonite, which is granular, iridescent and resinous, Melanite or black Garnet. Melanite is sometimes of volcanic origin. Some varieties contain $\ddot{T}i$. **G.**=3.64–4. Some of these varieties have part of the $\dot{F}e$ replaced by $\dot{M}n$. Such are the varieties called Rothoffite, Polyadelphite and Aplome.

OUVAROVITE.

Lime-Chromium Garnet.

Syn.—Uwarowit.

Color, emerald-green. **H.**=7.5. **G.**=3.41–3.52. Infusible. In some varieties part of the $\dddot{C}r$ is replaced by $\dddot{A}l$. Composition, $(\frac{1}{3}\dot{C}a^3+\frac{1}{3}\dddot{C}r)^2 \dddot{S}i^3$. It is a very rare variety, found only in the Urals in fine crystals. The crystals from Canada are very small.

Garnet has sometimes a peculiar relation to its gangues. It is often found in schists or in schistose and micaceous formations. In these cases it appears soldered to the mass, the latter being bent around the crystals, which are usually isolated. The rocks present the appearance then of being covered with tubercles, which are sometimes very large. On breaking them, the mass is usually found to be composed only of a thin crust of Garnet, which has the form of a rhombic dodecahedron. The interior is composed of the same substance that envelops it, either schist or Mica. We shall have occasion to refer to this property again, under Andalusite and Leucite. It is also found in large granular masses, formed of uncrystallized grains of a variable color, which are often iridescent. The natural faces of these grains have a lustre, which is in no way remarkable, but their fracture has a resinous lustre, which is quite easy to distinguish. Their usual color is red, inclining to brown. It is also found in large masses having a resinous lustre, which appear to be almost compact. It also occurs in irregular grains, associated with Calcite. These grains may pass insensibly into crystals, which become scattered through the limestone. They usually have a resinous lustre, and are frequently iridescent. They are called Colophonite, and are found particularly at Willsborough, N. Y. These granular masses can be easily broken up with the fingers. They might be confounded with Vesuvianite, as their colors are about the same; their lustre will however distinguish them. It is sometimes found in amorphous masses, which are generally fragments of large crystals. These masses are usually very much fissured, with traces of a cleavage parallel to the rhombic dodecahedron; their lustre will serve to distinguish them. The transparent varieties are much used for ornamental purposes. The coarse varieties are sometimes pulverized and used as a substitute for emery. In some cases, when it has been found abundantly, it has been used as a flux. It is a very common mineral, and is abundantly found in the U. S.

FORMULÆ OF THE CRYSTALS.

Pl. IV.

Fig. 21. ∞O; characteristic and most frequent form. *Fig.* 22. $2O2$. *Fig.* 23. $3O\frac{3}{2}$. *Fig.* 24. $2O2$. ∞O. *Fig.* 25. ∞O. $3O\frac{3}{2}$. *Pl.* V. *Fig.* 1. ∞O. $2O2$. $3O\frac{3}{2}$. *Fig.* 2. ∞O; lengthened diagonally. *Fig.* 3. ∞O; lengthened vertically. *Fig.* 4. ∞O. $2O2$; distorted and only having four planes of the dodecahedron. *Fig.* 5. ∞O; flattened diagonally.

Vesuvianite Group.

Zircon. $\ddot{Z}r\ \dddot{S}i$. TETRAGONAL.

SYN.—Hyacinth, Jacinth, Jargon.

Zircon is always crystallized. Its usual form is composed of the two prisms and octahedron, combined sometimes with the dioctahedron, *Pl.* V. *Figs.* 6–16. All of these forms with several different octahedra on the same crystal are sometimes found, *Fig.* 12. It has two quite difficult cleavages, one parallel to the faces of the prism, and the other parallel to the faces of the octahedron. When the crystals are partially decomposed these cleavages become easier. Its fracture is conchoidal and brilliant. Its lustre is adamantine, inclined to be resinous. It may even be entirely dull. Transparent, translucent, opaque. Color, colorless, pale yellow, brownish yellow, yellowish-green, reddish-brown, gray or blue. In the crystalline rocks there is frequently found a brown variety, which shows the prism with the principal octahedron and others, in which the prism is no longer present, but three octahedra, one above the other, at different angles, *Fig.* 12. The varieties found in the volcanic rocks may be white, brown, or red. These are called by the jewelers Hyacinths (to be distinguished from the Garnet also called Hyacinth), Jacinths and Jargons. These generally have the prism of the second order with the principal octahedron of the first order, *Fig.* 7. The prism is sometimes reduced so as to become rhombic in shape; the crystal then resembles a rhombic dodecahedron, but it can be distinguished from it both by the angles and the striations parallel to the base. Streak, colorless. **H.**=7.5. **G.**=4.05–4.75. Composition, $\ddot{Z}r$ 67, $\dddot{S}i$ 33. It sometimes contains 1–2% of $\ddot{F}e$, which perhaps replaces the $\ddot{Z}r$. It scratches Quartz, but the partially decomposed varieties have the same hardness as Quartz.

Pyr. &c. B. P. It is infusible. The red varieties lose their color without losing their transparency, and the dark colored varieties become white. It is possible, therefore, that the color is due to organic matter. It is not acted upon by acids, but is decomposed by fusion with alkaline carbonates.

It is frequently found as crystals disseminated in some of the older rocks. It is also found as little crystals disseminated in the rocks and sands of modern volcanoes, particularly of Mt. Somma, where it is frequently accompanied by Nephelite and the vitreous variety of Feldspar. It is also found in the beds of streams in volcanic countries as rolled crystals, having been detached from the rock by the action of water. The red varieties might be mistaken for the ruby Spinel or Garnet. Its form and hardness will distinguish it from either. It is found abundantly in New York, New Jersey, Pennsylvania, and N. Carolina.

FORMULÆ OF THE CRYSTALS.

Pl. V.

Fig. 6. ∞ P. P; most frequent form, P sometimes predominating. *Fig.* 7. ∞ P ∞. P; usual form of the Hyacinth. *Fig.* 8. ∞ P. ∞ P ∞. P. *Fig.* 9. ∞ P ∞. ∞ P. P. *Fig.* 10. The combination, *Fig.* 6, with 3 P 3. *Fig.* 11. The combination *Fig.* 8, with 3 P and 3 P 3. *Fig.* 12. P. 2 P. 3 P. ∞ P. 3 P 3. ∞ P ∞; from Miask. *Fig.* 13. ∞ P ∞. P. 2 P. 3 P. ∞ P. *Fig.* 14. The combination, *Fig.* 7, with 3 P 3. *Fig.* 15. P ∞. 3 P 3. P. *Fig.* 16. ∞ P. ∞ P ∞. P. P ∞. 3 P 3. 4 P 4. 5 P 5; from the Saualpe.

Vesuvianite. $(\frac{3}{5}(\dot{C}a, \dot{M}g, \dot{F}e)^3 + \frac{2}{5}(\dddot{A}l, \dddot{F}e))^2 \ddot{S}i^3$. TETRAGONAL.

SYN.—Idocrase, Vesuvian, Egeran, Wiluite, Cyprine.

It crystallizes as a square prism, with traces of cleavage parallel to the prisms of the two orders and the base. The simple forms of Vesuvianite are the prism of the first order, with traces of the second, the octahedron of the first order and the base, *Pl.* V. *Figs.* 17, 18. Complete octahedra are very rare. They often show the octagonal prism, *Fig.* 19. The dioctahedron is frequently found, *Pl.* VI. *Figs.* 1, 2; and also as many as five dioctahedra of different orders. The number of prismatic faces becomes so great at times, as to make the crystal appear almost round. They almost always show the octahedron of the first order. Very often the crystal is made up of a number of single crystals juxtaposed, so that they are striated in the direction of the prism. The faces of the octahedron are without lustre, but those of the base are brilliant. The crystals frequently become very complex; as many as 130 faces have been counted on a single crystal, *Pl.* VI. *Fig.* 3. Its fracture is smooth and conchoidal, but sometimes uneven, and the edges of the scales are sharp. The lustre of the fracture is bright. On the natural faces it is unequal and varies with the faces. The lustre of the base is the brightest, and is even like a mirror. Generally it is vitreous or resinous. Translucent, opaque. Colors, green, reddish-brown and yellow, which latter color is always of a peculiar tint; sometimes it is pale blue. The varieties which are colored brown with $\dddot{F}e$, do not show dichroism; the other varieties usually do. Perpendicular to the base, they are brownish-green; parallel to the base, bottle-green. Streak, white. **H.**=6.5. **G.**=3.49–3.45. Composition, $\dot{C}a$ 27–38, $\dot{M}g$ 0–10, $\dot{F}e$ 0–16, $\dddot{A}l$ 10–26, $\ddot{S}i$ 35–39. In all the varieties of Vesuvianite, whether altered or not, the analysis shows a small quantity of water varying from 1-2 %. Magnus has shown that this water is perfectly pure. It is evidently in the mineral in a state of combination, for it does not separate, except at a very high temperature, about that of the fusion of silver, but it is not known in what condition it is. It is possible that it plays the part of a base and replaces some one of the $\dot{R}$.

Pyr. &c. B. P. Fuses at 3, with intumescence, to a more or less colored globule. It is only partially attacked by acids, except when it has been fused. This is true of a large number of silicates, and to a certain extent of $\dddot{S}i$ itself.

Some varieties of Vesuvianite are bacillary and very much resemble Epidote, but they can always be distinguished because the bases of the *baguettes* in Epidote are curved, while in Vesuvianite they are plane and brilliant. It is also found in granular masses, made up of grains of unequal size interlaced. They might be confounded with the granular varieties of Chrysolite and Pyroxene. The blowpipe distinguishes it from Chrysolite, which is infusible, while Vesuvianite is fusible. For Pyroxene and Epidote the distinguishing marks are vague, and drawn from the distinction of colors. Pyroxene is dark bottle-green and Vesuvianite inclines more to yellow; Epidote is more of a pistachio-green color. Its colors may be brown, or reddish also, and it then becomes difficult to distinguish it from Garnet. A test must then be made to determine the presence or absence of water, which is found in Vesuvianite, but not in Garnet. Vesuvianite in mass is always difficult to determine in a single specimen. When the rock is found in place, however, its determination becomes easy, because there is always some part which is sufficiently well crystallized to distinguish the form. When it is in small indistinct crystals it resembles Garnet, but it can always be distinguished because it has square or rectangular faces which very rarely occur in the latter, the faces of which are almost always triangular or rhombic. A blue variety of Vesuvianite containing copper is called Cyprine; it is found associated with a rose-colored Epidote. It is bacillary and lamellar, and is one of the mineralogical curiosities. In this country Vesuvianite has been found in Maine, Massachusetts, New York, New Jersey, and elsewhere.

FORMULÆ OF THE CRYSTALS.

Pl. V.

Fig. 17. ∞P. ∞P∞. P. 0P; form of Wiluite. *Fig.* 18. The preceding, with P∞. *Fig.* 19. ∞P∞. ∞P. ∞P2. P. 0P. *Pl.* VI. *Fig.* 1. ∞P∞. ∞P. ∞P3. P. P∞. 0P. 3P. 3P3. *Fig.* 2. ∞P. ∞P∞. ∞P2. ∞P3. 0P. P. 3P. P∞. 2P∞. $\frac{3}{2}$P3. 3P3. 4P2. *Fig.* 3. ∞P. ∞P∞. ∞P2. ∞P3. 0P. P. 2P. 3P. P∞. $\frac{3}{2}$P3. 2P2. 4P2. 3P3. 5P5. *Fig.* 4. ∞P. ∞P∞. 0P. P. 3P. P∞. 3P3; from the Tyrol. *Fig.* 5. ∞P. ∞P∞. ∞P2. 0P. $\frac{1}{3}$P. $\frac{1}{3}$P. *Fig.* 6. ∞P. ∞P∞. 3P. P; from the Urals.

Epidote Group.

Epidote. $\left(\frac{1}{3}\dot{C}a^3+\frac{2}{3}(\dddot{\bar{F}}e, \dddot{\bar{A}}l)\right)^2 \dddot{S}i^3$. MONOCLINIC.

SYN.—Pistacite, Zoisite, Piedmontite, Thulite, Bucklandite pt.

It crystallizes as an inclined rhombic prism, with an easy cleavage parallel to the base, and also one parallel to the orthopinacoid, which gives it a fracture that is lamellar in one direction and sometimes fibrous.

Owing to the extension of the ortho- and clinopinacoids, the crystals are usually rectangular. The usual forms are shown in *Pl.* VI. *Figs.* 7–11. They are often rounded, owing to the presence of a number of hemi-orthodomes, *Fig.* 9. Twins formed by a hemitropy parallel to the orthopinacoid often occur, and are easily recognized by the reëntrant angle on the base. Sometimes, however, the base predominates, and then there is only a line in the place of the reëntrant angle; the striations will serve to place the crystal. Fracture, uneven. Lustre, vitreous, but variable on the different faces, sometimes pearly or resinous. Transparent, translucent, opaque. Color, pistachio-green, yellowish or brownish-green, black or greenish-black, gray, greenish-white; sometimes violet and sometimes clear red. Generally Epidote is of a more or less dark-green color, usually of a peculiar pistachio-green slightly yellowish. Some yellowish varieties are dichroic. They are reddish-yellow in one direction and greenish-yellow in the other. There are white varieties, but there is some doubt whether they are really Epidote. There are also violet varieties colored by manganese. **H.**=6–7. **G.**=3.25–3.5. Composition, $\dot{\mathrm{C}}\mathrm{a}$ 16–30, $\ddot{\mathrm{A}}\mathrm{l}$ 14–28, $\ddot{\mathrm{F}}\mathrm{e}$ 7–17, $\ddot{\mathrm{S}}\mathrm{i}$ 36–57. The formula which has been given refers to the varieties which are distinctly crystallized. When it has begun to decompose it often contains a certain proportion of water.

Pyr. &c. B. P. Fuses at 3–3.5, with intumescence, to a radiated scoria, which is often magnetic. It is difficult of fusion, however, especially in the clear varieties. With borax, it melts more easily, giving generally a dark green bead and the reactions for iron. Manganese varieties react for Mn. Partially soluble in acids. When ignited gelatinizes with acids.

It might at first sight be confounded with Pyroxene, as it has a similar form; but the type is different as it is generally lengthened in the direction of the orthodiagonal, while the crystals of Pyroxene are lengthened parallel to the faces of the prism. It is sometimes difficult to place the prism, but there is always such a want of symmetry in the crystals of Epidote as readily to distinguish it from Pyroxene. Instead of the faces of the orthodome, it may have a series of hemipyramids, which do not exist in Pyroxene and which make the want of symmetry perfectly evident. It is frequently found in bacillary masses, which may also be distinguished from Pyroxene by their form. The *baguettes* of Epidote seem to be made up of a large number of faces, so that they generally appear curved. The striations parallel to the intersection of the faces of the crystal may be easily seen and are all the more remarkable, as the crystals are often hemitrope parallel to the orthopinacoid. Frequently the bacillary masses are not terminated. It is then more difficult to determine them and to distinguish them from Pyroxene, but this last mineral has a number of cleavages, while Epidote has but two. However, with a little practice, the distinction can be easily made with the shades of color, the green of Pyroxene being generally bottle-green, while that of Epidote is more yellow. Epidote is also found in granular masses, having a granular fracture with very fine grains, when it can be easily distinguished by its color.

The violet variety of Epidote called Piedmontite is only found in mines of manganese associated with other minerals of this metal, and more especially with Braunite. Its ordinary gangue is Quartz. It is almost always in bacillary masses, showing some distinct crystals which are rarely terminated. Its peculiar color is generally sufficient to distinguish it, which is a very dark violet, inclining towards brown. When the manganese is in smaller proportion it is paler, but it is always violet. A special species is often made of the white variety of Epidote, which is called Zoisite. It is probably not the same substance as Epidote. As complete crystals of this substance are especially rare, it has been very little studied. It is generally found in large bacillary masses in quartzose gangues. It somewhat resembles Amphibole, Pyroxene and white Cyanite. It is distinguished, however, because it has but one cleavage, which is in the direction of the length and because it is lamellar, while Amphibole has two easy cleavages and Pyroxene four in the direction of its length and one across. Cyanite has one very easy cleavage in length, and one difficult across, but quite distinct. Its association with Quartz is also characteristic.

A rose variety called Thulite, is found associated with Cyprine: it is very rare. It is not known to what the rose color is owing. It is always crystallized or crystalline. Its color might cause it to be confounded with some of the cobalt minerals, but its hardness will distinguish it.

Large crystals of Epidote, which are generally hemitropes parallel to the vertical axis, have been found at Warren and Franconia, N. H. Crystals are found at Haddam, Conn; it is found in acicular masses, or massive, in New York, New Jersey, Pennsylvania and Lake Superior.

FORMULÆ OF THE CRYSTALS.

Pl. VI.

Fig 7. $0P$. $\infty\bar{P}\infty$. $\bar{P}\infty$. P; from the Urals. *Fig.* 8. $\infty\bar{P}\infty$. $\bar{P}\infty$. $-\bar{P}\infty$. ∞P. $\grave{P}\infty$. *Fig.* 9. $0P$. $\infty\bar{P}\infty$. $-\bar{P}\infty$. $2\bar{P}\infty$. $\bar{P}\infty$. ∞P. P. $\grave{P}\infty$. *Fig.* 10. ∞P. P. $\grave{P}\infty$; this combination is remarkable for not being lengthened in the direction of the orthodiagonal. It was formerly called Bucklandite. *Fig.* 11. Twin crystal; composition-face $\infty\bar{P}\infty$.

Iolite Group.

Iolite. $2(\dot{Mg}, \dot{Fe})\ddot{Si} + \ddot{\bar{Al}}^2\ddot{Si}^3$. ORTHORHOMBIC.

SYN.—Cordierite, Dichroite, Steinheilite, Hard Fahlunite, Luchssapphir, Wassersapphir, Saphir d'eau.

It crystallizes in the right rhombic prism of 119° 10′. It is therefore one of the limit forms, the modifications of which resemble those of the hexagonal system. It is generally found in prisms, which show the brachypinacoid and are consequently hexagonal in form, *Pl.* VI. *Fig.* 12. Sometimes the prism, a brachyprism and both pinacoids give it the appearance of two hexagonal prisms, *Figs.* 13, 14. Sometimes the pseudo-hexagonal prism and pyramid are found together. As many as three pseudo-hexagonal pyramids of different angles have been found on the same crystal, *Fig.* 14. If the octahedron and brachydome have the same apparent inclination, as is often the case in these limit forms, it is at first sight difficult to distinguish the crystal from an hexagonal pyramid. It has one cleavage parallel to the brachypinacoid, and another which appears to be parallel to the base. But this last cleavage may be owing to sheets of Mica, which are often interposed parallel to the base. The fracture is unequal, sometimes lamellar in appearance, owing probably to the pseudo-cleavage parallel to the base. The lustre is not very bright on the natural faces, but on the fracture it is vitreous, inclining to resinous.

Color, various shades of blue. The large crystals appear black, or almost black, on their natural faces. The fracture is blue, with greenish reflections, but the color is not usually uniform through the entire mass. In cut stones the color appears to be dark blue perpendicular to the base; in a direction perpendicular to this, it is a fallow-brown. In the other direction perpendicular to both, it appears to be brown, passing to green. From its having two colors it is often called Dichroite, but as it shows trichroism it should have been called Trichroite. Streak, colorless. **H.**=7–7.5. **G.**=2.56–2.67. Composition, $\dot{M}g$ 8.8, $\dot{F}e$ 7.9, $\ddot{\bar{A}}l$ 33.9, $\ddot{S}i$ 49.4.

Pyr. &c. B. P. Loses its transparency, and fuses at 5–5.5. Partially decomposed by acids.

It is found in crystalline masses, which may sometimes be quite large. These are generally formed by the union of a large number of crystals. On account of its crystalline form, which appears to be an hexagonal prism terminated by an hexagonal pyramid, it might be sometimes confounded with Quartz, as it has about the same hardness, but a little attention paid to its change of colors will easily distinguish it, and also the fact that it is fusible on the edges. Iolite may become altered without changing its form. These crystals are often covered over with an exterior crust, which is more or less white and less hard than the mineral. The interior, however, is sometimes not decomposed. One of these products of decomposition is Pinite. It is the result of its almost complete alteration. It has a stony look on the outside, but the interior usually contains a little core or undecomposed mineral. It is generally believed, though it is not positively certain, that all Pinites are the result of the decomposition of Iolite, the reason being that the forms are the same, as well as its associations. It is often found in cylindrical crystals in Granite, and it is not certainly known whether these are the result of the alterations of Iolite, or of some other silicate. Another variety in large hexagonal-shaped prisms is called Gigantolite. This variety shows in the highest degree, the pseudo-cleavage due to the presence of Mica. Iolite is occasionally used as an ornamental stone by jewelers. It has been found in large crystals in Haddam, Conn.

FORMULÆ OF THE CRYSTALS.

Pl. VI.

Fig. 12. ∞P. $\infty\breve{P}\infty$. $0P$. $\frac{1}{2}P$. $\breve{P}\infty$. *Fig.* 13. ∞P. $\infty\breve{P}\infty$. $\infty\bar{P}\infty$. $\infty\breve{P}3$. $\breve{P}\infty$. $0P$. *Fig.* 14. ∞P. $\infty\breve{P}\infty$. $\infty\bar{P}\infty$. $\infty\breve{P}3$. P. $\frac{1}{2}P$. $\frac{1}{3}P$. $2\breve{P}\infty$. $\breve{P}\infty$. $\frac{2}{3}\breve{P}\infty$. $0P$.

Mica Group.

Under the name of Mica a number of minerals are included, which have very analogous characteristics, but which have variable compositions. They contain $\ddot{\bar{S}}i$, $\dot{R}$, $\ddot{\bar{R}}$, $\ddot{\bar{E}}r$ and a little Fl and $\dot{H}$, which varies from .5–3%, but which does not appear to be essential to the composition. Some varieties contain $\ddot{T}i$ and $\ddot{Z}r$. All of these substances are however in very variable proportions. It is impossible to represent them, except approximately by formulæ, although it is necessary to make this approximation, in order to distinguish the varieties. Among the $\dot{R}$, $\dot{L}i$, $\dot{C}s$ and $\dot{R}u$ frequently occur in some of the varieties. Seven different varieties have been recognized. We shall have occasion to discuss only three.

Biotite. $(\frac{1}{3}(\dot{K}, \dot{M}g, \dot{F}e)^3+\frac{1}{2}(\dddot{A}l, \dddot{\bar{F}}e))^2 \ddot{S}i^3$. HEXAGONAL.

It is usually found in hexagonal plates, the crystals showing three or more hexagonal pyramids and sometimes a rhombohedron, *Pl.* VI. *Figs.* 15, 16. Cleavage, parallel to the base. Lustre, splendent, more or less pearly on the base. When black, submetallic. The prismatic faces are vitreous. Transparent, translucent, opaque. Color, generally green to black; in thin lamellæ, green, red or brown. Streak, colorless. **H.**=2.5–3. **G.**=2.7–3.1. Composition, $\dot{M}g$ 4–25, $\dot{F}e$ 0–20, $\dddot{A}l$ 11–21, $\dddot{\bar{F}}e$ 4–25, $\ddot{S}i$ 36–44.

Pyr. &c. B. P. In a closed tube, generally gives a little $\dot{H}$. Some varieties give a reaction for fluorine. Whitens and fuses on the edges. Gives reactions for iron. Decomposed by $\ddot{S}$.

It is found in large crystals at Greenwood Furnace, N. Y., and elsewhere.

FORMULÆ OF THE CRYSTALS.

Pl. VI.

Fig. 15. 0 R. ∞ P 2. $\frac{4}{3}$ P 2; the latter form has a part of the faces wanting. *Fig.* 16. 0 R. R. ∞ R. $\frac{1}{3}$ P 2. $\frac{4}{3}$ P 2. 2 P 2.

Muscovite. $(\dot{K}^3(\dddot{A}l, \dddot{\bar{F}}e))^2 \ddot{S}i^3+1\frac{1}{2} \ddot{S}i$. ORTHORHOMBIC.

SYN.—Common Mica, Biaxial Mica, Potash Mica, Kaliglimmer, Phengit Chrome Mica, Fuchsite, Chromglimmer.

It is found in the crystalline rocks in crystals, which usually have a rhombic or an hexagonal contour, *Pl.* VI. *Figs.* 17, 18, 19. It was thought for a long time that these crystals belonged to the hexagonal system, but they are orthorhombic. The primitive form is a rhombic prism, which has an angle of nearly 120°, producing limit forms. When the crystals are complete, they have six-sided terminations. Octahedra alone, or with a brachydome are rarely found, *Figs.* 20, 21. The edges of the crystals are never perfect, as they are made up of a large number of lamellæ superposed. Complete crystals are very rare. All the varieties of Mica have a very easy basal cleavage. The lamellæ may be separated with the nail, for which reason the crystals are usually flattened in the direction of the cleavage, which is parallel to the base. Lustre, pearly, sometimes submetallic. Transparent, translucent, opaque. Color, white, gray, brown, green, violet, yellow; sometimes rose-red. All of these colors pass into black when they become intense. Streak, colorless. It is always transparent, but it may absorb the greater part of the light or perhaps the whole of it if the color is very dark and if it is not in thin plates. Some varieties show the phenomenon of Dichroism. They are generally those which contain $\dddot{C}r$. These Micas have been very little studied. Muscovite possesses double refraction of one and sometimes of two axes, which has given rise to the distinction of uniaxial and biaxial Micas. This classification has no particular interest, and has a very remote connection with the chemical composition. It has been recently discovered, that by superposing several plates the same species may be made uniaxial or biaxial. **H.**=2–2.5. **G.**=2.75–3.1. Composition, $\dot{K}$ 5–12, $\dddot{A}l$ 31–39, $\dddot{\bar{F}}e$ 1–8, $\ddot{S}i$ 43–50.

Pyr. &c. B. P. In a closed tube, gives off water and reacts for fluorine. Whitens and fuses at 5.5 to a grayish or yellowish glass. Not attacked by acids.

When the plates are very thin they are exceedingly flexible and elastic.

When the Mica has been altered this elasticity is lessened and sometimes entirely lost. These plates are unctuous to the touch, which is probably owing to the presence of Magnesia. In the crystalline rocks it is found in plates, which are often very small, disseminated throughout the entire mass. These little plates do not show any crystalline characters and are probably the result of the destruction of larger crystals. It is usually found in large thin sheets, which have probably been produced by the cleavage of very large crystals. Besides the easy cleavage it has others that are more difficult. Cleavages are found parallel to the vertical faces of the prism and sometimes parallel to the brachypinacoid, so that when a strip of Mica is held up to the light a large number of striations are seen in the direction of the cleavages cutting up the plate, sometimes into rhombs and sometimes into hexagons. Besides the varieties more or less well crystallized, the mineral is found in some curiously distorted shapes. A variety known as Palm Mica shows the plates, which are usually very small, in the position of palm leaves. This position is caused by an apparently systematic arrangement around certain curved lines as axes. It is sometimes found in granular masses, which are almost saccharoidal. This is generally the condition of the Micas, which contain chromium.

It is used in Russia instead of glass. Its most extensive use is for the fronts of stoves and lanterns. It is abundantly found in the U. S.

FORMULÆ OF THE CRYSTALS.

Pl. VI.

Fig. 17. 0 P. ∞ P. $\infty \check{P} \infty$. *Fig.* 18. 0 P. ∞ P. $\infty \check{P} \infty$. P. *Fig.* 19. 0 P. $\infty \check{P} \infty$. 4 P. 2 P. $6 \check{P} 3$. *Fig.* 20. P; a rare form. *Fig.* 21. P. $2 \check{P} \infty$; apparently an hexagonal pyramid.

Lepidolite. $((\dot{K}, \dot{L}i)^3 (\dddot{A}l, \dddot{F}e))^2 \dddot{S}i^3 + 2 \dddot{S}i$. ORTHORHOMBIC.

SYN.—Lithia Mica, Lithionglimmer, Zinnwaldit, Lithionit.

It crystallizes as a right rhombic prism of 120°. It has a perfect basal cleavage. It is sometimes found in large imperfect crystals and in little plates almost granular, and appearing saccharoidal. This variety can be polished. Lustre, pearly. Translucent. Color, rose-red, violet, gray, lilac, grayish-white, white, or yellow. It is to these brilliant colors, which resemble the wings of certain Lepidoptera, that it owes its name. Streak, colorless. **H.**=2.5-4. **G.**=2.84-3. Composition, $\dot{K}$ 4-14, $\dot{L}i$ 1-5, $\dddot{A}l$ 14-38, $\dddot{F}e$ 0-11, $\dddot{S}i$ 42-54.

Pyr. &c. B. P. In a closed tube, gives off water and reacts for fluorine. Fuses at 2-2.5, with intumescence, to a grayish glass, coloring the flame red. Attacked by acids. Gelatinizes after fusion.

It is found in large quantities in Paris and Hebron, Maine.

Scapolite Group.

Wernerite. $(\frac{1}{3}(\dot{C}a, \dot{N}a)^3 + \frac{2}{3} \dddot{A}l)^2 \dddot{S}i^3 + \dddot{S}i$. TETRAGONAL.

SYN.—Scapolite, Paranthine, Glaucolite.

It crystallizes in a right square prism, with cleavages parallel to the faces of both prisms. The prism and octahedron, or the two prisms and octahedra and basal pinacoid are usually found in the small colorless crystals,

Pl. VII. *Figs.* 1, 2, 3. These crystals can generally be distinguished by their form from the substances which accompany them. They are usually found with Orthoclase, Nephelite and vitreous Feldspar. The earthy varieties have generally the same form, the prism of the second order being usually more developed than the others, so as to give the appearance of an octagonal prism. The octagonal prism is also found on the edges of the prism of the first order, *Fig.* 3. It sometimes happens that the prism is surmounted by a six-sided pyramid, formed of four faces of an octahedron and two of a sphenoid, which may be mistaken for an hexagonal form. This hemihedry is frequently seen not on the prism itself, but on the octagonal pyramid. This gives an hexagonal termination on a square prism. The termination is then not symmetric with regard to the prism. This form is quite frequent. Its fracture is usually lamellar on account of its numerous cleavages; otherwise it is conchoidal or uneven. Its lustre is vitreous, pearly or resinous and is very variable. In small colorless crystals it is vitreous; in the large crystals and in lamellar masses, which are sometimes translucent on the edges, it is generally somewhat resinous, but in many varieties it is earthy and dull. Transparent, translucent, opaque. Color, white, gray, bluish, greenish or reddish. The little vitreous crystals are generally colorless and unaltered. In the stony varieties it may be colorless, green or gray. The red color is generally an indication of an advanced state of decomposition. The rose-colored masses however, appear to be natural. Streak, colorless. **H.**=5–6. **G.**=2.63–2.8. Composition, $\dot{N}a$ 5, $\dot{C}a$ 18.1, $\dddot{A}l$ 28.5, $\dddot{S}i$ 48.4. Some varieties contain 2–3 % of water.

Pyr. &c. B. P. Fuses easily, with intumescence and gives usually a spongy scoria, especially if it has not been altered. When it has undergone a commencement of decomposition, the intumescence is much greater, the fusion is easier, and a blebby glass is obtained. It is decomposed by acids.

The earthy varieties are generally decomposed on the surface, and may be covered over with a whitish or reddish coating. When it is well crystallized, it can readily be determined by its crystalline form and by the fact that it is sometimes covered with a partially decomposed coating, which is more or less pulverulent. In cleavable crystalline masses, it is not always easy to distinguish it from the Feldspars. If the specimen is well chosen, it may be recognized by its lamellar and scaly fracture, and resinous and waxy lustre. When, however, these characters are not very well marked, the determination is difficult. The blowpipe then must be used. The intumescence will distinguish it. It has been found in some localities as bacillary radiations and fibrous, but these specimens are very difficult to determine and are only mineralogical curiosities. It has been found in Vermont, Massachusetts, Connecticut, New York and New Jersey.

FORMULÆ OF THE CRYSTALS.

Pl. VII.

Fig. 1. ∞P. ∞P∞. P. *Fig.* 2. ∞P∞. ∞P. P. P∞. *Fig.* 3. ∞P. ∞P∞. ∞P2. P. 0P. *Fig.* 4. ∞P. ∞P∞. P. P∞. 3P. 3P3; from Siberia.

Nephelite Group.

Nephelite. $(\dot{N}a^3, \dot{K}^3)^2 \dddot{S}i^3 + 3 \dddot{A}l^2 \dddot{S}i^3 + 3 \dddot{S}i$. HEXAGONAL.

SYN.—Elæolite, Davyne, Sommite, Eläolith.

Its form is the regular hexagonal prism. It has traces of cleavage parallel to the faces of the prism and base. It is very difficult, however, to produce them, as the crystals are almost always small. Its usual form is the hexagonal prism, with one or more hexagonal pyramids and the base, *Pl.* VII. *Figs.* 5, 6. Its fracture is conchoidal and brilliant. Its natural faces have a vitreous lustre. It is generally transparent and colorless, white or yellowish, but sometimes opalescent. The massive varieties are dark green, greenish, bluish, brownish and brick-red. Streak, colorless or same as color. **H.**=5.5–6. **G.**=2.5–2.65. Composition, $\dot{N}a$ 16.9, $\dot{K}$ 5.2, $\dddot{\bar{A}}l$ 33.7, $\ddot{S}i$ 44.2.

Pyr. &c. B. P. Fuses at 3.5 to a colorless glass. It gelatinizes with acids.

It is generally found in modern lavas, in the Dolomite of Mount Somma, and in the products of a few other modern volcanoes. It is usually accompanied by a number of other crystalline minerals, almost all of which are colorless as it is. These may be Zircon, Anorthite and the vitreous variety of Orthoclase. It can however be distinguished from all these, because its form is hexagonal.

FORMULÆ OF THE CRYSTALS.

Pl. VII.

Fig. 5. ∞P. 0P. P; usual form. *Fig.* 6. ∞P. ∞P2. ∞P$\frac{3}{2}$. 6P. 4P. 2P. P. $\frac{2}{3}$P. $\frac{1}{2}$P $\frac{2}{5}$P. 0P. 4P2.

Leucite Group.

Lapis-Lazuli. $\dot{N}a$, $\dot{C}a$, $\dddot{\bar{A}}l$, $\dddot{\bar{F}}e$, $\ddot{S}i$, $\dddot{S}$, S. ISOMETRIC.

SYN.—Lasurstein, Lasurit, Pierre d'Azur.

It generally occurs in amorphous masses, but crystals have been found which are rhombic dodecahedra. They come from Baikal in Siberia. It shows a trace of cleavage parallel to the face of a rhombic dodecahedron. Fracture, uneven. Lustre, vitreous. Translucent, opaque. Color, azure-blue, violet-blue, red, green; colorless. Streak, same as color. **H.**=5–5.5. **G.**=2.38–2.45. Composition, $\dot{N}a$ 0–12, $\dot{C}a$ 1–23, $\dddot{\bar{A}}l$ 11–43, $\dddot{\bar{F}}e$ 0–4, $\ddot{S}i$ 40–66, $\dddot{S}$ 0–5, S 0–4. Lapis-Lazuli is a mineral of very complex composition for which it is impossible to give any formula. It contains $\dot{N}a$, $\dot{C}a$, $\dddot{\bar{A}}l$, $\dddot{\bar{F}}e$, $\ddot{S}i$, $\dddot{S}$ and some free sulphur. It has been artificially produced and is the ultramarine of commerce. The presence of $\dddot{S}$ and S in this silicate has not yet been explained and it is not known how the iron exists. It is more than probable that it is a mixture. The color is different shades of blue, according to the locality from which it comes. Its hardness is variable and seems to be owing to the substances which accompany it. The hardest variety scratches glass with difficulty.

Pyr. &c. B. P. In a closed tube, gives a little water. Fuses easily at 3, with intumescence and gives a bluish bead. In acids it is more or less easily attacked and gelatinizes, evolving at the same time a little HS. The action of acids is frequently to decolorize it; sometimes it is not attacked by acids, except after calcination.

For a long time Lapis-Lazuli was only found in veins of limestone and was filled with Pyrite, as in Siberia, but within a few years some places have been found in this country where it is associated with Quartz and is consequently much harder than the other varieties. It is much used by jewelers, especially when it contains Pyrite. It was formerly used to make ultramarine, but is now superseded by a cheap, artificial preparation.

Haüynite. $(\frac{1}{4}\,\dot{N}a^3+\frac{3}{4}\,\dddot{\bar{A}}l)^3\,\dddot{S}i^3+\dot{C}a\,\dddot{S}$. ISOMETRIC.

SYN.—Hauyn, Berzeline.

It crystallizes in octahedra and rhombic dodecahedra, *Pl.* VII. *Figs.* 7, 8. Hemitropes parallel to a face of the octahedron, and twins with all the faces parallel also occur, *Figs.* 9, 10. It has a cleavage parallel to the rhombic dodecahedron. Fracture, conchoidal or uneven. Lustre, vitreous. Translucent, opaque. Color, various shades of blue or green. Streak, colorless or bluish. **H.**=5.5-6. **G.**=2.4–2.5. Composition, $\dot{N}a$ 16.5, $\dddot{\bar{A}}l$ 27.4, $\dddot{S}i$ 32, $\dot{C}a$ 9.9, $\dddot{S}$ 14.2. It is remarkable, like Lapis-Lazuli, for the $\dddot{S}$ which it contains.

Pyr. &c. B. P. In a closed tube, suffers no change. Fuses at 4.5 to a white glass. Reacts for sulphur. Gelatinizes with acids.

It is found in basalts and volcanic rocks and is easily recognized by its blue color. Vivianite resembles it somewhat, but is readily distinguished from it by its associations and its easy cleavage.

FORMULÆ OF THE CRYSTALS.

Pl. VII.

Fig. 7. O. *Fig.* 8. ∞O. *Fig.* 9. Twin crystal; composition-face O. *Fig.* 10. Hemitrope; combination O. ∞O.

Leucite. $\dot{K}\,\dddot{S}i+\dddot{\bar{A}}l\,\dddot{S}i^3$. ISOMETRIC.

SYN.—Amphigène.

It is always found crystallized as the tetragonal trisoctahedron. The one usually found has the notation 2 O 2, *Pl.* VII. *Fig.* 11. Very indistinct cleavages are sometimes found parallel to the rhombic dodecahedron. Fracture, conchoidal. It has very little lustre on the natural faces; on the fracture it is vitreous in the transparent varieties. In the others it is opaque and dull. Its crystals are sometimes transparent, but generally opaque as it has often undergone a decomposition towards porcelain clay. It is generally white or grayish, yellowish or reddish white. Streak, colorless. **H.**=5.5–6. **G.**=2.44–2.56. Composition, $\dot{K}$ 21.5, $\dddot{\bar{A}}l$ 23.5, $\dddot{S}i$ 55.

Pyr. &c. B. P. Infusible, Soluble in acids, without gelatinizing. Reduced to powder and treated with water, it gives an alkaline reaction.

It is a volcanic mineral, generally found in lava. It can generally be recognized by its characteristic crystalline form. It is often found in lava, when it will be known from its being white or reddish in a dark colored rock. In the granular varieties, it is rare to find an entire crystal or even the place occupied by one. When the large crystals are broken, it frequently becomes apparent that the outside only is Leucite and that the whole of the interior is made up of lava. This is a phenomenon very analogous to the one we shall presently speak of under Andalusite. The crystalline form will generally be sufficient to distinguish it, but it might be confounded with Analcite which has the same form. This however is fusible and contains water.

Feldspar Group.

Like Amphibole and Pyroxene, the feldspars form a group. Unlike them, however, the name feldspar applies to the group and the varieties are real species, while with Amphibole and Pyroxene the varieties are only peculiar conditions of the same species. The minerals generally known as the feldspars are Anorthite, Labradorite, Oligoclase, Albite and Orthoclase. The entire group, however, comprises a much larger number than this.

All the feldspars are made up of Ṙ, R̈, S̈i; but in the different species the relative proportions of these substances vary. Thus we have for the oxygen ratio in

	Bases.	Ṙ	R̈	S̈i	System of Crystallization.
Anorthite,	Ċa	1 :	3 :	4	5
Labradorite,	Ṅa, Ċa	1 :	3 :	6	5
Oligoclase,	Ṅa, Ċa	1 :	3 :	9	5
Albite,	Ṅa	1 :	3 :	12	5
Orthoclase,	K̇	1 :	3 :	12	4

Thus it will be seen, that these species differ always in the relation of the silicic acid to the bases, the relation between the bases being always 1 : 3. They are very analogous the one to the other and all crystallize in the triclinic system, except Orthoclase.

	∞'P ∧ ∞P'	0P ∧ ∞'P	0P ∧ ∞P'	0P ∧ ∞P̆∞
Anorthite,	120° 31′	110° 40′	114° 6½′	94° 10′ and 85° 50′
Labradorite,	121° 37′	110° 50′	113° 34′	93° 20′ " 86° 40′
Oligoclase,	120° 42′	110° 55′	114° 40′	93° 50′ " 86° 10′
Albite,	120° 47′	110° 50′	114° 42′	93° 36′ " 86° 24′
Orthoclase,	118° 48′	112° 13′	112° 13′	90°

It thus appears that these species of feldspar may be mistaken the one for the other, unless the measurements taken are very exact. The different varieties are analogous in their cleavage also.

	0P	∞P̈∞	∞'P	∞P'
Anorthite,	easy	imperfect	none	none.
Labradorite,	"	difficult	traces.	"
Oligoclase,	very easy	less easy	difficult	none.
Albite,	easy	"	"	traces.
Orthoclase,	"	easy	"	difficult.

All of these slight differences appear to be owing to the fact that the bases Ṙ, although isomorphous, are not entirely in the same conditions. For these variations in the cleavage correspond to differences in the bases that make up the different species, as we shall presently see. The densities of these different species are about the same, but have slight variations.

Anorthite,	2.66—2.78.
Labradorite,	2.67—2.76.
Oligoclase,	2.65—2.69
Albite,	2.59—2·65.
Orthoclase,	2.44—2.62.

In general, it may be said that Orthoclase has potash for its base and Albite, soda; but there is nothing absolute in this, for the analyses of a certain number of specimens of Orthoclase show that it may contain more $\dot{N}a$ than $\dot{K}$, and that Albite may contain more $\dot{K}$ than $\dot{N}a$. But in general the rule holds good and it is probably owing to this, that there is a difference in the forms of the two species. Oligoclase contains principally $\dot{N}a$, with $\dot{K}$ and a little $\dot{C}a$; Labradorite contains principally $\dot{C}a$. Anorthite resembles Oligoclase, but contains more $\dot{C}a$. Oligoclase, Albite and Orthoclase are not attacked by acids, probably on account of the very large proportion of $\ddot{S}i$. Labradorite and Anorthite are completely soluble in acids, gelatinizing. It is difficult to distinguish the symmetrical and dissymmetrical forms in simple crystals, as *Pl.* VII. *Figs.* 14, 15 and *Pl.* VIII. *Figs.* 11, 12, but when the crystals are composite or macled, it becomes easy. In *Pl.* VIII. *Fig.* 14 is shown a crystal of Orthoclase, macled by a hemitropy parallel to the clinopinacoid; *Pl.* VII. *Fig.* 18 is a crystal of Albite, macled in the same way. In the first case the two crystals are perfectly symmetrical and the angle between 0 P and $\infty \check{P} \infty$ is a right angle. In the second case they are not; the angle not being a right angle, where the faces come together, a reëntrant angle is formed on the line of junction. On the opposite faces below there will be a dome. Besides this, the cleavages parallel to the base go through the entire crystal of Orthoclase, which is not the case in Albite, since the two crystals do not correspond. The same is true of the other macled species.

Anorthite. $(\frac{1}{4}\dot{C}a^3 + \frac{3}{4}\dddot{A}l)^2 \ddot{S}i^3$. TRICLINIC.

SYN.—Indianite, Christianite, Amphodelite.

The primitive form is a doubly inclined rhombic prism of 120° 31′. Among the products thrown out by certain volcanoes, especially in the blocks from Dolomite of Mt. Somma, the vitreous variety of Orthoclase is found in small transparent crystals, which look like melted glass. If they are carefully examined with a glass, other crystals are sometimes found which resemble them, but which are not symmetrical and have reëntrant angles. This mineral is the feldspar Anorthite. These crystals of Anorthite are always very small, and often cannot be distinguished except with a glass. They are usually complete and so evidently of a different form, that they are easily distinguished from the crystals of vitreous Orthoclase, which are generally flattened in the direction of the brachypinacoid or are even reduced to the thinness of paper, while the crystals of Anorthite are about equally developed, and appear sometimes almost spherical. The usual forms are shown in *Pl.* VII. *Figs.* 12, 13. Cleavage, parallel to the base and brachypinacoid. Fracture, conchoidal. Lustre, vitreous, but pearly on the cleavage. Transparent, translucent, opaque. Color, white, grayish or reddish. Streak, colorless. **H.**=6–7. **G.**=2.66–2.78. Composition, $\dot{C}a$ 20, $\dddot{A}l$ 36.9, $\ddot{S}i$ 43.1.

Pyr. &c. B. P. Fuses at 5 to a colorless glass. Soluble in HCl, gelatinizing.

FORMULÆ OF THE CRYSTALS.

Pl. VII.

Fig. 12. $\infty P'$. $\infty' P$. $0 P$. $2_{,}\check{P}'\infty$. $2'\check{P}_{,}\infty$. $2'\bar{P}'\infty$. $2_{,}\bar{P}_{,}\infty$. $_{,}\bar{P}_{,}\infty$. $'P$. $P_{,}$. $_{,}P$. $2_{,}P$. *Fig.* 13. $\infty P'$. $\infty' P$. $0 P$. $\infty\check{P}\infty$. $2'\bar{P}'\infty$. $2_{,}\check{P}'\infty$. $2'\check{P}_{,}\infty$.

Labradorite. $(\dot{Na}, \dot{Ca})\,\dddot{Si} + \dddot{Al}\,\dddot{Si}^2$. TRICLINIC.

SYN.—Labrador, Pierre de Labrador.

Labradorite has the same forms and cleavages as the other species of feldspar. Crystals, however, are very rare; it is generally found in lamellar masses, which are highly striated and semi-translucent. Cleavage, easy, parallel to the base, less distinct, parallel to the brachypinacoid and prism. Lustre on the base pearly, elsewhere vitreous. Translucent, opaque. Color, gray, brown or greenish, sometimes colorless. It usually has a bright play of colors on the cleavage faces. Streak, colorless. **H.**=6. **G.**=2.67–2.76. Composition, $\dot{Na}$ 4.5, $\dot{Ca}$ 12.3, $\dddot{Al}$ 30.3, $\dddot{Si}$ 52.9.

Pyr. &c. B. P. Fuses at 3 to a colorless glass. Partially soluble in strong acids, and, when it has been reduced to a fine powder, it gelatinizes.

The lustre of its fracture is very bright, and somewhat resembles that of Orthoclase, but the striations and the solubility with acids distinguish it. Its chief characteristic is its play of colors. It is readily distinguished from a similar variety of Orthoclase by its striations.

Oligoclase. $(\frac{1}{4}(\dot{Na}, \dot{Ca})^3 + \frac{3}{4}\dddot{Al})\,\dddot{Si}^3 + 3\frac{3}{4}\,\dddot{Si}$. TRICLINIC.

SYN.—Oligoklas.

Oligoclase is sometimes found in perfect crystals, which resemble Orthoclase, but differ from it by want of symmetry. These crystals are frequently lengthened in the direction of the macrodiagonal, *Pl.* VII. *Fig.* 14. They are generally stony and do not have much lustre. Cleavage, parallel to the base and brachypinacoid. Fracture, conchoidal or uneven. Lustre, pearly, waxy or vitreous. Transparent, translucent, opaque. Color, white, grayish, greenish; in Gold Stone (Aventurine) reddish. It is usually greenish. Streak, colorless. **H.**=6–7. **G.**= 2.56–2.72. Composition, $\dot{Na}$ 2–12, $\dot{Ca}$ 0.5–5, $\dddot{Al}$ 19–24, $\dddot{Si}$ 59–64.

Pyr. &c. B. P. Fuses at 3.5 to a clear glass. Not acted upon by acids.

Crystals of Oligoclase are quite rare, but lamellar masses are found in a large number of rocks. They can be easily distinguished from Orthoclase by the striations, but it is quite difficult to distinguish them from Albite, since their external characters are nearly the same. When the mass is greenish it is probably Oligoclase, but when the mass is red or white it will be doubtful. The only way to distinguish them will be a chemical analysis. For a long time the massive specimens of Oligoclase were confounded with Albite. It can sometimes be recognized by the minerals which accompany it. Thus, from Arendal in Norway, it is usually associated with green Epidote. It can be distinguished most satisfactorily by a determination of the $\dddot{Si}$, which need not be very exact, for the proportions are very different in the two minerals. Albite contains 68–72%, while Oligoclase has only about 60%. It has been found in New Hampshire, Massachusetts, Connecticut, Pennsylvania, Delaware and elsewhere.

FORMULÆ OF THE CRYSTALS.

Pl. VII.

Fig. 14. $0P$. $\infty\breve{P}\infty$. $\infty'\breve{P}3$. $\infty'P$. $\infty P'$. ${}_{,}P$. $P_{,}$. $2'\breve{P}_{,}\infty$. $2_{,}\breve{P}'\infty$. ${}_{,}P_{,}\infty$. $2_{,}P_{,}\infty$.

Albite. $(\frac{1}{4}\dot{N}a^3 + \frac{3}{4}\dddot{A}l)^2 \ddot{S}i^3 + 6\ \ddot{S}i$. TRICLINIC.

SYN.—Tetartine, Pericline.

The crystalline forms of Albite are very nearly the same as those we shall describe under the head of Orthoclase. Simple crystals are quite rare, *Pl.* VII. *Figs.* 15, 16, 17. There is a very remarkable form of Albite, which shows a great development of the base and hemi-orthodome, *Figs.* 22, 23. It constitutes the variety called Pericline, which is generally milk-white and opaque, with a very bright lustre. They also sometimes contain the faces of a hemi-brachydome. These crystals appear at first sight to be symmetrical, but it will be remarked that the striations on the surface are not continuous, which shows that the crystals are formed of the union of several others. Hemitropes parallel to the base often occur, *Pl.* VIII. *Figs.* 1, 2. The exterior of the crystals of Pericline are often covered with iron and Chlorite. The ordinary form of Albite is the twin parallel to the brachypinacoid, *Figs.* 18, 19. The obliquity between the base and the brachypinacoid produces reëntrant angles on one side, and a dome on the other. This disposition is characteristic of Albite. In the hemitrope crystals the cleavage is never perfect across the crystal. Reëntrant angles are also made by a hemitrope around the normal to the brachypinacoid, sometimes parallel to the base. These crystals are rarely ever formed alone. Albite is generally found in collections of crystals, united to form one large one. A transverse section of a crystal often shows a series of reëntrant angles, analogous to those described under Quartz. Such crystals are composed of others which are generally hemitrope, sometimes in one direction, and sometimes in another. The faces of junction produce striations which are parallel to each other, and which often produce a pearly lustre. Albite has easy cleavages, one parallel to the base, and one parallel to the brachypinacoid. The fracture is lamellar and sometimes conchoidal or uneven. It is often granular in the imperfectly crystallized varieties. Its lustre on the faces is vitreous, but on the fracture it is pearly. Transparent, translucent, opaque. The color is generally white, either clear or milk-white. It is sometimes rosy-red, yellowish, greenish, bluish, grayish. Sometimes it has a bluish opalescence. Streak, colorless. It is lamellar and sometimes even fibrous. **H.**=6–7. **G.**=2.59–2.65. Composition, $\dot{N}a$ 11.8, $\dddot{A}l$ 19.6, $\ddot{S}i$ 68.6.

Pyr. &c. B. P. Fuses at 4 to a clear bead, coloring the flame yellow ($\dot{N}a$). Insoluble in acids.

Albite is found in a number of rocks; associated with Hornblende it forms Dioryte. It is quite common in Gneiss. It is found also in Trachytes and Phonolites. It is frequently found in Porphyries. When it is associated with Orthoclase in Granite it can be readily distinguished by its greater whiteness. Rare minerals and gems are often found in veins of albitic Granite. The lamellar masses of Albite are distinguished by the visible striations produced by hemitropy, which are generally less distinct in the other varieties of feldspars, except Oligoclase. The striations in pencil may also be seen sometimes by turning the crystal a little in the light. It can sometimes be distinguished by the feathered and fibrous disposition of the fracture. The fracture of Albite is also more generally granular than that of Orthoclase. Such characters are however often difficult of observation.

FORMULÆ OF THE CRYSTALS.

Pl. VII.

Fig. 15. ∞'P. ∞P'. ∞'P̆3. ∞P̆'3. ∞P̆∞. 0P. P̄∞. 2P̄∞. P̆∞. P'. ½P'. *Fig.* 16. The preceding, without ½P', but with 'P. *Fig.* 17. Twin crystal. *Fig.* 18. Hemitrope; composition-face ∞P̆∞. *Fig.* 19. Twin crystal, with the combination ∞P̆∞. ∞'P'. ∞'P̆'3. 0P. P̆∞; characteristic form of Albite. *Fig.* 20. Hemitrope as in *Fig.* 19, but the two halves by interpenetration crossing on a vertical line, so that the right quarter in front is continued in the left quarter behind, and the left in front in the right behind. *Fig.* 21. Double twin, formed by juxtaposition, parallel to the vertical axis, of two twins like *Fig.* 20, according to the law of the orthoclase twins, *Pl.* VIII. *Figs.* 21, 22. *Fig.* 22. 0P. ,P,∞. ∞'P. ∞P'. ∞P̆∞; form of Pericline. *Fig.* 23. The preceding, with ,P and ⅔P∞. *Pl.* VIII. *Fig.* 1. Hemitrope; composition-face 0P. As the part revolved in this case is the lower half and the salient angle is on the left, this twin is *left-handed.* *Fig.* 2. The same as the preceding, but *right-handed*, because the upper half is the part revolved.

Orthoclase. $(\frac{1}{4}\dot{K}\,3+\frac{3}{4}\dddot{Al})^2\,\ddot{Si}^3+6\,\ddot{Si}$. MONOCLINIC.

SYN.—Feldspar, Feldspath, Adularia, Adular, Eisspath, Orthose, Pegmatolith, Loxoclase, Ryacolite, Sanidine, Microcline, Amazon Stone.

It crystallizes as an inclined rhombic prism of 118° 48'. The primitive form of Orthoclase, *Pl.* VIII. *Fig.* 3, without modifications is comparatively rare. At first sight it might easily be mistaken for a rhombohedron, but the direction of the striations shows that it is not. The striations of the base are parallel to the orthodiagonal, while those of the prism are parallel to the edges of the base. There is another form made up of the prism and a hemi-orthodome, *Fig.* 4, which might still more easily be mistaken for the rhombohedron, but it can be distinguished from the first by the striations. These crystals are sometimes quite large and their surface is generally covered with Chlorite, which sometimes masks the striæ. In order to find out what form it is, it is sufficient to effect a cleavage, which shows the disposition of the striæ. Besides these forms which belong to detached crystals of Orthoclase, others are found in eruptive rocks. They are generally long crystals as in *Figs.* 9–11. Those of the first type are very often vitreous. They are then very thin in the direction of the clinopinacoid, In the second form, *Fig.* 10, the faces of a clinoprism appear. These faces are generally small and without lustre, which distinguishes them from the others, which are usually brilliant. It is often found in Granite with a lengthened form, *Fig.* 12, in which the rectangular prism predominates. Beside these forms, it is susceptible of macles and hemitropes, which may be produced according to different laws. Sometimes the hemitropy takes place around the face of the clinopinacoid, and then the form *Fig.* 14 is produced. As we have seen before, there is generally a cleavage parallel to the base, but sometimes it happens that the striations do not continue in the same line in the two crystals either parallel to the base or the hemi-orthodome, which shows that the crystal is hemitrope. Sometimes, particularly in the trachytic rocks, the crystal is hemitrope around the clinopinacoid and is then displaced laterally, but the parts penetrate each other very nearly half and half, *Figs.* 20–22. The very lengthened forms, *Fig.* 12, may give rise to two hemitropes. The first, *Fig.* 13, takes place around a normal to the base. In this case, as the angle between the base and the clinopinacoid is 90°, the cleavage parallel to the clinopinacoid goes through the whole crystal. The second hemitrope takes place around a diagonal, *Fig.* 15. The two faces of the clinopinacoid are almost at right angles. In this form the faces of a clinoprism are frequently found. These forms are easily distinguished from Albite; for when the hemitropy is in the first case the cleavage parallel to the clinopinacoid does not go through the crystal. It is sometimes found in very large crystals. It is rarely simple, but is formed of macles of several crystals, according to one or more of the laws just described. These clusters of macles can readily be distinguished by their striations. It has an easy cleavage parallel to the base and clinopinacoid and a less easy one parallel to the orthopinacoid and prism.

On account of the two very easy cleavages, Orthoclase is almost always more or less split, so that it has the appearance of having brilliant faces or mirrors in its interior. Its fracture is generally lamellar; across the crystal it is sometimes conchoidal and sometimes smooth. The lustre of Orthoclase is vitreous, in the transparent crystals; on the cleavage faces it is often pearly. The faces of the base and hemi-orthodome are always less brilliant than the faces of the prism; and the faces of the clinoprism are always without lustre. Transparent, translucent, opaque. The transparent varieties are generally whitish and show mirrors caused by cleavage. Color, flesh-red, white, gray, greenish or bright green. Streak, colorless. **H.**=6–6.5. **G.**=2.44–2.62. Composition, $\dot{K}$ 16.9, $\dddot{Al}$ 18.8, $\dddot{Si}$ 64.6. Some varieties contain as much as 10% of soda. Generally the largest amount is 6%; most varieties contain only a trace.

Pyr. &c. B. P. The colored varieties whiten. In thin scales it is fusible between 4 and 5 to a white glass. With borax, it gives a transparent glass and with S. Ph., a silica skeleton. Not acted on by acids.

The colors of Orthoclase are in general not very decided; they are white, grayish and flesh-color, more rarely rose-red and green. This last variety is used for ornaments under the name of Amazon Stone. A transparent, vitreous variety of Orthoclase is found among volcanic products, generally in Lavas or Trachytes. This variety has been called vitreous feldspar or Sanidine and was for a long time known as Ryacolite, which was supposed to be a separate species of feldspar. They are generally characterized by rounded exterior surfaces, which gives them the appearance of having been melted. It appears, however, to be owing to the circumstances which prevented them from forming completely. Some of these varieties show the hemitropy described, where the crystals have been displaced laterally. They are found in the granitic, trachytic and porphyritic rocks. The lengthened form is also found and the hemitropes to which it gives rise. A variety is found which changes its color, sometimes showing the phenomenon of asteria; the light falling upon it is decomposed and shows a pencil of rays. These varieties always have the cleavage parallel to the clinopinacoid and the base, which form fine parallel striations on the surface, but this phenomenon is exceptional. This variety is called Microcline. Generally the crystals which show a change of color are Labradorite. Beside these crystalline varieties, there are those which are only imperfectly crystallized, and this is generally the case in the Granitic rocks. These varieties are generally in lamellar masses having the rectangular cleavages. They cannot be scratched with the knife, but they scratch glass. Their color is generally darker than that of the crystals. They are red, brown, yellow, green, but generally flesh-color. It is also found in granular and earthy masses. This is owing to a commencement of decomposition, the mineral passing into porcelain clay. Massive Orthoclase is frequently found associated with crystallized Quartz, both crystallizing together. The crystallization of the feldspar has been hindered and has assumed the shape of irregular thin plates. The Quartz is frequently smoky and gives to the mass the appearance of having hebrew characters written upon it. The rock is called Graphic Granite. Geodes filled with perfect crystals are frequently met with in massive feldspar. Sunstone is the name of a variety which is transparent in thin lamellar and has gold-colored Mica scattered through it. Part of the mineral known as Sunstone is Oligoclase.

Orthoclase is the essential constituent of a large number of rocks. The feldspar of the porphyritic and trachytic rocks is generally Orthoclase; sometimes, however, it is Albite or Oligoclase.

FORMULÆ OF THE CRYSTALS.

Pl. VIII.

Fig. 3. ∞P. 0P. *Fig.* 4. ∞P. P̀∞. *Fig.* 5. ∞P. P̀∞. 0P; usual on Adularia. *Fig.* 6. The preceding, but with the base and hemi-orthodome equally developed. *Fig.* 7. The preceding with ∞P̀∞. *Fig.* 8. The preceding, with ∞P∞. P. *Fig.* 9. ∞P̀∞. ∞P. 0P. 2P̀∞; usual form of the Orthoclase in Granite and Porphyry. The same form is shown in *Fig.* 19, in a different position. *Fig.* 10. The preceding with ∞P̀3. *Fig.* 11. ∞P. ∞P̀∞. 0P. 2P̀∞. P̀∞. P. 2P̀∞. *Fig.* 12. 0P. ∞P̀∞. ∞P. 2P̀∞. 2P̀∞; usual form of the rectangular crystals. *Fig.* 13. Twin crystal; composition-face 0P. *Fig.* 14. Hemitrope; composition-face ∞P̀∞. *Fig.* 15. Hemitrope; composition-face 2P̀∞. *Fig.* 16. Projection of a twin crystal similar to the preceding. *Fig.* 17. Double twin crystal, formed in the same way as *Fig.* 15. *Fig.* 18. ∞P̀∞. ∞P̀3. ∞P. 0P. 2P̀∞. P̀∞. ⅔P̀∞. P. 2P̀∞. *Fig.* 19. Usual form of Orthoclase. *Fig.* 20. Twin crystal by juxtaposition upon the face ∞P̀∞. *Figs.* 21, 22. Twin crystals by interpenetration; composition-face ∞P̀∞. *Fig.* 21, is a *left-handed* and *Fig.* 22 a *right-handed* twin, as the left and right halves are respectively the parts revolved.

PETROSILEX.

Syn.—Felsite.

Petrosilex, or compact feldspar, as it is often called, is for the feldspars what Corneine is for Amphibole, and Lherzolyte for Pyroxene. It is a rock of analogous composition with the feldspars and is sometimes their matrix. The composition is variable. $\ddot{S}i$ 70—80, sometimes less, $\dddot{A}l$ 18—20, Alkalies 11—16. According to the proportion of the different elements, Petrosilex has given rise to different feldspars. It is never crystallized and never has any well defined character. Its hardness is about the same as feldspar. It is not acted on by acids. Fusible to an enamel, intumescing slightly, which is owing to 1 or 2% of $\dot{H}$, which it sometimes contains. It has an intermediate form and structure. Its fracture is scaly with sharp edges, which are sometimes translucent. It is often smooth and conchoidal like a coarse Agate. It has a fatty lustre and its color is variable, but generally light. By the hardness it might be confounded with compact Quartz, but its fusibility distinguishes it. It is found in nature under similar conditions with the feldspars of which it sometimes contains crystals. Some varieties of Petrosilex approach Albite, others Orthoclase or Oligoclase. They rarely approach Labradorite. They are often found associated with the Granitic Porphyries, either as nodules or veins.

PITCHSTONE.

SYN.—Pechstein.

Pitchstone, Pechstein, and resinous feldspar are names given to a rock the composition of which is analogous to the feldspars, but which contains a considerable amount of water. These rocks are quite fragile and not so hard as Petrosilex. They are not attacked by acids and only partially attacked by alkalies. They fuse with intumescence and generally whiten, as their color is usually due to organic matter. Their composition is analogous to that of Petrosilex, except that they contain an amount of water varying from 4–8 and even 10%. They have a peculiar, vitreous, lustrous or resinous appearance, which cannot be described. Their fracture is irregular, though generally smooth, but rarely conchoidal or plane. They are usually more translucent than Petrosilex and their colors are generally darker. They are often found associated with feldspar. Owing to their very peculiar and resinous lustre, they can easily be distinguished from the varieties of Opal called resinous Quartz. Their fusibility will however distinguish them in case of doubt.

OBSIDIAN.

Obsidian is a volcanic glass, which has a variable composition, but which approaches that of feldspar. $\ddot{S}i$ 70-80, $\dddot{A}l$ 2-10, $\dot{N}a$ and $\dot{K}$ 3-10, $\dot{C}a$ and $\dot{M}g$ 1-4, $\dddot{F}e$ 0-3, $\dot{M}n$ traces, Volatile matter 0-4. The volatile matter may be water or combustible material. Obsidian is generally of a dark green color, which is owing to the presence of iron, or of a dark brown, owing to organic matter. Its fracture is conchoidal and its lustre vitreous. Parts of it sometimes present the appearance of becoming devitrified; such parts usually contain more lime than the others and are thus more stony, and then are often found as distinct globules in the mass. It is sometimes found stratified in beds of vitreous and stony matter; such specimens have not been thoroughly melted. Obsidian is found in great abundance in volcanic products and in lavas, through which it is often scattered in the shape of globules. *Pele's Hair* is the name given by the natives of the Sandwich Islands, to the volcanic glass from Mauna Loa, which has been spun out into fine filiaments by the wind. A similar product is sometimes found in blast furnaces.

PUMICE.

SYN.—Bimstein, Pearlstone, Spherulite, Perlstein, Perlite.

Pumice and Perlite are two products which are analagous in composition, and in their mode of formation to Obsidian. Some silicates when they are in a fluid state have the property of becoming porous and swelling, so as to resemble sponge. This is the case with some artificial slags. The contact of such silicates thrown out by volcanoes with water produces Pumice. The name Perlite has been given to the same substance, when it is in little globules soldered together. Pumice resembles a very bulby and spongy glass; sometimes it is drawn out into fine fibers, so that the mass looks fibrous, but sometimes it exists as real fibers. It is generally of a white color, but frequently passes into Obsidian, which has about the same mean composition and differs from it only in being compact.

PRODUCTS OF DECOMPOSITION.

Under the influence of atmospheric agencies, the silicates undergo a peculiar decomposition, especially when they contain any bases susceptible of a higher degree of oxidation, such as Fe. In such cases the alkalies, the $\dot{C}a$ and $\dot{M}g$, and even the silica, either by the action of pure water or water charged with $\ddot{C}$, are at times partly dissolved. Besides the action of water, other causes, such as the action of acid vapors, may have produced this decomposition. These vapors sometimes consist of HCl, which appears to be given off in abundance in certain formations, since chlorides are found in many rocks. At other times the decomposition is produced by $\ddot{S}$ or HS, which fill the ground of certain regions. More often however, the decomposition is produced by damp air or by water, which may be water from thermal or other springs, or may be in the state of saturated vapor. There is no silicate, which will not after a certain time be affected by such action. It is generally the feldspathic rocks which are affected in this way, which is the reason why we treat of this subject here. Sometimes, when the decomposition has not gone very far, the structure and cleavage of feldspar can still be seen. Generally however, they are in a more or less earthy condition, quite distinct from the parts which have not been acted upon. The products of decomposition can thus be generally separated. They are usually composed of silica, alumina and water. If these materials are not in perfectly defined proportions, and their variable composition seems to prove it, they are not mixtures of $\ddot{\bar{A}}l$, $\ddot{S}i$ and $\dot{H}$, but rather mixtures of definite combinations of these elements; for by the action of the alkalies, only a very little $\ddot{S}i$ is dissolved, and without acids only a very little alumina. Although a part of the water may be driven off at a low temperature, the greater part can only be separated at a very high heat. These products of decomposition, not having a definite chemical composition, cannot have any very well defined characters. They are generally in earthy masses. When dry, they absorb water and attract the tongue; in contact with water, they fall to pieces and make mud. This mud makes a paste which is more or less plastic, but cannot generally be drawn out without breaking. With heat, they become hard and compact, but do not lose their water, except at a very high temperature. They are infusible, unless mixed with some metallic oxides. When the decomposition has taken place in a rock, the elements of which are well separated, as large grained Granites and Pegmatites, the Quartz is unaltered and the Mica is not decomposed; the feldspar only has undergone decomposition. The Mica, however, undergoes certain changes and takes a silvery look, which it did not have in the unaltered rock. We shall separate these kinds of products of decomposition as follows:

1. Kaolins, or porcelain clays, resulting from the decomposition of rocks in place.
2. Ordinary Clays, formed as sediments.
3. Clays, produced by chemical deposition.

KAOLIN.

Syn.—Kaolinite, Porcelain Clay, Porzellanerde, Argiles à porcelaine.

Kaolin is generally produced by the decomposition of the feldspar of granitic rocks, and is generally found in place. It is usually white and somewhat plastic, not very coherent, earthy and without argillaceous odor when breathed upon. By crushing and washing, it is very easy to separate it from the undecomposed materials accompanying it. It is very much sought for when free from iron, for the manufacture of porcelain. For this purpose it is indispensible that all the Mica should be washed out.

Brongniart analyzed a great number of Kaolins used in the arts and arrived at the following limits:

$\dddot{Si}$ 23–46, Metallic oxides 0.5–1, $\overline{\dddot{Al}}$ 21–43, $\dot{C}a$, $\dot{M}g$ 0–6, Alkalies 0–0.5, $\dot{H}$ 5–12, residue not argillaceous 0–3.

Its general formulæ may be expressed as $\overline{\dddot{Al}}\ \dddot{Si}^2+2\ \dot{H}$; $\overline{\dddot{Al}}$ 39.8, $\dddot{Si}$ 46.3, $\dot{H}$ 13.9

ORDINARY CLAYS.

Clays seem to have been formed from the product of decomposition carried off by water and deposited in beds in the stratified formations. They do not have any well-defined characters. When dry they rapidly absorb water, which they lose easily and then contract and crack in every direction. They have an earthy and sometimes a lamellar aspect. When taken from the earth they are sometimes somewhat translucent on the edges and have a soapy look and a slight lustre. When breathed upon they give a peculiar odor called argillaceous, like the smell of ground after a rain. Their colors are very variable; so remarkable that those found with salt are frequently called *variegated clays*. They seem to have undergone a sort of metamorphism at a very high pressure, under the influence of which the iron and other metals in different degrees of oxidation have given different tints to different parts of the mass. These different colors are frequently arranged in regular stratified layers. The chemical composition of clays is very variable but they can all be arranged around two types, represented by the following compositions;

	I.	II.
$\dddot{Si}$	45—50	60—66
$\overline{\dddot{Al}}$	34—38	18—25
$\dot{H}$	9—15	9—15

These may be represented by the formulæ

$\overline{\dddot{Al}}^3\ \dddot{Si}^5+4\ \dot{H}$; $\dddot{Si}$ 51.83, $\overline{\dddot{Al}}$ 35.36, $\dot{H}$ 12.46 and
$\overline{\dddot{Al}}\ \dddot{Si}^5+3\ \dot{H}$; $\dddot{Si}$ 65.64, $\overline{\dddot{Al}}$ 22.54, $\dot{H}$ 4.82.

These clays are generally plastic enough to allow their use in moulding and for pottery. When they contain but little iron they can be used for fire brick. They absorb water rapidly and have a very distinct argillaceous odor, and are only partially acted on by acids.

CHEMICAL CLAYS.

SYN.—Smectite, Terre à foulon, Steinmark, Lithomarge, Walkerde.

The third variety constitutes the varieties known as fuller's earth or smectic clays. They can be almost entirely attacked by acids and alkalies, and are scarcely at all plastic. They do not fall to pieces in water and make a paste hardly capable of being drawn out, or, as it is said, is very *short*. In the fracture their lustre is quite bright; they may even be translucent on their edges. This translucidity disappears and re-appears, when they are placed in water. This may be done several times. When rubbed with the nail they sometimes acquire a very high lustre. They do not absorb water as easily as the first two varieties, but they unite with fats, even when cold, and saponify. They are largely used for soap in the countries where they are found.

The chemical clays differ materially in their characteristics from the ordinary clays, and also in their formation. They generally accompany certain metallic oxides and have the general appearance of definite compounds. They are not entirely amorphous, and when just taken from the earth are quite translucent on their edges and even sometimes transparent. They do not always become entirely opaque afterwards. Their fracture is smooth and sometimes scaly and bright. When rubbed they take a bright polish. In water they do not fall to pieces, and are not at all plastic, but they absorb water and become much more transparent. Like the smectic clays, this translucidity may be made to appear and disappear several times. They are all attacked by acids and gelatinize. They have no absolutely definite composition and therefore are not species. Certain authors have regarded them as species, but they should not have names given to them, except to serve a sort of pseudo-scientific purpose in distinguishing them. Their chemical composition is generally $\ddot{S}i$ 44–50, $\ddot{\bar{A}}l$ 17–23, $\dot{H}$ 22–27. They sometimes contain a little $\dot{M}g$ and $\dot{F}e$. Some varieties of these chemical clays which have a schistose structure are called Lithomarge and also those which form the fluccan of certain metallic veins.

III. Subsilicates.

Chrondrodite. $\dot{M}g^8\ \ddot{S}i^3$. ORTHORHOMBIC.

SYN.—Brucite, Humite.

It crystallizes as a right rhombic prism of 94° 26′. The crystals show a very large number of faces and present several types some of which show hemihedral forms *Pl.* VIII. *Fig.* 23. The volcanic varieties have generally the most complicated forms. Cleavage indistinct, parallel to the base. Fracture, conchoidal or uneven. Lustre, vitreous or resinous. Transparent, translucent, opaque. Color, white, yellow, brown, red, green, gray or black. Streak, white or yellowish. **H.**=6–6.5. **G.**=3.118–3.24. Composition, $\dot{M}g$ 54.50, $\dot{F}e$ 6.75, $\dddot{S}i$ 33.19, Fl 5.56. Part of the oxygen is always replaced by fluorine.

Pyr. &c. B. P. Infusible. Some varieties blacken and then become white. With fluxes, gives the reactions for iron. With S.Ph. or $\dddot{S}$, gives the reactions for Fl. Gelatinizes with acids.

It is almost always found as rounded crystals, or in grains in limestone, sometimes associated with Graphite. Humite is generally found in the feldspathic rocks of Mt. Somma. It might resemble certain varieties of Garnet or Tourmaline, but its gravity is less and its infusibility and action with acids will also distinguish it.

It is found abundantly in Massachussetts, New York, New Jersey and Pennsylvania.

FORMULÆ OF THE CRYSTALS.

Pl. VIII.

Fig. 23. ∞P. $\bar{P}\infty$. $\frac{4}{3}P$. $\frac{8}{7}\bar{P}2$. $4\breve{P}\infty$; some of the faces are hemihedral.

Tourmaline. $((\dot{N}a, \dot{K}, \dot{C}a, \dot{M}g, \dot{F}e)^3 (\ddot{\overline{F}}e, \ddot{Al}, \ddot{B}))^8 \dddot{S}i^9$. HEXAGONAL.

SYN.—Schörl, Rubellite, Indicolite.

It crystallizes in forms derived from a rhombohedron of 103°, parallel to which it has traces of cleavage. The forms of Tourmaline are all hemihedral, with inclined faces. Thus the trigonal prism is frequently found, *Pl.* VIII. *Figs.* 24–27. This same hemihedry shows itself in the terminations. The ditrigonal prism is also found, *Pl.* IX. *Fig.* 5. Crystals are very often found hemimorphic, *Pl.* VIII. *Figs.* 25, 26. *Pl.* IX. *Figs.* 2, 3, 4, 5. Generally the crystals have a very large number of modifications, which show themselves by striations on the trigonal prism, making the section of the crystal a spherical triangle, which is generally equilateral. Fracture, conchoidal or uneven. Lustre, vitreous. Transparent, translucent or opaque. Even the opaque crystals are translucent when cut into very thin plates. Color, black, brown, blue, green, red, sometimes colorless. The black color may be intense green, blue, violet or reddish-brown, which colors are also found lighter. The red variety of Tourmaline is sometimes called Rubellite and the blue variety Indicolite. Very often the same crystal has two or more of these colors together. The Tourmalines from Elba often show two or even three colors. Streak, colorless. The varieties which contain the alkalies and the alkaline earths, are generally of a light color, while those containing the metallic oxides are darker. It sometimes has a remarkable dichroism. In the direction of the vertical axis, some crystals are yellowish-brown, and perpendicular to this asparagus green; or respectively brownish-violet and greenish-blue; or purple and bluish. It is pyroelectric. **H.**=7–7.5. **G.**=2.94–3.3. Composition, $\dot{N}a$ 0–5, $\dot{K}$ 0–4, $\dot{C}a$ 0–2, $\dot{M}g$ 0–15, $\dot{F}e$ 0–17, $\ddot{\overline{F}}e$ 0–11, $\ddot{Al}$ 30–44, $\ddot{B}$ 4–11, $\dddot{S}i$ 35–40. It is possible that a part of the $\ddot{B}$ may replace a part of the $\dddot{S}i$. A small portion of fluorine is almost always found, which probably replaces some of the oxygen. It is impossible to make anything more than a general formula, on account of the difficulties which the simultaneous presence of $\dddot{S}i$, $\ddot{B}$ and Fl causes in the analysis.

Pyr. &c. B. P. Most of the varieties fuse to a blebby glass, but some are infusible. With fluxes, it generally gives the reactions for iron and manganese. Fused with bisulphate of potash and Fluorite, it gives the reaction for boracic acid. Heated, loses weight, giving off fluoride of silicon or perhaps fluoride of boron. Not decomposed by acids, unless after fusion.

Those varieties which contain metallic oxides have different characters. The darkest are fusible, with intumescence, but the clear varieties are generally infusible. They all color the flame green, owing to the presence of $\dddot{B}$. None of them are attacked by acids. The red variety called Rubellite has been used to imitate Ruby, but it possesses only simple refraction. Tourmaline is very remarkable for certain physical properties; it is much used in optics, particularly in the study of polarization. A bacillary variety is also found made up of *baguettes* more or less fine, generally placed together in a parallel position, but sometimes radiated. It might be confounded with the radiated varieties of Andalusite, but can be distinguished by the complete absence of cleavage. From Amphibole and Pyroxene, they are distinguished by the reaction for $\dddot{B}$. Generally the cylindrical disposition of the crystals may be distinguished even in the *baguettes;* when the crystals are acicular, the glass will have to be used. Very fine crystals of Rubellite and Indicolite have been found in Maine and Massachusetts. The common varieties are found abundantly in the U. S.

FORMULÆ OF THE CRYSTALS.

Pl. VIII.

Fig. 24. $\infty P2. \frac{\infty R}{2}. -\frac{1}{2}R$; a very frequent form. *Fig.* 25. $\infty P2. \frac{\infty R}{2}. -\frac{1}{2}R$ above, below $-\frac{1}{2}R. 0R$; a usual hemimorphic crystal. *Fig.* 26. $\infty P2. -\frac{\infty R}{2}. R. R5$ above, R below. *Fig.* 27. $\infty P2. -\frac{\infty R}{2}. -\frac{1}{2}R3. R.$ *Pl.* IX. *Fig.* 1. $\infty P2. \frac{\infty R}{2}. R$; a common form. The prisms are often oscillatory combined, so that the crystal becomes triangular in shape. *Fig.* 2. $\infty P2. \frac{\infty R}{2}. -2R. R.$ above, $-2R$ alone below. *Fig.* 3. $\infty P2. R. -2R$ above, $0R$ below. *Fig.* 4. $\infty P2. \frac{\infty R}{2}. R. -\frac{1}{2}R$ above, $-\frac{1}{2}R. \frac{1}{4}R$ below. *Fig.* 5. $\infty P2. \frac{\infty R}{2}. -\frac{\infty R}{2}. -\frac{1}{2}R. R. -2R. \frac{1}{2}R5$ above, $R. -\frac{1}{2}R.$ below; from St. Lawrence Co., N. Y.

Andalusite. $\dddot{\bar{\text{Al}}}\ \dddot{\text{Si}}$. ORTHORHOMBIC.

SYN—Chiastolite, Macle, Hohlspath.

It crystallizes as a right rhombic prism of 94° 48′. The crystals usually affect the form of the prism, sometimes with the faces of the macro- and brachydome, *Pl.* IX. *Figs.* 6–8; and sometimes this form is associated with the brachydome and a macroprism, *Fig.* 9. These prisms are easily mistaken for the square prism, as their angle varies so slightly from a right angle. A particular disposition of Andalusite is called Chiastolite or Macle. This name has however no connection with the crystallographical property known as macle. The mineral has generally been found in black schistose rocks. In crystallizing it frequently envelops them, as we have seen that it did under Mica. This absorption of the schist is, however, made in accordance with a law; the foreign substances are regularly placed in the crystal, following a law which varies from one crystal to another and may even vary in the different parts of the same crystal. The color of the mineral is a reddish-white and the schist is black, so that the appearance of the crystal when cut perpendicularly to its greatest axis will show the disposition of the mineral and the schist, *Figs.* 10–16; at the two extremities of the same crystal it is almost always different, *Figs.* 12, 13, 16. These masses are also characterized by the interpenetration of Mica. They are often partially decomposed. It has a very easy cleavage, parallel to the prism, which is very marked in the transparent varieties from Brazil, but less so in the opaque ones. It has two other cleavages less easy, parallel to the macro- and brachypinacoid. Fracture, unequal. The fracture of the bacillary varieties is granular. It has a vitreous lustre which is quite apparent in the transparent varieties, but not so much so in the opaque ones. It is often earthy. Transparent, translucent, opaque. Color, whitish, rose-red, flesh-red, violet, gray, reddish-brown, olive-green. The transparent varieties have a clear greenish color in one direction; in the other by transmitted light it is dark blood-red. It thus presents the phenomenon of dichroism. The compact masses have almost always a rosy tint. The starry varieties are whitish-gray or sometimes reddish. Streak, colorless. **H.**=7.5 in the transparent, and 3.1–3.2 in the opaque varieties. **G.**=3.05-3.35. Composition, $\overset{\cdots}{\overline{\rm Al}}$ 63.2, $\overset{\cdots}{\rm Si}$ 36.8. The earthy varieties of Andalusite, which are not pure, do not correspond perfectly to the formula; they sometimes contain as high as 5–7% of $\overset{\cdots}{\overline{\rm Fe}}$, and a small quantity of $\dot{\rm Ca}$ and $\dot{\rm Mg}$, which is sometimes substituted for the $\overset{\cdots}{\overline{\rm Al}}$ with which it is isomorphous. It is this iron that gives the greenish tint to the crystals.

Pyr. &c. B. P. Infusible. With cobalt solution, gives a blue color ($\overset{\cdots}{\overline{\rm Al}}$). With borax, usually reacts for iron. Not acted on by acids.

The opaque and stony varieties are found in the metamorphic rocks roughly crystallized. They generally occur in groups of large crystals almost always covered with, or penetrated by Mica. This structure gives an apparently easy cleavage in certain directions. It is also frequently found in bacillary masses, composed of fibers placed together either parallel, or more rarely somewhat diverging from a common center. It may be confounded, when in bacillary masses, with certain red Tourmalines; the form of the crystal, however, is different. Tourmaline reacts for boracic acid, and Andalusite does not. It can also be distinguished by its macles. The Mica, which is almost always found with it, may be used as an empirical character. It might also be confounded with some varieties of Pyroxene, Wernerite, Spodumene or Feldspar; all of these are however fusible. It is found in most of the New England states and Pennsylvania.

FORMULÆ OF THE CRYSTALS.

Pl. IX.

Fig. 6. ∞P. $0P$. $\breve{P}\infty$. *Fig.* 7. ∞P. $\breve{P}\infty$. $0P$; the brachydome predominating. *Fig.* 8. The combination *Fig.* 6, with $\bar{P}\infty$. *Fig.* 9.

∞P. ∞P̄2. 0P. P̄∞. 2P̆2. *Figs.* 10–16. Sections of crystals of Chiastolite, showing the most usual markings. *Fig.* 12. Section from opposite ends of the same crystal. *Fig.* 13. The same. *Fig.* 15. Apparently a twin crystal. *Fig.* 16 shows the variation of the markings in the successive portions of a single crystal.

Fibrolite. $\ddot{\text{A}}\text{l}\ \ddot{\text{S}}\text{i}$. MONOCLINIC.

SYN.—Bucholzite, Sillimanite, Xenolite, Monrolite, Bamlite, Faserkiesel.

It crystallizes as an inclined rhombic prism of 90° to 98°. The crystals are generally rough, so that the angle cannot be determined exactly. In the rough varieties the angles are large. The crystals are generally long, slender, sometimes rounded and very much striated, and rarely terminated. It has an easy cleavage, parallel to the orthopinacoid. Fracture, uneven. Lustre, vitreous. Transparent, translucent, opaque. Color, brown, grayish-white or green. Streak, colorless. **H.**=6–7. **G.**=3.2–3.3. Composition, $\ddot{\text{A}}\text{l}$ 63.2, $\ddot{\text{S}}\text{i}$ 36.8. The $\ddot{\text{A}}\text{l}$ may be replaced by 2 % of $\ddot{\text{F}}\text{e}$ or 0.8 % of $\dot{\text{M}}\text{g}$; even $\dot{\text{H}}$ is also present in very small proportions.

Pyr. &c. B. P. Infusible. Not acted on by acids.

The crystals are usually found imbedded in Gneiss or Mica Schist. As they are fibrous or very highly striated, they might be mistaken for varieties of the Amphibole group, but these are fusible while Fibrolite is not. It might be mistaken for Cyanite, but this shows flat crystals which are bluish and nearly rectangular, while Fibrolite is fibrous and rhombic in form. It is found in Massachusetts, Connecticut, New York and Pennsylvania.

Cyanite. $\ddot{\text{A}}\text{l}\ \ddot{\text{S}}\text{i}$. TRICLINIC.

SYN.—Kyanite, Disthen, Disthène, Rhætizite.

It crystallizes as a doubly inclined rhombic prism of 97°4′. Its angles are ∞P̄∞ ∧ ∞P̆∞=106° 16′, 0P ∧ ∞P̄∞=100° 50′, 0P ∧ ∞P̆∞=93° 15′. Sometimes it is found in isolated bacillary crystals in the rock, which are easily distinguished by the characteristic bluish color. These crystals are generally the primitive form alone, or combined with the macro- and brachypinacoids and a hemi-clinodome. *Pl.* IX. *Figs.* 17–20. Generally the mineral is found as lamellar masses, or as fibrous crystals joined together either parallel to each other or radiating from a common center. Hemitropes parallel to the vertical axis occur quite frequently, *Fig.* 20 The color of the masses is generally blue. It has three cleavages; one very easy and the other two less so. The first of these cleavages parallel to the macropinacoid, which is the one oftenest seen, gives a brilliant surface which is polished and has a vitreous lustre. The second, parallel to the brachypinacoid is less brilliant and is less easily produced, but still quite pronounced; it is smooth and brilliant with a vitreous lustre. The third, parallel to the base is more difficult, has no lustre and is fibrous. In this last cleavage there are so many striations, as to render the measurement of the angles very difficult. Lustre, vitreous or pearly. Transparent, translucent or opaque. Color, blue, white, or a combination of white and blue, gray, green, black. Streak, colorless. **H.**=5–7.25. **G.**=3.45–3.7. Composition, $\ddot{\text{A}}\text{l}$ 63.2, $\ddot{\text{S}}\text{i}$ 36.8. Its hardness varies on the different faces.

Pyr. &c. B. P. Infusible. The colored varieties become white.

With borax, it gives a colorless transparent bead. Not acted on by acids.

White varieties are also found and are called Rhætizite; the blackish and yellowish colors are generally accidental, and are caused by organic matter or iron. It can almost always be recognized by its crystalline forms, and its association with Mica and Quartz, but it is especially characterized by its bluish color, whence the name Cyanite, from *Kuanos*, blue. When blue and perfectly transparent, it is cut as a gem and used to imitate Sapphire. It might sometimes resemble some of the varieties of Amphibole, but it can be easily distinguished from it by its infusibility. The colorless varieties might be mistaken for Diaspore, but Cyanite does not give off water when heated. Short crystals may sometimes resemble Staurolite; they differ however, in their terminations and cleavage. It is found abundantly in the U. S. White, blue and green varieties are found on New York Island.

FORMULÆ OF THE CRYSTALS.

Pl. IX.

Fig. 17. 0 P. ∞'P. ∞P'. ∞P̆∞; usual form, but the base is rarely observed. *Fig.* 18. The preceding, with ∞ P ∞. *Fig.* 19. The preceding, with 'P̆,∞. *Fig.* 20. Twin crystal.

Topaz. $\dddot{A}l$ $\ddot{S}i$. ORTHORHOMBIC.

SYN.—Topas, Topaze, Pycnite, Pyrophysalite, Physalite.

It crystallizes as a right rhombic prism of 124° 17′. There are three types which it usually takes, especially in the bluish varieties. The first type is shown in *Pl.* IX. *Figs.* 21–23. They frequently however, have a much greater number of faces. These crystals appear to be rectangular, the angle between the brachydome being 92° 42′. In the yellow or slightly colored varieties which are found in Saxony, the faces of the brachydome are generally greatly developed, and replace the base completely, or almost completely, *Figs.* 24, 25. The last type is that of the orange varieties from Brazil, in which the octahedron replaces the base, *Pl.* X. *Fig.* 2. Cleavage very easy, parallel to the base. On account of the cleavage it is rare to find perfect crystals. Fracture, conchoidal or uneven. Lustre, vitreous, bright but unequal on the different faces. The vertical faces are generally striated and dull. Transparent, translucent, opaque. Color, blue, green, yellow, orange-yellow, red and colorless. The colors vary with the locality and crystalline form, and appear to be generally owing to organic substances. Streak, colorless. **H.**=8. **G.**=3.4–3.65. Composition, Si 15.17, Al 29.58, O 34.67, Fl 28.58; one-fifth of the oxygen of $\ddot{S}i$ is replaced by Fl.

Pyr. &c. B. P. Infusible. The yellow varieties when heated take a pink or red color, and are then known as burnt Topaz. If the heating is continued, it becomes colorless. Partially attacked by $\ddot{S}$.

Nebular masses are often distinguished in Topaz. On examining them with a glass they are found to be made up of very small cavities, which are often quite regular and seem to have the shape of a crystal of Topaz. These cavities usually contain a drop of liquid, which generally shows a little bubble of gas. These cavities have been found large enough to admit of the chemical examination of the liquid, which was found to be one of the hydrocarbons. This liquid is extraordinarily volatile. If the crystal is heated to 20° or 30° C. the liquid is volatilized, which, with the change of color, would appear to show that the mineral was formed at a low temperature. In the very rare cases in which it has been possible to examine it, it has been found to be a hydrocarbon which must have been liquified under an enormous pressure. In the colored varieties of Topaz the color in not only variable, but the shades of color are often arranged in zones which are sometimes concentric. In some of the streams of Brazil, pebbles of Topaz which are perfectly colorless are found. These varieties are often called Water Drops. Their hardness might cause them to be confounded with Quartz or Beryl, but their density is quite different. That of Topaz is 3.5, while that of Quartz and Beryl is about 2.5. The cleavage is also different, being very easy in the direction of the base, which would easily distinguish it from Quartz. The varieties called Water Drops (*gouttes d'eau*) are frequently used to imitate the Diamond; they have very much less fire and possess double refraction, while the Diamond is only simple. The earthy variety of Topaz is known under the name of Pycnite. It is found in a rock associated with Mica. The general appearance of this rock is usually very characteristic. The Mica is almost always of a dark color, and the Topaz is in *baguettes* placed together in parallel lines and separated by Mica. These *baguettes* can thus be very easily separated the one from the other by a stroke of the hammer, and show the peculiar cleavage of Topaz. The color of this variety is always yellow and slightly greenish, and is not liable to be mistaken for anything else. In Sweden, large crystals of earthy Topaz have been found analogous to Beryl. These varieties are called Pyrophysalite. The difference in the ease of cleavage and in density will readily distinguish them. It is easily distinguished from other minerals which it resembles by the great ease and brilliancy of its cleavage. It is found in the U. S., at Trumbull, Ct., in N. Carolina and in Utah.

FORMULÆ OF THE CRYSTALS.

Pl. IX.

Fig. 21. ∞P. $\infty\check{P}2$. $0P$. $2\check{P}\infty$. $\frac{2}{3}P$. $\frac{4}{3}P2$; from Saxony. *Fig.* 22. ∞P. $\infty\check{P}2$. $\infty\check{P}3$. $0P$. $2\check{P}\infty$. $4\check{P}\infty$. $\frac{2}{3}P$. *Fig.* 23. ∞P. $\infty\check{P}2$. $0P$. $\frac{2}{3}P$. P. $2P$. *Fig.* 24. ∞P. $\infty\check{P}2$. $2\check{P}\infty$. P. *Fig.* 25. The preceding, with $0P$. *Pl.* X. *Fig.* 1. ∞P. $\infty\check{P}2$. P. $\frac{4}{3}\check{P}2$. $2\check{P}\infty$. *Fig.* 2. ∞P. $\infty\check{P}2$. P; from Brazil. *Fig.* 3. ∞P. $\infty\check{P}2$. $\infty\check{P}\infty$. $0P$. P. $\frac{3}{2}P$. $\frac{3}{2}\check{P}2$. $4\check{P}\infty$. $\frac{3}{2}P\infty$.

Euclase. $(\frac{1}{6}\dot{H}^3+\frac{2}{6}\dot{B}e^3+\frac{3}{6}\ddot{A}l)\dddot{S}i$. MONOCLINIC.

SYN.—Euklas.

It crystallizes as an inclined rhombic prism of 115°. Its ordinary form is the prism with clino- and orthopinacoids; the prism is generally terminated by a clinopyramid which is very much inclined, so that the crystal is terminated by two very sharp points, *Pl.* X. *Figs.* 4,5. The crystals are also often striated in the direction of their length. It has a very easy cleavage parallel to the clinopinacoid; on account of the great ease with which this cleavage is produced, it is very difficult to find perfect crystals. Fracture, conchoidal. Lustre, vitreous and somewhat pearly on the cleavage faces. It is always transparent. Color, greenish-blue, yellow and colorless; the color is not always the same at the two extremities of the crystal. Streak, colorless. Very brittle. **H.**=7.5. **G.**=3.098. Composition, $\ddot{B}e$ 17.4, $\bar{\ddot{A}}l$ 35.3, $\ddot{S}i$ 41.1, $\dot{H}$ 6.2. Fl is present in variable proportions. It appears to replace the oxygen, so that it is impossible to refer this mineral to any very well-defined formula. Otherwise it very much resembles the composition of Phenacite.

Pyr. &c. B. P. In a closed tube yields water; fuses, with intumescence, at 2. Not acted on by acids.

The gangue of Euclase appears to be massive Hematite, but it is rarely found in place. As yet it has only been found in Brazil in the same localities as Diamond. It is sometimes used by jewelers on account of its very great lustre. It is too brittle however, to be much used as an ornament. It is a very rare mineral and has only been found in the Urals and Brazil.

FORMULÆ OF THE CRYSTALS.

Pl. X.

Fig. 4. ∞P. $\infty \mathbf{P} 2$. $\infty \mathbf{P} \infty$. $\infty \grave{P} \infty$. $\frac{3}{2} \grave{P} \frac{3}{2}$. $\grave{P} \infty$. $-\mathbf{P} 2$. $-P$. *Fig.* 5. The same crystal in a different position.

Datolite. $(\dot{C}a^3, \dot{H}^3, \dddot{B}) \ddot{S}i$. MONOCLINIC.

SYN.—Datholite, Botryolite, Humboldtite.

It crystallizes in an inclined rhombic prism of 115° 3′. It is very near a right rhombic prism however, the angle between the base and the prism being 90° 5′ only. Its ordinary crystalline forms are shown in *Pl.* X. *Figs.* 6–8. The faces are generally striated in the direction of their length. Cleavage easy, parallel to the base. The fracture of Datolite is conchoidal and smooth, and at the same time lamellar and scaly. On the lamellar fracture the lustre is very vitreous and bright, and sometimes even pearly. The crystals often show mirrors. On the conchoidal fracture the lustre is generally resinous, but is sometimes also vitreous and bright. Transparent, translucent, opaque. On the natural faces the lustre is vitreous and very bright. Color, slightly green or white, grayish, yellow and amethystine. Streak, white. Brittle. **H.**=5–5.5. **G.**=2.8–3. Composition, $\dot{C}a$ 35.0, $\dot{H}$ 5.6, $\dddot{B}$ 21.9, $\ddot{S}i$ 37.5.

Pyr. &c. B. P. In the flame without the aid of a blowpipe, the transparent varieties become opaque, friable and lose water. Fuses, with intumescence, at 2 to a clear bead, which colors the flame green on account of $\dddot{B}$; this is especially apparent with bisulphate of potash or Fluorite. Gelatinizes with acids.

A mineral, the composition of which is almost the same as Datolite and which has been called Botryolite, is usually placed with Datolite. It is not crystallized, but is found in little spheroidal concretions. The white compact variety is found at Lake Superior associated with Copper. It was formerly found in great abundance at Bergen Hill, N. J. It is also found at Lake Superior. Until these discoveries were made the mineral was exceedingly rare.

FORMULÆ OF THE CRYSTALS.

Pl. X.

Fig. 6. ∞P. $\infty \grave{P} 2$. $0P$. $2P$. $-2 \mathbf{P} \infty$. $2 \grave{P} \infty$. $-2 \grave{P} 2$. *Fig.* 7. ∞P. $\infty \mathbf{P} \infty$. $0P$. $-4P$. $2P$. $\frac{4}{3} P$. $\frac{4}{5} P$. $-2 \mathbf{P} \infty$. $\frac{3}{2} \grave{P} \infty$. $2 \grave{P} \infty$. $4 \grave{P} \infty$. *Fig.* 8. ∞P. $\infty \mathbf{P} \infty$. $0P$. $\frac{3}{4} P$. P. $\frac{4}{3} P$. $2P$. $-4P$. $-8 \grave{P} 2$. $-2 \mathbf{P} \infty$. $-6 \mathbf{P} \infty$. $-6 \mathbf{P} 3$. $2 \grave{P} \infty$. $4 \grave{P} \infty$.

Titanite. ($\dot{C}a+\ddot{T}i$) $\dddot{S}i$. MONOCLINIC.

SYN.—Sphene, Greenovite, Spinthère, Semeline, Castellite, Lederite, Pictite.

It crystallizes in an inclined rhombic prism of 113° 31′. The light-colored varieties have generally two types, *Pl.* X. *Figs.* 9, 10. The crystals are often very flat in the direction of the vertical axis. These crystals are usually very small and detached, and their color may vary from white to brown. The central part of the crystal is usually of a clearer color than the rest. Other varieties, usually of a honey-yellow color, are generally lengthened in the direction of the orthodiagonal and have fewer modifications, *Fig.* 14. The acute angles are usually of darker color than the rest of the crystal. This form is susceptible of a very remarkable macle around the face of the orthopinacoid giving rise to a very deep reëntrant angle which is characteristic of this variety. A twin by interpenetration also occurs frequently, *Fig.* 15. The color of this variety is mixed and variable. The varieties which are of a dark maroon-brown have a type, *Fig.* 11, which is very different from the preceding. They are usually flattened crystals disseminated in a mass of Quartz, Scapolite or other minerals. This form sometimes becomes a flattened octahedron. This variety of Titanite might be confounded with Zircon or Garnet. The rose varieties are called Greenovite and have a much more resinous lustre, *Fig.* 12. The color is somewhat inclined to violet. It is found in lamellæ and its color is owing to manganese. This variety is quite rare and is only found in manganese mines, often associated with violet Epidote and other oxides of manganese. The crystals of Titanite are often hemimorphic. It has a cleavage more or less easy, parallel to the prism and another less so parallel to the orthopinacoid. Fracture, imperfectly conchoidal or uneven. Lustre, resinous and sometimes very bright, almost adamantine, except in the rose variety which is only resinous. Transparent, translucent, opaque. Color, nearly white, black, greenish-yellow, orange-yellow, and all the shades of green and brown, maroon and rose. Streak, white, reddish in Greenovite. **H.**=5–5.5. **G.**=3.4–3.56. Composition, $\dot{C}a$ 21–28, $\ddot{T}i$ 33–43, $\dddot{S}i$ 30–35. The relation of the $\dot{R}$, $\ddot{R}$ and $\dddot{S}i$ is 1 : 2 : 2. It will be seen that the $\dot{R}+\ddot{R}=\ddot{R}$, or that the formula might be written $\ddot{R}$ $\dddot{S}i$. The relation would then be 3 : 2. The $\dot{C}a$ may be replaced by $\dot{M}g$ or $\dot{F}e$.

Pyr. &c. B. P. The transparent varieties are fusible with difficulty; the dark colored ones change color, becoming yellow, and fuse at 3, with intumescence. Gives reactions for $\ddot{T}i$. Decomposed by $\dddot{S}$ and HCl.

It is sometimes found in brown crystalline masses which show an apparently rhombohedral cleavage. Titanite occurs in Syenite, Gneiss, Mica, Slates and limestones and sometimes in volcanic rocks associated with Pyroxene, Hornblende and Nephelite. It is found in very large crystals at Natural Bridge and elsewhere in New York. Also in Maine, Massachusetts, New Jersey and Pennsylvania.

FORMULÆ OF THE CRYSTALS.

Pl. X.

Fig. 9. ∞P. ∞P∞. 2P. P∞. -5P∞; form of Spinthere. *Fig.* 10. ∞P. ∞P∞. 2P. 0P; form of Semeline. *Fig.* 11. 2P. 0P. ∞P∞. *Fig.* 12. 2P. -2P. -4P. $\frac{4}{5}$P∞; Greenovite. *Fig.* 13. ∞P. ∞P∞. 0P. 2P. -2P; Lederite. *Fig.* 14. ∞P. P∞. $\frac{1}{2}$P∞. 0P; a very frequent form. *Fig.* 15. Twin crystal by interpenetration parallel to 0P.

Staurolite. $(\frac{1}{5}(\frac{3}{4}\dot{H}+\frac{2}{4}\dot{M}g+\frac{1}{4}\dot{F}e)^3+\frac{4}{5}\dddot{A}l)^4\ddot{S}i^3$. ORTHORHOMBIC.

SYN.—Staurotide.

It crystallizes as a right rhombic prism of 129° 20′; the crystals are sometimes simple, *Pl.* X. *Figs.* 16–18. It is more frequently macled in the shape of a greek or St. Andrew's cross, *Fig.* 19, 20, and *Pl.* XI. *Figs.* 1, 2. When they are at right angles, the crystals are placed together by the faces $\frac{3}{2}\breve{P}\infty$, so that the faces of the brachypinacoid are almost at right angles with each other; this is the most frequent form, *Fig.* 20. When the macle is inclined, the axes intersect at an oblique angle, *Fig.* 19 and *Pl.* XI. *Fig.* 2. Even the crystals which appear simple, are usually macled, as shown in *Fig.* 3. It has a cleavage parallel to the brachypinacoid. Its fracture is conchoidal and unequal, sometimes lamellar when the cleavage is distinct. Lustre, vitreous, resinous when not altered, otherwise earthy. Translucent, opaque. Color, dark reddish-brown, brownish-black or yellowish-brown. Streak, colorless or grayish. It shows a remarkable dichroism. The brown transparent varieties appear blood-red by transmitted light. **H.**=7–7.5. **G.**=3.4–3.8. Composition, $\dot{H}$ 1.7, $\dot{M}g$ 2.5, $\dot{F}e$ 15.8, $\dddot{A}l$ 51.7, $\ddot{S}i$ 28.3. A variety from Sweden shows 11.61 % of $\dot{M}n$.

Pyr. &c. B. P. Infusible. The dark varieties become fritted in the R. F. and sometimes give a magnetic globule. The manganesian varieties fuse easily to a magnetic mass. With borax, the reactions for iron are shown. Insoluble in acids.

It is frequently found associated with Cyanite. At St. Gothard the two minerals are almost always found together. It is usually easily distinguished by its color, form, and association with Cyanite. It is found abundantly in the U. S.

FORMULÆ OF THE CRYSTALS.

Pl. X.

Fig. 16. ∞P. $\infty\breve{P}\infty$. $0P$; primitive form. *Fig.* 17. The preceding, with $\bar{P}\infty$. *Fig.* 18. The same, but with $\infty\breve{P}\infty$ predominating. *Fig.* 19. Oblique twin crystal. *Fig.* 20. Twin crystal; composition-face $\frac{3}{2}\breve{P}\infty$. *Pl.* XI. *Fig.* 1. The same as the preceding. *Fig.* 2. Oblique twin crystal. *Fig.* 3. Section of a crystal showing the striations from twinning.

HYDROUS SILICATES.

I. Bisilicates.

Pectolite Group.

Pectolite. $(\frac{1}{6}\dot{H}+\frac{1}{6}\dot{N}a+\frac{4}{6}\dot{C}a)\ddot{S}i$. MONOCLINIC.

SYN.—Pektolith.

It crystallizes in an inclined rhombic prism, and is isomorphous with Wollastonite. Distinct crystals are rarely found. It is almost always in columnar or fibrous masses, which are divergent. The fibers are sometimes 5–6 c. m. in length, and are often in very sharp crystals easily detached. It has an easy cleavage parallel to the orthopinacoid. Lustre, vitreous or silky. Transparent, opaque. Color, white, gray, or brown when altered. Streak, colorless. Tough. **H.**=5. **G.**=2.68–2.78. Composition, $\dot{H}$ 2.7, $\dot{N}a$ 9.3, $\dot{C}a$ 33.8, $\dddot{S}i$ 54.2.

Pyr. &c. B. P. In a closed tube yields water. Fuses at 2 to a white enamel. Gelatinizes with acids. Sometimes phosphoresces.

It is generally found in trap rock, in cavities and thin seams. A white, compact variety is found at Lake Superior. It resembles Tremolite, but is distinguished by gelatinizing with acids. It can be distinguished from Natrolite by its mode of occurrence. The compact varieties resemble Wollastonite; the presence of $\dot{N}a$, $\dddot{A}l$ and $\dot{H}$ distinguishes it. It has been found in beautiful radiated masses at Bergen Hill, N. J.

Laumontite. $(\frac{1}{4}\,\dot{C}a^3+\frac{3}{4}\,\dddot{A}l)\,\dddot{S}i^3+3\,\dot{H}$. MONOCLINIC.

SYN.—Laumonite.

It crystallizes as an inclined rhombic prism of 86° 16′. It has a cleavage parallel to the clinopinacoid and another parallel to the prism. The usual forms are given in *Pl.* XI. *Figs.* 4, 5, 6. The crystals are rarely complete; they are generally broken, or simply *baguettes* without definite form. Fracture, uneven. Lustre, vitreous, or pearly. Transparent, translucent, opaque. Becomes opaque from loss of water. Color, white, somewhat red or yellow. Streak, colorless. All of its external characters are however variable as the mineral is generally more or less decomposed by loss of water. **H.**=3.5–4. **G.**=2.25–2.36. Composition, $\dot{C}a$ 11.9, $\dddot{A}l$ 21.9, $\dddot{S}i$ 50.9, $\dot{H}$ 15.3. The $\dot{H}$ may vary. It is always lost by exposure to the air, so that the substance falls to pieces and can only be preserved under water or coated with gum.

Pyr. &c. B. P. Swells and melts at 2.7–3 to a white enamel. Gelatinizes with acids.

It is usually found associated with Calcite and Quartz. It occurs at Bergen Hill, N. J., and is found abundantly at Lake Superior and elsewhere.

FORMULÆ OF THE CRYSTALS.

Pl. XI.

Fig. 4. ∞P. 2P∞; the most usual form. *Fig.* 5. ∞P. ∞P∞. 2P∞. 6P∞. -2P∞. -P. *Fig.* 6. ∞P. ∞P̀∞. 2P∞. 0P. P. -P.

Dioptase Group.

Dioptase. $\dot{C}u\,\dddot{S}i+\dot{H}$. HEXAGONAL.

SYN.—Achirite.

The primitive form is a rhombohedron of 126° 24′, but it is found as an hexagonal prism terminated by one or more rhombohedra, *Pl* XI. *Figs.* 7, 8. It has an easy cleavage parallel to the rhombohedron, Fracture, conchoidal or uneven. Lustre, vitreous. Color, emerald-green, which is sometimes bluish. Streak, green. **H.**=5. **G.**=3.278–3.48. Composition, $\dot{C}u$ 50.4, $\dddot{S}i$ 38.2, $\dot{H}$ 11.4.

Pyr. &c. B. P. Decrepitates, loses water, becomes blackened and gives a half-melted scoria. In the R. F. with Soda, gives a globule of copper. Soluble in ammonia. Gelatinizes with acids.

On account of its color it might be mistaken for Emerald; it is however easily distinguished by its hardness and blowpipe reactions.

It is found in the Steppes of Siberia and is a very rare mineral.

FORMULÆ OF THE CRYSTALS.

Pl. XI.

Fig. 7. $\infty P 2$. $-2R$; the most usual form. *Fig.* 8. $\infty P 2$. $-2R$. $-\frac{12}{7} R \frac{10}{9}$. $-2 R \frac{7}{6}$; the last two forms are hemi-scalenohedral.

Chrysocolla. $\dot{C}u \ddot{S}i + 2 \dot{H}$.

SYN.—Kupfergrün, Kieselkupfer, Kieselmalachit, Kupferblau, Cuivre hydraté silicifère.

It is amorphous and has a composition varying around the formula. Fracture, smooth and conchoidal. Lustre, resinous, vitreous, shining, earthy. Translucent, opaque. Color, bluish-green, sky-blue or turquoise-blue; sometimes brown or black, when impure. Streak, when pure, white, **H.**=2–4. **G.**=2–2.38. Composition, $\dot{C}u$ 45.3, $\dot{H}$ 20.5, $\ddot{S}i$ 34.2.

Pyr. &c. B. P. In a closed tube, blackens and gives off water, Decrepitates and colors the flame green, but is infusible. With fluxes, reacts for copper. Soluble in acids without gelatinizing.

The resinous fracture makes it resemble the resinous varieties of Quartz and Feldspar, but these substances rarely have blue colors. Chrysocolla is distinguished from them by being soluble in acids, and many varieties even in ammonia, giving the characteristic colors of copper. The blowpipe also shows copper. It is sometimes found in large masses with other ores, especially in the upper parts of veins, but it is rarely rich in copper; when pure it contains about 30% and is a valuable ore of copper.

II. Unisilicates.

Calamine Group.

Calamine. $\dot{Z}n^2 \ddot{S}i + \dot{H}$. ORTHORHOMBIC..

SYN.—Electric Calamine, Hemimorphite, Galmei, Zinksilicat, Kieselzinkerz, Zinc silicaté.

It crystallizes as a right rhombic prism of 104° 13′, with an easy cleavage parallel to the prism, and a less easy one parallel to the base. The simple crystals are usually short and thick, and show a great want of symmetry, *Pl.* XI. *Figs.* 9–17. It is quite easy to distinguish Calamine from Willemite, because the terminal modifications are arranged by 2 and 4 and not by 3 and 6. It is usually hemimorphic. Crystals which are thin, parallel to the brachypinacoid are also found, which sometimes become real lamellæ, having a silky lustre and filled with little open spaces in every direction, *Fig.* 11. Twin crystals, joined together on the base also occur, *Fig.* 18. Fracture, sometimes lamellar, sometimes uneven or conchoidal. The transparent crystals have usually a vitreous lustre, which is sometimes adamantine. On the base it is pearly. Sometimes, on account of impure reflections, it appears silky. Transparent, translucent, opaque. When crystallized, it is almost always colorless and transparent. When concretionary, it is yellow, and sometimes blue or green when mixed with salts of copper. Streak, white. When heated, it becomes electric, which gave it the name of Electric Calamine. Its forms are hemihedral, with inclined faces, as is usual with crystals which become electric. **H.**=4.5–5. **G.**=3.16–3.9. Composition, $\acute{Z}n$ 67.5, $\dot{H}$ 7.5, $\ddot{S}i$ 25.

Pyr. &c. B. P. Heated in a tube, decrepitates and gives off water and becomes milky-white. Fuses at 6. Gives a green color with nitrate of cobalt. With soda on Ch., a coating is formed, which is yellow while hot and white when cold. Gelatinizes with acids.

It is found as crystalline incrustations and also concretionary in little mamelons covered with crystals, showing concentric bands of different colors. The fracture of these masses is fibrous which distinguishes them from the carbonate (Smithsonite), which is scaly. It is found compact and is then distinguished from the carbonate by its hardness. Its fracture is rough. Its color is then whitish, sometimes yellow or brown when it is impure, and it is filled with cavities which contain crystals of Calamine, Willemite or Smithsonite. It is sometimes called Electric Calamine. The name Calamine is sometimes wrongly given to carbonate of zinc, which is Smithsonite. It was changed for no good reason and tends to create confusion. It is, however, the industrial name in many districts and it is frequently impossible to know without trial what the ore is, though the distinction Electric Calamine for the silicate and simply Calamine for the carbonate is made.

The lamellar variety resembles Stilbite. Its want of fusibility, gelatinizing with acids, and the coat given off before the blowpipe distinguish it. It is also electric when heated. It is sufficient to heat it, and to bring it near a light body to show its repelling action. It is easily distinguished from Smithsonite by its action with acids; by its infusibility, from ores of lead; and from certain Chalcedonies which it resembles, by its inferior hardness and gelatinizing with acids. Calamine is a valuable ore of zinc. It is found abundantly in Pennsylvania, Virginia and Missouri.

FORMULÆ OF THE CRYSTALS.

Pl. XI.

Fig. 9. $\infty\breve{P}\infty$. ∞P. $0P$. $3\bar{P}\infty$; below, $2\breve{P}2$. *Fig.* 10. The preceding, with $3\breve{P}\infty$. *Fig.* 11. $\infty\breve{P}\infty$. ∞P. $3\bar{P}\infty$. $\breve{P}\infty$; below, $2\breve{P}2$. *Fig.* 12. $\infty\breve{P}\infty$. ∞P. $3\bar{P}\infty$. $\bar{P}\infty$. $0P$; below, $2\breve{P}2$. *Fig.* 13. $\infty\breve{P}\infty$. ∞P. $\bar{P}\infty$. $0P$; below, $2\breve{P}2$. *Fig.* 14. $\infty\breve{P}\infty$. ∞P. $\breve{P}\infty$. $3\bar{P}\infty$. $\bar{P}\infty$; below, $2\breve{P}2$ and $\breve{P}\infty$. *Fig.* 15. $\infty\bar{P}\infty$. $\infty\breve{P}\infty$. ∞P. $3\bar{P}\infty$. $3\breve{P}\infty$. $\bar{P}\infty$. $\breve{P}\infty$. $0P$; below, $2\breve{P}2$. *Fig.* 16. The preceding, with $2\breve{P}2$, $2\bar{P}2$ and $4\breve{P}4$ above. *Fig.* 17. $\infty\bar{P}\infty$. $\infty\breve{P}\infty$. ∞P. $\infty\breve{P}5$. $7\breve{P}\infty$. $2\breve{P}2$. $4\bar{P}\frac{4}{3}$; below, $2\breve{P}2$. *Fig.* 18. Twin crystal; composition-face $0P$. The two crystals are joined together by the lower extremities.

Prehnite. $(\frac{1}{6}\dot{H}^3+\frac{2}{6}\mathring{C}a+\frac{3}{6}\ddot{A}l)^2\ddot{S}i^3$. ORTHORHOMBIC.

SYN.—Koupholite, Chrysolite du Cap.

It crystallizes in the right rhombic prism of 99° 56′, with an easy cleavage parallel to the base, which is perhaps less a cleavage than a tendency of composite crystals to separate, as the apparently simple crystals are often made up of others. It is sometimes found as simple crystals, *Pl.* XI. *Figs.* 19, 20, which are usually very small, only slightly colored and which might be mistaken for Gypsum, but they are very much harder. This variety which is called Koupholite, is usually associated with Asbestus and Epidote. Another form resembles Barite, but is distinguished with a little attention, as all its characters are different. The simple forms of Prehnite, *Figs.* 19–22, are rare; it is almost always found in complex crystals. Sometimes single crystals are placed together by their bases, but not exactly parallel the one to the other, so that while the prismatic faces are straight the bases are curved. At other times the crystals are fan-shaped, placed together parallel to the brachypinacoid. Very large masses are sometimes formed in this way. It is also found in mamelonated masses showing the ends of crystals. The fracture of these masses is fibrous and diverging. They are easily recognized by their greenish color, their semi-transparency and their resinous appearance. Fracture, unequal, often fibrous, sometimes scaly. It has a vitreous lustre, which is often resinous; pearly on the base. Transparent, translucent. Color, yellowish-green passing to white or gray, often faded from exposure. Streak, colorless. Pyroelectric. **H.**=6–6.5. **G.**=2.8–2.95. Composition, $\dot{C}a$ 27.1, $\dot{H}$ 4.4 $\dddot{Al}$ 24.9, $\dddot{Si}$ 43.6.

Pyr. &c. B. P. In a closed tube, gives off water and becomes white. Fuses at 2, with intumescence, to a blebby glass. Some varieties blacken and give an organic odor. With acids, it is decomposed without gelatinizing.

It is usually found with the Zeolites, but does not belong to that family. It sometimes resembles Beryl, green Quartz and Chalcedony. It is distinguished by its fusibility and solubility. It is much harder than any of the Zeolite family. It is sometimes polished and used for ornamental work and is found in many localities of the U. S.

FORMULÆ OF THE CRYSTALS.

Pl. XI.

Fig. 19. 0 P. ∞ P. *Fig.* 20. 0 P. ∞ $\breve{P}$ ∞. ∞ P. *Fig.* 21. ∞ P. 0 P. ∞ $\breve{P}$ ∞. 3 $\breve{P}$ ∞. *Fig.* 22. ∞ P. 0 P. ∞ $\bar{P}$ ∞. P. $\frac{3}{4}$ $\bar{P}$ ∞.

Chlorastrolite. $(\dot{C}a^3, \dot{N}a^3)^2\ \dddot{Si}^3+2\ (\dddot{Al}, \dddot{Fe})^2\ \dddot{Si}^3+6\ \dot{H}$.

It is never crystallized, but is found in radiated, globular masses, sometimes in a gangue of sandstone, but generally detached from it in the sand. Fracture, uneven. Lustre, vitreous or pearly. Opaque. Color, light bluish-green, showing different shades in rings. Slightly chatoyant on the fracture. **H.**=5.5–6. **G.**=3.18. Composition, $\dot{N}a$ 5.2, $\dot{C}a$ 18.7, $\dddot{Fe}$ 6.4, $\dddot{Al}$ 24.6, $\dddot{Si}$ 37.6, $\dot{H}$ 7.5.

Pyr. &c. B. P. In a closed tube, gives off water and becomes white. Fuses easily, with intumescence, to a grayish glass. Reacts with fluxes for iron and alumina. Soluble in acids, with separation of flocculent silica.

It is found only on the shores of Lake Superior.

Apophyllite Group.

Apophyllite. $(\frac{1}{2}\dot{H}+\frac{1}{2}(\frac{1}{9}\dot{K}+\frac{8}{9}\dot{C}a))^2\ \dddot{Si}+\dot{H}\ \dddot{Si}$. TETRAGONAL.

SYN.—Ichthyopthalmite, Albine.

It occurs in several different types of crystals, *Pl.* XI. *Figs.* 23–25 and *Pl.* XII. *Figs.* 1–5. The prism of the second order, with or without the octahedron of the first order, and the base, *Pl.* XI. *Figs.* 23, 24, is found. This form is sometimes so much flattened as to resemble a cube, but the vertical faces are always striated and the base pearly. The prism is often surmounted by the octahedron of the first order, which has a vitreous lustre, while the prism is striated and curved, *Fig.* 25 and *Pl.* XII. *Figs.* 1, 2. Sometimes the octahedron alone is found, *Fig.* 3. These crystals are usually more or less broken, on account of the cleavage, and show the usual pearly lustre of the base. The next variety is very much flattened, the base predominating, *Fig.* 4; or the octahedron has almost obliterated the faces of the prism, *Fig.* 5. These forms might be mistaken for Barite, but the cleavage will distinguish it. It has a very easy cleavage parallel to the base, and a less distinct one parallel to the prism. Lustre, pearly on the base and on the cleavage parallel to it; on the other faces it is vitreous. It is almost always transparent, or at least translucent. It sometimes has an earthy appearance, owing to commencement of decomposition. Often colorless, sometimes slightly yellowish, greenish, bluish or flesh-red. In the Hartz Mts. and in India, a rose variety is found, which is very much esteemed. At Bergen Hill, N. J., large white and transparent crystals were found. A white, opaque variety from Bohemia is called Albine, *Fig.* 2. The translucent, lamellar varieties having the lustre of a fish's eye are called Icthyopthalmite. Streak, colorless. **H.**=4.5–5. **G.**=2.3–2.4. Composition, $\dot{H}$ 16.7, $\dot{K}$ 4.8, $\dot{C}a$ 23, $\dddot{S}i$ 55.5.

Pyr. &c. B. P. In a closed tube, it exfoliates, loses water which is acid, and becomes white. Fuses at 1.5, giving at first a blebby glass, but afterwards a vitreous globule. Reacts for fluorine. Soluble in acids, giving slimy silica.

It is also frequently found in imperfectly crystalline, lamellar masses, made up of a large number of crystals wedged together, thus making a very fragile mass. These varieties usually have an earthy aspect; they are whitish or bluish and sometimes rosy. This opacity is owing to commencement of decomposition, the mineral having lost part of its water. It is easily distinguished by its lustre, cleavage and hardness. It might be mistaken for Barite or Celestite, but the hardness and peculiar lustre, density and action of acids distinguish it. Very large crystals have been found at Bergen Hill, N. J. Some of these have been altered to Pectolite.

FORMULÆ OF THE CRYSTALS.

Pl. XI.

Fig. 23. $\infty P \infty$. 0 P. *Fig.* 24. $\infty P \infty$. P. 0 P. *Fig.* 25. $\infty P \infty$. P. *Pl.* XII. *Fig.* 1. $\infty P \infty$. $\infty P 2$. P. *Fig.* 2. P. $\infty P \infty$. 0 P. *Fig.* 3. P. *Fig.* 4. 0 P. $\infty P \infty$. P. *Fig.* 5. The preceding, with P further developed.

ZEOLITE SECTION.

A certain number of the hydrous silicates, which generally contain for $\dot{R}$ the alkalies and $\dot{C}a$, for $\ddot{R}$, $\ddot{A}l$, have been placed together and called the *Zeolite Group.* The greater part of them have a composition which is almost identical. They are however distinct in crystalline form and present sufficient variation in the chemical composition to be regarded as species and not as varieties. They are all, or nearly all, found under the same circumstances and many of them together, and even associated on the same specimen. Their physical characters are also frequently very much alike. To be able to determine all of them requires practice. They can generally, however, be distinguished by their crystalline form and by a few tests. They are usually found in the amygdaloidal rocks. These are rocks of volcanic origin, having a large number of bases in their composition, and are filled with cavities in which the Zeolites are found lining the sides. They also occur in veins, serving as the vein-rock for metallic minerals. The formation of these Zeolites is posterior to the formation of the rock, which contains them. They owe their formation probably to the infiltration of water charged with silica, in which the different bases were dissolved. These waters were probably under great pressure or at a high temperature and, finding cavities where they could penetrate, deposited crystals of the different Zeolites, accordingly as one base or the other was present in the right proportion. Some of them have been artificially formed. Wöhler has reproduced some by keeping water, charged with the powder of the mineral, for a long time at a temperature of 200° C. At Plombières, Daubry found the Zeolites in the very act of formation. The Romans were in the habit of using the thermal springs that are found there and had constructed different works of art for their use and preservation. The masonry, like all that of this epoch, was made of concrete and of bricks, and the waters of these thermal springs contain silica. The works of the Romans are now abandoned, but the water still flows through them to a certain extent and deposits Zeolites on the sides and in the cavities. It is remarkable that the minerals differ in their composition, according to the part of the masonry where they are deposited. Thus in the concrete, Apophyllite, which contains $\dot{\text{Ca}}$, is always found. In the bricks, however, it is Stilbite which forms, as this contains principally $\dddot{\text{Al}}$. The group of the Zeolites contains a large number of species; some of them are very rare, and others are not very well defined and may be only varieties of the species. We shall therefore not mention all of them but only such as are well defined and are most generally found.

I. Unisilicates.

Natrolite. $\dot{\text{Na}}\,\dddot{\text{Al}}, 3\,\dddot{\text{Si}}, 2\,\dot{\text{H}}$. ORTHORHOMBIC.

SYN.—Natrolith, Spreustein, Natron-Mesotype.

It crystallizes in a right rhombic prism of 91°, and has a cleavage parallel to the prism. It is almost always crystallized. Its usual form is the prism surmounted by the octahedron, *Pl.* XII. *Figs.* 6, 7. This octahedron is easily distinguished from that of Apophyllite, as it is very much flatter. It is rarely ever found in large crystals; they are generally acicular. It is usually found in the Phonolites and Amygdaloids, and at Bergen Hill, in Greenstone. It may be associated with Fluorite, Calcite and Wavellite. Its fracture is conchoidal and uneven; sometimes fibrous, in which case the crystals are placed together in the direction of their length. Transparent, translucent. Its lustre is vitreous, sometimes pearly in the fibrous varieties. Color, white or colorless, sometimes reddish, grayish or yellowish. Streak, colorless. **H.**=5-5.5. **G.**=2.17-2.25. Composition, $\dot{\text{Na}}$ 16.3, $\dddot{\text{Al}}$ 27, $\dddot{\text{Si}}$ 47.2, $\dot{\text{H}}$ 9.5. The oxygen ratio is 1 : 3 : 6 : 2. The $\dot{\text{Na}}$ is sometimes associated with $\dot{\text{Ca}}$.

Pyr. &c. B. P. Loses its water at 90° C. and whitens, becoming opaque. Fuses at 2 to a colorless glass. It is fusible in the flame of a candle. Gelatinizes with acids.

Besides the well crystallized varieties, it is also found in fibrous masses, made up of a large number of crystals placed together, and radiating from a center. Sometimes these radiations show the octahedron, and sometimes they are penetrated by crystals, which have formed on the other side of the rock in the same way. These masses of Natrolite often resemble those of Stilbite, but the latter are always lamellar, on account of an easy cleavage parallel to the brachypinacoid. As this cleavage is not found in Natrolite, the fracture is simply fibrous, and the lustre of Natrolite is not so uniformly pearly like that of Stilbite. One of the varieties of Natrolite has a very peculiar color. It is made up of radiated fibres, which are generally in the shape of spheres that show concentric rings of red, white and yellow. It has been found in fine crystals at Bergen Hill, N. J. and in Lake Superior.

FORMULÆ OF THE CRYSTALS.

Pl. XII.
Fig. 6. ∞P. P. *Fig.* 7. ∞P. ∞P̆∞. P.

II. Bisilicates.

Analcite. $\dot{N}a, \dddot{A}l, 4\,\dddot{S}i, 2\,\dot{H}$. ISOMETRIC.

SYN.—Analcime, Analzim, Kuboit.

It crystallizes in the isometric system, with traces of cleavage parallel to the faces of the cube. It is found in two different associations. In lavas and veins it occurs as the tetragonal trisoctahedron 2 O 2, *Pl.* XII. *Fig.* 8. The crystals are usually quite small, brilliant and transparent, with a decidedly vitreous lustre. In veins and in amygdaloidal rocks it has the same form, but the crystals are much larger, milk-white and opalescent or rosy. They are usually opaque, though sometimes translucent and the lustre is much less decided than in the volcanic variety. In modern lavas the cube predominates and the angles are only truncated with the tetragonal trisoctahedron, *Fig.* 9. These crystals are rarely ever large; they are sometimes transparent and line the sides of cavities. Its fracture is conchoidal, unequal and undulated. Lustre, vitreous. Transparent, translucent, opaque. It is generally white, often perfectly clear, sometimes gray, greenish, yellowish or reddish. Streak, white. **H.**=5–5.5 **G.**=2.22–2.29. Composition, $\dot{N}a$ 14.1, $\dddot{A}l$ 23.3, $\dddot{S}i$ 54.4. $\dot{H}$ 8.2.

Pyr. &c. B. P. Heated in a tube, it loses water and becomes opaque. Fuses at 2.5 to a colorless glass. Gelatinizes with acids.

It is sometimes found compact in red or white masses and is then difficult to determine. The form is the same as that of Leucite; it is easily distinguished however by the action of acids and by the blowpipe. From Quartz it is distinguished by its hardness; from Chabazite, by fusing without intumescence, and by its form. It is found at Bergen Hill, N. J. associated with Natrolite, Stilbite, Apophyllite and Datolite; and at Lake Superior, associated with Copper.

FORMULÆ OF THE CRYSTALS.

Pl. XII.
Fig. 8. 2 O 2; characteristic form. *Fig.* 9. ∞O∞. 2 O 2.

Chabazite. $(\frac{4}{5}\dot{C}a+\frac{1}{5}(\dot{N}a, \dot{K}))$, $\dddot{\bar{A}}l$, 4 $\dddot{S}i$, 6 $\dot{H}$. HEXAGONAL.

SYN.—Acadialite, Haydenite, Chabasit, Phakolith, Chabasie.

It crystallizes in rhombohedra of 94° 46, which very much resemble cubes, *Pl* XII. *Fig.* 10. It has traces of cleavage parallel to the faces of the rhombohedron. The crystals are often simply rhombohedra, without modifications and might be mistaken for cubes as the angle is 94°, but they are easily distinguished by the disposition of the striations. Although the crystals are simple, there is a tendency towards modifications on the terminal edges, so as to form a rhombohedron tangent to the primitive form, *Fig.* 11. This tendency shows itself by striations parallel to the terminal edges. If the form was a cube, the striations would be parallel to all of the edges. The crystals are sometimes twins of two primitive rhombohedra revolved 180° and interpenetrating, so that the edges of one come out of the faces of the other, *Figs.* 12, 13, 14, 15. This form resembles the hexagonal pyramid, with reëntrant angles on the base and pyramidal faces. These crystals resemble Calcite somewhat, but are distinguished by the blowpipe and action with acids. It is always crystallized. The crystals usually line cavities in amygdaloidal rocks. Sometimes, as in the variety called Phacolite, the hexagonal prism has been observed. The fracture is unequal, but this character is not very distinct, as the crystals are usually very small. Lustre, vitreous. Color, white, colorless, reddish, yellowish or greenish. A blood-red variety is found in Nova Scotia and is called Acadialite. The color is owing to the interposition of oxide of iron, which is a mechanical admixture. Streak, colorless. **H.**=4.5. **G.**=2.08–2.19. Composition, $\dot{C}a$ 4–11, $\dot{N}a$ 0–4, $\dot{K}$ 0.17–2.58, $\dddot{\bar{A}}l$ 17–21, $\dddot{S}i$ 45–52, $\dot{H}$ 19–22. Some varieties might be mistaken for Fluorite, but are distinguished by the want of cleavage and the action of acids.

Pyr. &c. B. P. In a closed tube, gives off water. Intumesces and fuses to a white spongy glass, which is nearly opaque. With acids, gives slimy silica.

It is found in Massachusetts, Connecticut, New York and New Jersey.

FORMULÆ OF THE CRYSTALS.

Pl. XII.

Fig. 10. R; usual form. *Fig.* 11. R. $-\frac{1}{2}$R. -2 R. *Fig.* 12. Twin crystal by interpenetration; composition-face 0 P. *Fig.* 13. Twin crystal. *Fig.* 14. Twin crystal, with the combination R. $-\frac{1}{2}$ R. -2 R. $\frac{1}{4}$ R 3. ∞P 2. *Fig.* 15. Twin crystal, $-\frac{1}{2}$ R; form of Phacolite. *Fig.* 16. $\frac{13}{16}$ R $\frac{5}{4}$; Haydenite.

Harmotome. $\dot{B}a$, $\dddot{\bar{A}}l$, 5 $\dddot{S}i$, 5 $\dot{H}$. ORTHORHOMBIC.

SYN.—Morvenite, Kreuzstein.

There are two varieties which differ in their chemical composition. Both of these varieties contain $\dot{K}$, but in one the $\dot{B}a$ is replaced by $\dot{C}a$. The one containing lime is called Phillipsite; the other, Harmotome. These two varieties are also distinct in their crystallographic characters.

It crystallizes in a right rhombic prism of 120° 47′, with easy cleavages parallel to the base and prism. The variety containing $\dot{B}a$ is in detached crystals, which are always twins, *Pl.* XII. *Figs.* 17–21. These crystals might be mistaken for Apophyllite, but are distinguished by the striations. The octahedron is also less acute than in Apophyllite. When the mineral has $\dot{C}a$ for a base, the crystals are twinned parallel to the prism, showing the characteristic cross. Sometimes the faces of the prism do not exist and in others they are very smalll. Simple crystals are unknown. Even the crystals containing $\dot{B}a$ which appear simple are twins, as can be seen by the optical characters and striations, *Fig.* 17. The macle is however as it were latent and does not affect the crystalline form. Fracture, conchoidal and uneven. Lustre, vitreous on the exterior faces, but less bright on the fracture. Transparent on the edges. Color, white, gray, yellow, red, brown or greenish. It possesses a certain degree of opalescence, which is owing to the presence of a certain amount of $\dddot{S}i$ in excess, which is in a gelatinous condition. Streak, white. **H.**=4.5. **G.**=2.44–2.45. Composition, $\dot{B}a$ 23.7, $\ddot{\bar{A}}l$ 15.9, $\dddot{S}i$ 46.5, $\dot{H}$ 13.9; when it contains lime, $\dot{C}a$ 7.4, $\ddot{\bar{A}}l$ 20.5, $\dddot{S}i$ 47.9, $\dot{K}$ 6 3, $\dot{H}$ 17.9.

Pyr. &c. B. P. Whitens and becomes opaque; exfoliates and melts at 3.5 to a white glass. Sometimes phosphoresces. Soluble with separation of pulverulent $\dddot{S}i$. This mineral is frequently found with Chabazite. Its usual localities are Scotland and the Hartz.

FORMULÆ OF THE CRYSTALS.

Pl. XII.

Fig. 17. $\infty \breve{P} \infty$. $\infty \bar{P} \infty$. P. $\breve{P} \infty$. *Fig.* 18. Twin crystal; composition-face ∞P. *Fig.* 19. Twin crystal by interpenetration. *Fig.* 20. The same. *Fig.* 21. Horizontal projection of a twin crystal like the preceding.

Stilbite. $\dot{C}a$, $\ddot{\bar{A}}l$, 6 $\dddot{S}i$, 6 $\dot{H}$. ORTHORHOMBIC

SYN.—Desmine, Strahlzeolith.

It crystallizes as a rhombic prism of 94° 16′. It has an easy cleavage parallel to the brachypinacoid and a less easy one parallel to the orthopinacoid. The usual forms are shown in *Pl.* XII. *Figs.* 22, 23. The faces of the prism and octahedron have a feeble and vitreous lustre, while that of the brachypinacoid is pearly and very bright. The base is sometimes found on these crystals, but it is very small.. They are not generally found detached, but as a large number of crystals attached to one another parallel to the brachypinacoid. This face is in this case the most brilliant, the others being striated, rounded and bent. It frequently happens, and this is the general case, that the crystals are not placed together exactly parallel, but diverging slightly from the center of the crystal, so that it looks somewhat fan-shaped. The faces of the brachypinacoid are then curved, but still pearly and very brilliant, while the others are striated and dull. Fracture, uneven. Lustre vitreous, pearly on the brachypinacoid. Color, usually white, but often colored yellow by $\ddot{\bar{F}}e$ $\dot{H}$, or blood-red by $\ddot{\bar{F}}e$. This iron is interposed between the lamellæ of the crystal and is not a part of its chemical composition. It is sometimes found of a gray color, which is due to organic matter. This color generally affects the point of the crystal. Streak, colorless. **H.**=3.5–4. **G.**=2.094–2.205. Composition, $\dot{C}a$ 8.9, $\ddot{\bar{A}}l$ 16.5, $\dddot{S}i$ 57.4, $\dot{H}$ 17.2. The hardness is not the same on the different faces, it being less on the brachypinacoid than on the other.

Pyr. &c. B. P. Swells, becomes fan-like and vermicular and melts to a white enamel. Soluble in acids, with separation of pulverulent silica.

Besides the distinctly crystallized varieties, it is found in masses which are spheroidal and made up of a large number of crystals grouped together. These masses have a radiated, fibrous structure and are at the same time lamellar, on account of cleavage parallel to the brachypinacoid which is very easy. They have a characteristic pearly lustre. It is quite rare in these masses not to find a distinct crystal showing the fan-like shape. They frequently have a gray, rose or yellow color. It might sometimes be confounded with Gypsum, but is distinguished from it by its pearly lustre and its hardness. The intensity of its lustre will generally distinguish it from Apophyllite, which is usually less pearly. It might also be confounded with one of the oxides of antimony, but a blowpipe trial distinguishes it immediately. It is found in Massachusetts, Connecticut, New York, New Jersey and Lake Superior.

FORMULÆ OF THE CRYSTALS.

Pl. XII.

Fig. 22. $\infty \breve{P} \infty$. $\infty \bar{P} \infty$. P. *Fig.* 23. $\infty \breve{P} \infty$. $\infty \bar{P} \infty$. ∞P. P. 0 P.

Heulandite. $\dot{C}a, \dddot{A}l, 6\,\dddot{S}i, 5\,\dot{H}$. MONOCLINIC.

SYN.—Blätterzeolith, Euzeolith.

It crystallizes as an inclined rhombic prism of 136° 4′. The ordinary forms are given in *Pl.* XII. *Figs.* 24–28. These crystals might at first sight be taken for Stilbite, as they, like it, have a pearly lustre on the brachypinacoid and a vitreous lustre on the other faces. It is, however, quite easy to distinguish them by the form of the face of the brachypinacoid. That of Stilbite is symmetrical, while that of Heulandite is dissymmetrical, as may easily be seen in the figures. It has a very easy cleavage, parallel to the clinopinacoid. This cleavage is easier even than that of Stilbite. Fracture, conchoidal or uneven. Lustre on the clinopinacoid, pearly, elsewhere vitreous. Transparent, translucent, opaque. Colors, white, gray, yellow and often dark red. **H.**=3.5–4. **G.**=2.2. Composition, $\dot{C}a$ 9.2, $\dddot{A}l$ 16.9, $\dddot{S}i$ 59.1, $\dot{H}$ 14.8. The composition, with the exception of one atom of water, is the same as that of Stilbite. It differs from it, however, in its crystalline form, which is monoclinic.

Pyr. &c. B. P. Acts like Stilbite.

It is not always easy to appreciate this difference of crystalline form, since the two substances have the same associations and are often found together on the same specimen. The alkalies are often associated with the $\dot{C}a$.

It is found in Connecticut, Massachusetts, New York, New Jersey, Pennsylvania and Maryland.

FORMULÆ OF THE CRYSTALS.

Pl. XII.

Fig. 24. $\infty \grave{P} \infty$. $\infty P \infty$. $\grave{P} \infty$. 0 P. *Fig.* 25. The preceding form, lengthened in the direction of the orthodiagonal. *Fig.* 26. The combination, *Fig.* 24, with 2 P. *Fig.* 27. The preceding, with $\frac{4}{3}$ P. *Fig.* 28. $\infty \grave{P} \infty$. $\infty P \infty$. $\grave{P} \infty$. $2 \grave{P} \infty$. 2 P; usual form of the crystals from the Fassathal.

MARGAROPHYLLITE SECTION.

I. Bisilicates.

Talc Group.

Talc. $(\frac{4}{5}\dot{M}g+\frac{1}{5}\dot{H})\ddot{S}i$. ORTHORHOMBIC.

SYN.—Steatite, Soapstone, Potstone, Speckstein, Agalmatolite, Pagodite.

It crystallizes in a right rhombic prism of 120°. It is found more or less crystallized, but the crystals are never complete; they are usually nothing but lamellæ, having generally an angle near 120°, or lamellar masses. These masses show an easy cleavage parallel to the base. Fracture, uneven. Translucent, opaque. Lustre, argentine or pearly. Color, green, white, red, gray. Streak, white, or lighter than color. It is flexible but not elastic, which allows of its being distinguished from Mica. Its touch is unctuous and soapy, on account of the large quantity of magnesia which it contains. Its powder is even more unctuous than the mass. **H.**=1.15. **G.**=2.565–2.8. Composition, $\dot{M}g$ 33.5, $\ddot{S}i$ 62.8, $\dot{H}$ 3.7. The composition is quite variable.

Pyr. &c. B. P. It whitens, swells and sometimes decrepitates a little, fusing with difficulty on the thin edges. It does not give off water like the Micas, except at a very high heat. With nitrate of cobalt, it gives the characteristic color of magnesia. Not affected by acids.

In its physical characters it resembles the Micas, but it differs from them in its chemical composition, which, however, is not always fixed. It may be referred to a general formula, in which $\dot{M}g$ is the $\dot{R}$. It rarely contains $\dot{F}e$. It contains water and fluorine in varying proportions, which appear to be essential to its composition. These two substances seem to be present as a sort of witness to the circumstances under which it was formed. It is sometimes found in lamellar fibers, coating the sides of rocks and placed perpendicularly to them. When it is formed on the sides of narow fissures, these fibers meet and interlace and often twist and turn on each other. It is also found in lamellar, schistose masses, which are not very pure. It is also found earthy and compact when it is known under the name of Steatite. Its fracture is then compact. It attracts the tongue, which distinguishes it from the Kaolins and clays. It is soapy to the touch. These varieties are always very impure. Compact and earthy Talc often replaces by pseudomorphism a number of mineral species. In some places crystals of Quartz have been found completely transformed into Talc. There is no explanation to this phenomenon. Did the Talc fill up a cavity left by the Quartz, or was it substituted molecule for molecule, as Quartz itself has become substituted in organisms? These questions have not as yet been answered. This pseudomorphism appears, however, to be intimately connected with the decomposition of minerals which the Talc has replaced. A slaty variety is known as talcose schist. These schists have a very variable aspect and composition. Talc is also found in lamellæ, which are generally unctuous and very tender. To this variety belongs Pagodite and Agalmatolite, a rock from which the Chinese are in the habit of cutting figures. It is possible that some of the images are artificially made, as the Chinese are most skilled masters of falsification. Some of these earthy varieties are capable of supporting heat and are used, as the variety called Soapstone and Potstone, for culinary and heating purposes. The powdered mineral is often used as a lubricator. It is found abundantly in the U. S.

Sepiolite Group.

Sepiolite. $\dot{M}g^3 \ddot{S}i^3+2\dot{H}$.

SYN.—Meerschaum, L'Ecume de Mer.

It is never crystallized, but always found in amorphous masses, with a conchoidal fracture, which is sometimes granular when it is mixed with foreign substances. Its fracture is dull and without lustre, while its natural surfaces are slightly lustrous. Color, white, grayish-white, yellowish or reddish. Opaque. Its natural surfaces are smooth and unctuous to the touch, like all the magnesian compounds. When it is pure, it does not give an odor when breathed upon and does not attract the tongue. **H.**=2-2.5. When dry, floats upon water. Composition, $\dot{M}g$ 27.1, $\dot{H}$ 12.1, $\ddot{S}i$ 60.8. It sometimes contains $\ddot{C}$, and the proportion of $\dot{H}$ is quite variable.

Pyr. &c. B. P. Heated in a tube, it gives off water. It cracks and fuses with difficulty on the edges. With nitrate of cobalt, it gives the reaction for magnesia. With acids, it sometimes effervesces slightly, and gelatinizes.

It is sometimes found porous and sometimes quite compact. It is the common meerschaum of commerce. It is found in beds interstratified between lime or clay. It is then often lamellar and impure, often containing these substances. It is no longer a chemical combination but a sort of magma. It is generally found in Greece or Asia Minor.

II. Unisilicates.

Serpentine Group.

Serpentine. $(\frac{3}{4}\dot{M}g+\frac{1}{4}\dot{H})^2\ddot{S}i+\frac{1}{2}\dot{H}$. ORTHORHOMBIC.

SYN.—Verd antique, Chrysotile, Picrolite, Marmolite, Schiller Spar, Williamsite, Baltimorite.

Serpentine is found crystallized only as pseudomorphs after other minerals. Fracture, scaly sometimes conchoidal. Lustre, resinous, greasy, pearly or earthy. Translucent, in thin splinters; opaque. It has generally a more or less green color, inclining to yellow, but is sometimes darker and red. Its color is sometimes homogenous, and sometimes composed of many colors blended together without order, but sometimes in bands. Streak, white and shining. It may be easily cut and worked, and it has some appearance of ductility. It is however tenacious, and resists the stroke of the hammer somewhat like Corneine. Its touch is unctuous, as it is principally composed of magnesia. The composition varies as follows: $\ddot{S}i$ 42-46, $\dot{M}g$ 36-40, $\dot{H}$ 12-13. For the formula, $\dot{M}g$ 42.97, $\ddot{S}i$ 44.14, $\dot{H}$ 12.89. Its water appears to be in part hygrometric, for a part of it disappears under the air pump. It contains frequently as much as 1-2 % of organic matter. Its composition is very variable, but may be represented by the formula given above. Its composition is, however, never very well defined, so that it is rather a rock than a mineral.

Pyr. &c. B. P. Heated in a tube, it gives off water. Fuses on the edges, with great difficulty, at 6. With nitrate of cobalt, it gives the characteristic rose of magnesia. It is always somewhat attacked by acids, sometimes even entirely dissolved.

The highly colored varieties are sometimes called noble Serpentine, and are in use as ornaments. Like Talc, it is found in limestones, and gives a peculiar variety of marble, which is much sought for in the arts. Without being crystallized, it shows traces of crystallization. It is supposed to crystallize in the orthorhombic system. A brilliant and lamellar substance is sometimes found in Serpentine, which is usually referred to Diallage; it has, however, the same general composition as, and is undoubtedly Serpentine. Other fibrous substances are also found in Serpentine. They are not very well characterized, but are quite near it in composition. Without being very well defined, they have optical characters that bring them near to crystallized substances. Thus, they possess dichroism, being yellowish-green in one direction, and red in another. These fibrous substances are sometimes found in veins and resemble Asbestus. They are called Chrysotile. Serpentine is found abundantly in the U. S.

III. Subsilicates.

Chlorite Group.

Chlorite.

CHLORITE is the name given to a family of minerals. They are analogous to Mica and Talc in their exterior characters, and also in the variety of their composition. The $\dot{R}$, is $\dot{M}g$ and $\dot{F}e$; the $\bar{R}$, $\bar{F}e$, $\ddot{\bar{A}}l$, $\ddot{\bar{C}}r$. It is more than probable that a part of the $\dot{H}$ is associated with the $\bar{R}$.

Prochlorite. $(\frac{4}{7}(\dot{M}g, \dot{F}e)^3+\frac{3}{7}\ddot{\bar{A}}l)\dddot{S}i+\frac{4}{3}\dot{H}$. HEXAGONAL.

SYN.—Chlorite.

It crystallizes in an hexagonal prism and has a very easy cleavage parallel to the base, which allows of its being exfoliated, but less easily than Mica. The crystals are not very well formed. They are generally in hexagonal or triangular lamellæ. Generally it is imperfectly crystallized, the lamellæ being fastened to each other on the edges and radiated, so as to form a spheroidal mass, a section of which would be fan-shaped. This variety is usually called Ripidolite. Its lustre is very bright and sometimes almost metallic on the edges. It is also somewhat vitreous. Transparent, in thin lamellæ; opaque. Color, green, more or less dark. Some varieties show a remarkable dichroism. Parallel to the base in the direction of the greatest length of the prism, it appears somewhat dark-red and even black, while perpendicular to the base it is green. Streak, white or greenish. **H.**=1–2. **G**=2.78–2.96. Composition, $\dot{M}g$ 15.3, $\dot{F}e$ 27.5, $\ddot{\bar{A}}l$ 19.7, $\dddot{S}i$ 26.8, $\dot{H}$ 11.7. Its touch as well as its powder is unctuous, less so however than Talc. This difference is probably owing to the greater proportion of water.

Pyr. &c. B. P. In a closed tube, gives off water. Whitens, exfoliates and fuses, with difficulty, to a grayish-black glass. Decomposed by $\dddot{S}$.

Prochlorite is frequently found as a powder covering other minerals, as Quartz, Albite and Orthoclase. It is found in lamellar masses, so that it almost always has a spathic fracture. This variety is abundant in many localities, and is frequently found with Vesuvianite. It is also found in compact masses, somewhat analogous to Talc, from which it can be distinguished by the proportion of water. It is also found in little detached plates forming sand, resulting from the weathering action on rocks which contained it. These sands, some of them at least, were formed at a geological epoch anterior to our own, and have sometimes been transported considerable distances. It is thus found disseminated in different formations. It is found abundantly in the U. S.

POTASSIUM.

Nitre. $\dot{K}\,\ddot{N}$. ORTHORHOMBIC.

SYN.—Saltpetre, Kalisalpeter, Potasse nitratée.

The primitive form is a right rhombic prism of 118° 50′, consequently it is one of the limit forms whose combinations resemble the Hexagonal system. Crystals, when they are simple, have the forms *Pl.* XIII. *Figs.* 1,2,3,4,5. They are frequently hemitrope around the brachy-axis. The crystals are often twins, which are mostly the union of several crystals, attached by the faces of the prism, *Pl.* XIII. *Fig.* 8. generally the interior is left hollow. Crystals of Nitre, which are not hollow must be formed by the double decomposition of $\dot{K}\,\ddot{C}$ and $\dot{N}a\,\ddot{N}$ and crystallized in the same liquid. The crystals obtained from solution in pure water are usually twins and are made up of 2, 3, or even 4 crystals. Nitre is dimorphous; at a high temperature it crystallizes in rhombohedra of 105°30′. Calcite has a rhombohedron of 105°5′ and is isomorphous with Nitre; Calcite is also dimorphous with the right rhombic prism of 120°, which is isomorphous with Nitre. These two minerals are therefore isomorphous in their dimorphism. This property has been called isodimorphism. Nitre has difficult cleavages parallel to the prisms and to a brachy dome. Its fracture is conchoidal. It is usually white and transparent or at least translucent. Its lustre is vitreous and its streak white. It is generally found as thin crusts, efflorescences or as small crystals. **H.**=2. **G.**=1.937. The composition of Nitre is $\dot{K}$ 46.6, $\ddot{N}$ 53.54.

Pyr. &c. Deflagrates on Ch. colors the flame violet ($\dot{K}$). It is soluble in its weight of cold, and half its weight of warm water. It has a peculiar, cooling, saline taste. It is not altered by exposure.

Nitre is the saltpetre of commerce. It is found abundantly in the soils of some parts of Spain, Egypt and Russia. It is also manufactured from soils containing other nitrates. In this country it is found in caves in Madison Co., Ky., in Tennessee and in the Mississippi valley.

FORMULÆ OF THE CRYSTALS.

Pl. XIII.

Fig. 1. $\infty P.\ \infty\breve{P}\infty.\ P.\ 2\breve{P}\infty.$ This combination resembles the ordinary crystal of quartz, $\infty P.\ P.$ when the faces of the brachy-dome and brachy-pinacoid are respectively equal to those of the pyramid and prism. *Fig.* 2. The above with $\breve{P}\infty$. *Fig.* 3. $\infty P.\ \infty\breve{P}\infty.\ 2\breve{P}\infty$ *Fig.* 4. The combination *Fig.* 3, with $\breve{P}\infty$. but more tabular. *Fig* 5. The combination *Fig.* 4, with $4\breve{P}\infty$. *Fig.* 6, Section parallel to 0P, showing the mirrors of the prismatic cleavage. *Fig.* 7. The same, in a twin crystal. *Fig.* 8. Trilling crystal, hollow, formed by the juxtaposition of three prisms.

Aphthitalite. $\dot{K}\,\dddot{S}$. ORTHORHOMBIC.

SYN.—Glaserite, Kalisulphat, Potasse sulfatée.

It crystallizes in right rhombic prisms of 120° 30′. Its fracture is conchoidal and it scratches **Gypsum.** It is generally white or colored slightly blue or green; transparent, translucent or opaque; lustre, vitreous or resinous. It often appears like an hexagonal prism both in the simple crystals, which are quite rare, and in the crystals which are composite 3 by 3 or 6 by 6. When obtained by crystallization in a solution, where it has been produced by double decomposition with $\dot{N}a\ \ddot{C}$, it is dimorphous and gives an hexagonal prism. **H.**=3—3.5. **G.**=1.731. Composition, $\dot{K}$ 54.1, $\dddot{S}$ 45.9.

Pyr. &c. Before the blowpipe, it decrepitates and melts. In the red flame it is reduced to K S; the moistened globule then blackens silver. It is not very soluble, has a bitter taste.

Anhydrous sulphate of potash is found among volcanic products in crystalline crusts which appear to be produced by the vapors of $\dddot{S}$ on rocks containing alkalies.

SODIUM.

Soda Nitre. $\dot{N}a\ \overset{\cdot\cdot\cdot\cdot\cdot}{N}$. HEXAGONAL.

SYN.—Nitratin, Natron Salpeter, Soude nitratée.

It crystallizes as a rhombohedron of 106° 33′, it is isomorphous with Dolomite and Nitre, which under certain circumstances takes the same angle. When it is evaporated rapidly it takes the form of a rhombic prism of 117° or 118° and is thus isodimorphous with Nitre. It has cleavages parallel to the primitive rhombohedron. It is generally white or grayish, with a slight vitreous lustre; transparent, translucent or opaque. Its fracture is sometimes granular, sometimes lamellar or conchoidal. **H.**=2. **G.**=1.937. Composition, $\dot{N}a$ 36.5, $\overset{\cdot\cdot\cdot\cdot\cdot}{N}$ 63.5.

Pyr. &c. Deflagrates on Ch., but less than Nitre; colors the flame yellow. It is soluble in 3 parts of $\dot{H}$ and has a cooling, saline taste.

Soda Nitre is found in great abundance in certain countries, especially in Peru, where it is found in crystalline masses of large extent, which appear to be the residues of ancient lakes. It is also found in India, where it continually effloresces on the surface of the ground.

Thenardite. $\dot{N}a\ \dddot{S}$. ORTHORHOMBIC.

It is generally found in crystalline masses, with forms that approach the prism. They are usually rhombic octahedra, with very sharp edges. They have a cleavage parallel to the base and brachy-pinacoid. Lustre, vitreous; color, white to brown; translucent. **H.**=2—3. **G.**=2.5 to 2.7. Composition, $\dot{N}a$ 56.3, $\dddot{S}$ 43.7.

Pyr. &c. Colors the flame yellow; in R. F. gives Na S. If dissolved above 33°, on cooling it crystallizes as anhydrous and below it as hydrous.

It is found in deposits left by Thermal springs.

FORMULÆ OF THE CRYSTALS.

Pl. XIII.

Fig. 9. P. Surface, rough and with but little lustre. *Fig.* 10. P. 0P. ∞P.

Mirabilite. $\dot{N}a\ \dddot{S}+10\ \dot{H}$. MONOCLINIC.

SYN.—Glauber Salz, Mirabilite, Soude sulfatée.

The primitive form is an inclined rhombic prism 86° 31′. Lustre vitreous, color white; transparent, opaque, **H.**=1.5—2. **G.**=1.481. Composition, $\dot{N}a$ 19.3, $\dddot{S}$ 24.8, $\dot{H}$ 55.9.

Pyr. &c. Yields water in a tube, gives a yellow flame, gives the reactions for S. Tastes cool, afterwards salty and bitter. In the air, falls to pieces and becomes anhydrous.

In Spain, especially near Madrid, it is found in large masses not crystallized, but fibrous and bent. These masses are semi-transparent, with a slight pearly lustre, and appear to result from alteration. In the air, these masses effloresce rapidly; they might sometimes be mistaken for salt. It might be mistaken for $\dot{Mg}\,\dddot{S}$, but this is less soluble and less efflorescent.

FORMULÆ OF THE CRYSTALS.

Pl. XIII.

Fig. 11. $\infty\bar{P}\infty.\ \infty P.\ P.\ 0P.\ \infty\grave{P}\infty.\ \bar{P}\infty.\ \grave{P}\infty.$ *Fig.* 12. $\infty\bar{P}\infty.\ \infty\grave{P}\infty.\ \infty P.\ P.-P.\ \frac{1}{2}P.\ \bar{P}\infty.\ \frac{1}{2}P\infty.-\frac{1}{2}P\infty.\ P\infty.\ 2P\infty.\ 0P.$

Glauberite. $(\frac{1}{2}\dot{Na}+\frac{1}{2}\dot{Ca})\dddot{S}$. MONOCLINIC.

The form is a very oblique rhombic prism. If not crystallized, it is crystalline. It may have two types, which are formed by the same laws of derivation, but differ somewhat in their exterior. The first type, *Fig.* 13, *Pl.* XIII. is the ordinary one. It is the primitive crystal with the modification $-P$, but very flat parallel to $0P$, with sharp edges. The angle ∞P being very obtuse, 104°15′, the face $-P$ is very much striated, so much so as sometimes to appear step shaped. This type is that generally found in Spain. in the salt mines and in the variegated clays which accompany them. The same variety has been found in France, at Vic, colored red with iron. In Peru, at Iquique, Glauberite has been found with the second type, which shows the same faces, but the crystal is much longer from the extension of I; this variety is usually found in colorless crystals, with a very bright resinous lustre. It has a very easy cleavage parallel to the base. The fracture is conchoidal and brilliant. Its lustre is resinous and sometimes very bright. Its ordinary color is yellow, somewhat gray, but when $\ddot{Fe}$ is present, it is red. Streak, white. **H.**=2.5 to 3. **G.**=2.64 to 2.85. Composition, $\dddot{S}$ 57.5, $\dot{Ca}$ 20.1, $\dot{Na}$ 22.4.

Pyr. &c. B. P. Decrepitates and melts into a bead which is transparent, when hot, but opaline, when cold. By the action of $\dot{H}$ the two sulphates are isolated, the $\dot{Na}\,\dddot{S}$ being dissolved. At first it has no taste, then it gives successively a salt, bitter and astringent taste.

Near Madrid, a large mass of Glauberite was found, 14 to 15m. thick and several leagues square. It was from the decomposition of this mass that the fibers of $\dot{Na}\,\dddot{S}$ came, of which we have spoken. It is generally found in the same associations as salt.

FORMULÆ OF THE CRYSTALS.

Pl. XIII.

Fig. 13. $0P.-P.$ *Fig.* 14. $0P.-P.\ \infty P.\ -P\infty.$ *Fig.* 15. $\infty P.\ 0P.-P.\ -P\infty.$

Halite. Na Cl. ISOMETRIC.

SYN.—Salt, Rock Salt, Kochsalz, Steinsalz, Soude muriatée, Sel gemme, Halites.

It crystallizes in the isometric system, almost always in cubes. It has a very easy cubical cleavage, so that the crystals are generally filled with mirrors in three directions. Lustre, vitreous; streak, white. Its colors are very variable. When pure, it is colorless, but generally it is colored by some earthy or organic matter. It may be gray, red, violet, blue or green. The cause of all these colors is not very well understood. They may be owing to traces of $\dot{N}i$, $\dot{C}o$, $\dot{C}u$, or organic matter. It is soluble, and has its own peculiar taste. The fracture of salt is lamellar and sometimes conchoidal across the cleavage. **H.**=2.5. **G.**=2.1.-2.257. Composition, Na 39.3, Cl 60.7.

Pyr. &c. Heated, it first decrepitates and then melts. Some peculiar varieties do not decrepitate. It sometimes decrepitates, when it is dissolved, which is owing to the presence of cavities in the crystals. These cavities are often cubical in shape and contain water and $\ddot{C}$, which is under pressure and produces the phenomena of decrepitation, when by solution the sides of the cavities are no longer strong enough to bear the pressure.

Besides the crystallized and lamellar varieties, it is frequently found in fibrous masses, which sometimes adhere to the rock by one extremity only; the fibers are then very much bent. When it fills fissures, the fibers are usually straight, and the two opposite walls of the rock are often joined. This variety is frequently colored red by oxide of iron.

It is found in Europe in the deposits of Gypsum, and especially in the variegated clays (Marnes Irisées) of the Trias. In the U. S. it has been found forming beds associated with Gypsum in Virginia and Oregon. In Louisiana, in large beds of later origin. Most of the salt obtained in this country is from Salt springs.

FORMULÆ OF THE CRYSTALS.

Pl. XIII.

Fig. 16. ∞O∞. *Fig.* 17. Hopper-shaped crystal. *Fig.* 18. Cavernous cube.

Borax. $\dot{N}a\,\ddot{B}^2 + 10\,\dot{H}$. MONOCLINIC.

SYN.—Tinkal, Borsaures Natron, Soude boratée.

It crystallizes as an inclined rhombic prism of 87°, with the very simple forms shown in *Pl.* XIII. *Figs.* 19–21. It has a number of difficult cleavages. The crystals are mostly white and opaline, but when they are stained by organic matter they are colored gray, more or less dark, sometimes bluish or greenish, and then give off an empyreumatic odor when heated. It is isomorphous with Pyroxene. Streak, white; translucent, opaque; fracture, conchoidal. It is soluble, has a salt, alkaline soapy taste; boiling water dissolves double its weight. **H.**=2.25. **G.**=1.716. Composition, $\dot{N}a$ 16.2, $\ddot{B}$ 36.6, $\dot{H}$ 47.2.

Pyr. &c. B. P. Intumesces and fuses to a transparent glass. Fused with Fluorite and bisulphate of potash, gives the reaction for $\ddot{B}$.

Borax is found in nature either in solution in certain lakes, or as a residue of their evaporation. It is found most abundantly in India and California.

FORMULÆ OF THE CRYSTALS.

Pl. XIII.

Fig. 19. ∞P̄∞. ∞P. ∞P̀∞. P. 2 P. 0P. *Fig.* 20. ∞P̄∞. ∞P. P.2P 0P. *Fig.* 21. ∞P̄∞. ∞P̀∞. ∞P. P. 2 P. 4 P̀∞ 0P.

Natron. $\dot{N}a\ \ddot{C}+10\ \dot{H}$. MONOCLINIC.

SYN.—Nitrum, Soude carbonatée.

The usual form is given *Fig.* 22, *Plate* XIII.

Under this name a number of mixtures of different carbonates have been described, as for example the neutral and sesqui-carbonate with varying proportions of $\dot{H}$, which are found in certain African lakes. These masses are white or gray, from the interposition of foreign substances. They are generally earthy. In these masses crystals may be seen, which may be white or opaque and seem to belong to a partially decomposed sesqui-carbonate; the rest of the mass being a neutral carbonate, which is much less efflorescent. In the air, the crystals usually fall to powder, and have a very disagreeable alkaline taste. **H.**=1–1.5. **G.**=1.423. Composition $\dot{N}a$ 18.8, $\ddot{C}$ 26.7, $\dot{H}$ 54.5.

Pyr. &c. B. P. Heated, these masses give up water and fuse. Soluble in water. In acid, soluble with great effervescence.

FORMULÆ OF THE CRYSTALS.

Pl. XIII.
Fig. 22. P. ∞P. ∞$\grave{P}$∞.

AMMONIUM.

Mascagnite. $\dot{N}H^4\ \dddot{S}+\dot{H}$. ORTHORHOMBIC.

SYN.—Mascagnin, Ammoniaque sulfatée.

Right Rhombic Prism of 107° 40′.

It is found as an incrustation on volcanic rocks. It is similar to the same salt in the laboratories. It is isomorphous with Aphthitalite. Lustre, vitreous; translucent; color, yellowish gray to lemon yellow; taste, pungent and bitter. **H.**=2–2.5. **G.**=1.72–1.73. Composition $\dot{N}H^4$ 34.7, $\dddot{S}$ 53.3, $\dot{H}$ 12.

Pyr. &c. It is soluble and volatile, giving off very thick vapors. It has the peculiar flavor of Sal-Ammoniac, but is bitter, which distinguishes it.

It is generally found on volcanoes and in coal mines which have been on fire.

Sal-Ammoniac. $NH^4\ Cl$. ISOMETRIC.

Tetragonal-trisoctahedron and hexoctahedron. Sometimes lengthened as in *Fig.* 26, *Pl.* XIII. Lustre, vitreous. Color white, yellowish or grayish. Fracture, conchoidal. **H.**=1.5–2. **G.**=1.528. Composition $\dot{N}H^4$ 33.7, Cl 66.3.

Pyr. &c. Sublimes in a closed tube without fusion. Heated with lime or caustic alkali, gives ammonia reaction. It is soluble in three times its weight of $\dot{H}$ and has a peculiar saline and pungent taste.

If in the solution of Sal-Ammoniac a few milliems of acid are added, it crystallizes in cubes. If however salt is added, it crystallizes in tetragonal-trisoctahedra. They have an octahedral cleavage. It is isomorphous with the Sylvite and Halite, but these have a cubical cleavage. It is found in certain volcanic formations and has the same formula that it has in the laboratories. It is generally found as incrustations on the surfaces of lakes, or in coal mines on fire. These incrustations are white or colored yellow by Fe Cl which is volatilized with it, as frequently happens in the laboratories.

FORMULÆ OF THE CRYSTALS.

Pl. XIII.—*Fig.* 23. O. *Fig.* 24. 303. *Fig.* 25. $30\frac{3}{2}$. *Fig.* 26. 303, very much distorted.

BARIUM.

Barite. Ḃa S̈. ORTHORHOMBIC.

SYN.—Heavy Spar, Schwerspath, Baryte sulfatée.

It crystallizes as a right rhombic prism of 101° 40′, with three cleavages; two parallel to the prism and one parallel to the base. Crystals are found, according to localities, in four general types. The first is quite flat. It is composed of the macrodome and brachy-pinacoid. These crystals are usually not very large, and are often modified as *Figs.* 28, 29, 30, *Plate* XIII. The second is also flat, made up of the base and the macro and brachy-pinacoids, *Fig.* 31, *Plate* XIII. and shows a rectangular form. This may also be modified. It belongs to the varieties which are gangues of metals. The third type belongs to veins. The base has disappeared and the brachy-dome become very large, so that the crystal has a roof shape, *Fig.* 34, *Plate* XIII. The fourth type is much rarer. It shows a predominance of the macro-dome, so that it has the dome at right angles to the third type, *Fig.* 2, *Pl.* XIV. The disposition of the cleavage allows of the distinction being always made between this form and the previous one. Besides the forms which have been given, the crystal often has faces which have a much more complicated notation. These faces are generally small and are distinguished by their difference in lustre. Its colors are usually not very distinct. White, gray or yellow are the usual ones, but it is sometimes blue or reddish. The yellow is usually owing to F̈e, and the red often to Cinnabar. The lustre of the mineral is vitreous, generally very bright; sometimes it is resinous. **H.**=2.5–35. **G.**=4.3–4.86. Composition Ḃa 65.7, S̈ 34.3.

Pyr. &c. B. P. Decrepitates and fuses to a white enamel, which falls to pieces after some time. The flame, during the fusion, is colored yellowish-green. The globule obtained in the red flame is BaS, moistened, it blackens Ag. It is insoluble to all intents and purposes, but not completely so. It has the same composition as the Ḃa S̈ of the laboratories.

Besides the simple forms, it is often grouped. One of the most frequent is the crested form, made up of crystals which are usually small and which are radiated in the form of a sphere. It is also found in lamellar and bacillary masses which are striated in the direction of their lengths, and have mirrors in every direction, owing to easy cleavages. In concretionary masses, in stalactites, which are fibrous in their interior, having usually different varieties of brown for their colors, which seem mostly to be owing to organic matter. Saccharoidal, having the aspect of Italian marble. Sometimes it is entirely earthy. It may also be found in every intermediate stage. Compact, as the gangue of veins, looking quite like Petrosilex or Agate, but is easily distinguished by its inferior hardness. The character, which is the one depended on to distinguish it from Celestite and Aragonite, is its density, which is 4.5. It is distinguished from the various carbonates by not effervescing with acids: from metallic minerals, by giving no metallic reaction before the **B. P.**

Uses.—Its general use is as an adulteration for white lead. A mixture of equal parts of Barite and white lead is often called *Venice White.* Twice the weight in Barite is called *Hamburg White.* Three times, *Dutch*

White. The use of Barite is often advantageous, especially when the paint is likely to be exposed to sulphurous vapors. It then protects the lead from becoming immediately blackened. Barite used for this purpose must be very white, and is often purified, especially from iron, by digestion in dilute $\ddot{S}$.

FORMULÆ OF THE CRYSTALS.

Pl. XIII.

Fig. 27. $\infty\breve{P}\infty$. $\bar{P}\infty$. *Fig.* 28. $\infty\breve{P}\infty$. $\breve{P}\infty$. $\bar{P}\infty$. *Fig.* 29. $\infty\breve{P}\infty$. $\bar{P}\infty$. $\breve{P}\infty$. ∞P. *Fig.* 30. $\infty\breve{P}\infty$. $\bar{P}\infty$. $\breve{P}\infty$. P. *Fig.* 31. $\infty\breve{P}\infty$. $\infty\bar{P}\infty$. 0P. Rectangular form. *Fig.* 32. $\infty P\infty$. $\infty\breve{P}2$. $\breve{P}\infty$. *Fig.* 33. The same elongated in the direction of $\breve{P}\infty$. *Fig.* 34. $\bar{P}\infty$. $\breve{P}\infty$. *Fig.* 35. $\bar{P}\infty$. $\breve{P}\infty$. $\infty\breve{P}2$. $\infty\breve{P}\infty$. *Pl.* XIV. *Fig.* 1. The same, elongated in the direction of $\infty\breve{P}2$. *Fig.* 2. $\bar{P}\infty$. $\breve{P}\infty$. *Fig.* 3. $\infty\breve{P}\infty$. $\bar{P}\infty$. $\breve{P}\infty$. $2\breve{P}\infty$. *Fig.* 4. $\infty\breve{P}\infty$. $\bar{P}\infty$. $\breve{P}\infty$. $\infty\breve{P}2$. ∞P. *Fig.* 5. $\infty\breve{P}\infty$. $\infty\breve{P}2$. $\bar{P}\infty$. $\breve{P}\infty$. $2\breve{P}\infty$. *Fig.* 6. $\infty\breve{P}\infty$. $\infty\breve{P}2$. $\bar{P}\infty$. P. *Fig.* 7. $\infty\breve{P}\infty$. $\infty\breve{P}2$. $\infty\breve{P}4$. $\bar{P}\infty$. 0P. *Fig.* 8. $\infty\breve{P}\infty$. $\infty\breve{P}2$. $\infty\breve{P}5$. $\bar{P}\infty$. 0P.

Witherite. $\dot{B}a\ddot{C}$. ORTHORHOMBIC.

SYN.—Baryte carbonatée.

The primitive form is a right rhombic prism of 118° 30′. It is one consequence of the limit forms. It has a cleavage parallel to the prism. Its primitive form is rarely ever found. It is generally in hexagonal prisms.

These hexagonal crystals are rarely ever simple. They are generally macles made up of six other crystals. They often show an empty space in the centre, because, since the angles are not quite 60°, they do not exactly coincide. This juxtaposition takes place on the faces of the prism. When the prism disappears, as it frequently does, we have a pseudo-hexagonal pyramid made up of the brachy-dome and rhombic octahedron. These crystals are themselves frequently hemitrope, which does not change the form but makes the mass opalescent instead of transparent. Its fracture is unequal, scaly, with a resinous lustre, while the natural faces are usually covered with a pulverulent coating. Colors generally whitish, yellowish, or grayish; transparent or translucent. Streak, white. **H.**=3–3.75. **G.**=4.29–4.35. Composition $\dot{B}a$ 77.7, $\ddot{C}$ 22.3.

Pyr. &c. B. P. With acids, it effervesces slowly. It must be heated to dissolve it. It decrepitates, colors the flame yellowish-green and fuses to a limpid globule, which becomes opaque and is absorbed by the charcoal.

It is also found in crystalline masses, and is sometimes concretionary and fibrous. Its fracture then has a peculiar silky look. It is, at the same time, scaly. These masses are remarkably translucent. Their density will distinguish them from similar masses of Calcite. It is distinguished from Calcite and Aragonite by its fusibility; by giving a green flame, from Strontianite; by its effervescence, from similar minerals which are not carbonates and, by giving no metallic reaction, from metallic minerals. It is largely used in chemical manufactories for the preparation of Salts of

Baryta, which are used in analysis and, to some extent, in pyrotechny. Artificially prepared, it is used as a paint for water-color drawings. It is sometimes used as a poison.

FORMULÆ OF THE CRYSTALS.

Pl. XIV.

Fig. 9. P. 2 P̆ ∞. Both forms are equal and apparently make a hexagonal pyramid. *Fig.* 10. The former combination, with 0P. *Fig.* 11. ∞P. ∞P̆∞. P. 2 P̆∞. 0P. *Fig.* 12. ∞P. ∞P̆∞. P. 2P̆∞. $\frac{3}{2}$ P. 2 P. 3 P̆∞. 4 P̆∞. *Fig.* 13. Horizontal projection of a composite crystal, formed by six simple crystals, ∞P. being the composition face.

Barytocalcite. ($\frac{1}{2}$ Ḃa+$\frac{1}{2}$ Ċa) C̈. MONOCLINIC.

Its primitive form is an inclined rhombic prism of 106° 54′. The usual form is shown in *Fig.* 14. *Pl.* XIV. Cleavage, easy parallel to the prism, less easy, parallel to the base. Lustre, vitreous. Color white, gray, greenish or yellowish. Fracture, uneven. Streak, white. Composition Ḃa C̈ 66. 3, Ċa C̈ 33.7.

Pyr. &c. B. P. Fuses with difficulty on the edges, coloring the flame yellowish-green and becomes greenish. With Ṅa C̈ on Ch., the Ḃa is absorbed and the Ċa remains on the Ch. Soluble in dilute N̈. This mineral is comparatively rare. It is quite remarkable that it is not isomorphous with the variety of Calcite which contains Ṗb C̈, while the carbonates of Ṗb and Ḃa are both prismatic and isomorphous.

FORMULÆ OF THE CRYSTALS.

Pl. XIV.—*Fig.* 14. ∞ P. ∞ P̄ 3. P. P̄ ∞.

STRONTIUM.

Celestite. Ṡr S̈. ORTHORHOMBIC.

SYN.—Celestine, Cälestin, Strontiane sulfatée.

The primitive form is a right rhombic prism of 104° 2′. The bluish crystals have usually the first type of the crystals of Barite. The white Celestite from Sicily has another form, which is that of the last type, *Fig.* 18, *Pl.* XIV. It has also another form which seems to belong to the tertiary formations. This is from the round masses of Celestite which are found at Paris and in the silex of the chalk. These are produced as follows: During the tertiary epoch, while the formation of Gypsum and of chalk was going on, the silex of the chalk was more or less cracked. Into these cracks waters, charged with Ṡr S̈ in solution, infiltrated and soldered them together.

All of these crystals have a special form, from which the primitive type has entirely disappeared. They are often acicular, and it is not always easy to distinguish them as orthorhombic.

Celestite is isomorphous with Barite and Anglesite. The prismatic angle which in ḂaS̈ was 101° is 104° in Ṡr S̈. The cleavages are the same. The crystals are usually smaller and less distinct than those of Barite. It is oftener white than Ḃa S̈ and has frequently a bluish tinge, which is characteristic and from whence it received the name Celestite. Sometimes it is soluble, occasionally trichroic. Its lustre is very bright, often pearly, generally less resinous than that of Barite. Its fracture is lamellar and sometimes conchoidal. Streak, white. **H.**=3–3.5. **G.**=3.92–3.975. Composition, Ṡr 56.4, S̈ 43.6.

Pyr. &c. B. P. Decrepitates and fuses, coloring the flame red. If fused in R. F. and then dissolved in acid, it gives to the alcohol flame the characteristic red color. On Ch., acts like Barite. Insoluble in acids.

It is also found in lamellar masses like Barite, but the cleavage is less easy. It is also found fibrous, usually in parallel fibres filling up cracks in the rocks. They are usually perpendicular to the sides of the rock, sometimes on one side, sometimes on the other. These masses are generally blue. In regarding the cleavages, it will be seen that each fibre is a crystal which belongs to the type of the tertiary formation. It is often saccharoidal when it is found in marls and clays. It is also found in the tertiary formations in compact earthy masses. These are generally found in rounded masses, which often lie together in beds, like the silex of the chalk formation. These masses are often hollow or cracked in their interior, which is then filled with crystals of the tertiary type. The density of Celestite is great, it is however a little less than Barite never reaching 4. It is quite difficult to distinguish it from Barite, especially when it is in lamellar, saccharoidal or compact masses; a test must then be made. It is reduced to a sulphuret, dissolved in H Cl or $\overset{.....}{N}$ and mixed with alcohol to get the flame. The two bases might be distinguished by the crystals of Nitrate or Chloride.

Celestite is used in the arts for making Nitrate of Strontia which produces the red color in fireworks. An attempt was made, some years ago, to use the very large deposit on Strontian Island, in Lake Erie, for adulterating white paint, but it does not appear to have been successful, as only a few cargoes were used.

FORMULÆ OF THE CRYSTALS.

Pl. XIV.

Fig. 15. $\bar{P}\infty$. $\breve{P}\infty$. $\infty\breve{P}\infty$. *Fig.* 16. $\bar{P}\infty$. $\infty\breve{P}2$. $\breve{P}\infty$. $\infty\breve{P}\infty$.
Fig. 17. $\breve{P}3$. $\breve{P}\infty$. *Fig.* 18. $\bar{P}\infty$. $\breve{P}\infty$.

Strontianite. $\dot{Sr}\ddot{C}$. ORTHORHOMBIC.

Crystallizes as a right rhombic prism of 117° 19′. It has a cleavage parallel to the prism and to the brachy-pinacoid. The usual forms are shown in *Figs.* 19, 20, 21, 22. *Pl.* XIV. It is isomorphous with Witherite. The pseudo-hexagonal pyramids are however much rarer. The crystals are usually complete prisms, but are almost always small and not very distinct. They are most frequently needles, the details of which are only visible with a glass. Lustre, vitreous or resinous. Colors gray, white, yellow, brownish and pale green; transparent, translucent. Streak, white. Fracture, uneven. **H.**=3.5–4. **G.**=3.605–3.713. Composition $\dot{Sr}$ 70.2, $\ddot{C}$ 29.8.

Pyr. &c. B. P. Swells, arboresces and fuses on the thin edges. Moistened with HCl, gives the Strontia color. With $\dot{Na}\ddot{C}$ on Ch., fuses to a clear glass and all the $\dot{Sr}$ is absorbed by the Ch. Soluble in H Cl.

It is often found in concretionary masses which are fibrous and tufty, and are usually yellowish, with a very resinous lustre. These concretionary masses are often grouped round a crystal. It usually has a greenish color, which is characteristic, and which has caused the name of asparagus stone to be given to it. This name, however, properly belongs to Apatite. These masses are generally concretionary, translucent and analogous to Witherite. The density would confound it with $\dot{B}a\ \ddot{C}$, when the distinction must be made by the color of the flame. It is used for Pyrotechnics.

FORMULÆ OF THE CRYSTALS.

Pl. XIV.

Fig. 19. ∞P. ∞$\breve{P}$∞. 0P. P. 2$\breve{P}$∞. *Fig.* 20. The preceding, with ⅓ P. $\breve{P}$∞. *Fig.* 21. ∞P. ∞$\breve{P}$∞. 0P. P. ⅓ P. *Fig.* 22. ∞P. ∞$\breve{P}$∞. 0P. P. ⅓ P. 2$\breve{P}$∞. The last face predominant.

CALCIUM.

Anhydrite. $\dot{C}a\ \dddot{S}$. ORTHORHOMBIC.

SYN.—Muriazite, Karstenite, Chaux sulfatée anhydre.

It crystallizes as a right rhombic prism of 100° 30′, with three easy, but unequal cleavages which lead to a rectangular prism, usually the dominant form. Lustre, vitreous and bright on the cleavage faces; when massive, it is vitreous or pearly. It is translucent or transparent, and generally colorless. The colors are never very decided, but are light tints of gray, rose, blue, or violet. Fracture, uneven. Streak, white. **H.**=3-3.5. **G.**=2.899-2.895. Composition, $\dot{C}a$ 41.2, $\dddot{S}$ 58.8. It is found especially in association with salt, in fibrous masses, which are generally bent and show traces of cleavage in the direction of a normal to the curves. In this condition, it resembles some of the masses of Celestite, but its density is much lower, being 2.9 at the most. Like Celestite, it is found bluish in certain localities. It is also found in saccharoidal masses, analagous to Calcite, but it is usually more transparent, its density is greater and its cleavage is different. It sometimes occurs in large, compact masses, with a conchoidal fracture, which may also be rectangular or scaly. These masses can be distinguished by their transparency and hardness.

It is susceptible of decomposition, or rather of alteration, as in damp air it becomes hydrated, effloresces and changes to Gypsum. Masses thus altered have been found, retaining their transparency and their rectangular cleavage, but which can be scratched by the nail, and are Gypsum, at least near the surface of the crystal. At Bex, in Switzerland, the beds are sometimes altered to the depth of 20 to 30m. This alteration is also quite frequent in collections. It is the cause of difficulty in working certain salt mines. When one of the galleries of the mines passes through a mass of Anhydrite, at the end of a certain time the latter becomes hydrated on the surfaces and swells. This throws down the wood-work or supports of any kind that may have been used, so that the gallery has to be rebuilt, sometimes several times. until the alteration has taken its full effect and has entered for a considerable distance into the mass.

Anhydrite is distinguished from Gypsum by being harder; from the zeolites and the carbonates, by its action with acids. Some varieties are cut and polished and are called alabaster. It is found at Lockport, N. Y., associated with Dolomite and limestone.

Pyr. &c. B. P. Heated, it becomes white, coloring the flame reddish-yellow, but does not exfoliate; heated longer in the R. F., it melts at 3, with an enamel which breaks up and becomes pulverulent and which, when moistened, shows the usual reaction for sulphur. With

$\dot{N}a\,\dot{C}$, does not give a clear bead and is not absorbed by the charcoal. Soluble in HCl. 100 parts of $\dot{H}$ at 18° C dissolve 0.2 parts.

It is generally found in semi-crystalline, lamellar masses, quite analagous to Calcite, but more transparent and with rectangular instead of rhombohedral cleavages.

FORMULÆ OF THE CRYSTALS.

Pl. XIV.

Fig. 23. 0P. $\infty\bar{P}\infty$. $\infty\breve{P}\infty$. ∞P. $\breve{P}\infty$. P. $2\bar{P}2$. $3\bar{P}3$. *Fig.* 24. ∞P. $\breve{P}\infty$. Stassfurth. *Fig.* 25. 0P. ∞P. $\breve{P}\infty$. Stassfurth.

Gypsum. $\dot{C}a\,\dddot{S}+2\,\dot{H}$. MONOCLINIC.

SYN.—Gyps, Chaux sulfatée, Selenite, Satin Spar.

It takes the form of the oblique rhombic prism of 138° 28′, and has three cleavages. The one parallel to the clinopinacoid is very easy, the two others are perpendicular to this; one of them has a vitreous lustre, the other is fibrous. The lustre is pearly, owing to the fact that the mass is not perfectly coherent. It is rarely ever found in its primitive form, on account of the very easy cleavage parallel to $\infty\grave{P}\infty$. The two others are perpendicular to this one, one parallel to the base, which has the fibrous cleavage, and the other is parallel to $\infty\bar{P}\infty$, which is vitreous. They sometimes have the modifications $\infty\grave{P}\infty$, with the faces of the prism and hemioctahedron P. Sometimes the faces of the prism have the other hemioctahedron -P. on the lower acute angle. Sometimes the faces of the hemiorthodome $\bar{P}\infty$ appear on the obtuse angle.

These intermediate faces are almost always curved, which appears to take place in the middle rather than on the edges. The faces of the prism ∞P, frequently disappear and give rise to rounded figures, *Fig.* 34, *Pl.* XIV.

The mirrors of cleavage however are visible upon these rounded aces. These crystals are often hemitrope around one of the faces, which gives rise to arrow-shaped crystals, which, notwithstanding the rounding of the faces of the hemioctahedron, have the line of junction always straight. Sometimes a large number of these crystals are joined, forming a peculiar wedge-shaped crystal.

Other hemitropes also occur by a hemitropy parallel to $\infty\bar{P}\infty$. The same twins sometimes occur that we have noticed as occurring in the Orthoclase of the Trachytes by half penetration, *Fig.* 37. When this takes place in lenticular crystals, the form of the crystal is hardly recognizable. A great many crystals are frequently grouped together and have very complicated dispositions. Gypsum is quite soft and is easily scratched with the nail. Its colors are very variable, generally not very strong. It is genally colorless or whitish. It often has $\ddot{F}e$ interposed, when it is red. It is also yellow, blue, brown or black, which colors seem to be owing to organic matter. It is often transparent or translucent. It has a vitreous lustre, which, on some of the faces. may be adamantine. It is soluble in HCl, or in 400 to 500 parts of $\dot{H}$. **H.**=1.5-2. **G.**=2.314-3.28. Composition, $\dot{C}a$ 32.6, $\dddot{S}$ 46.5, $\dot{H}$ 20.9.

Pyr. &c. B. P. At the first touch of heat, it swells and exfoliates in direction of the easy cleavage, becomes white and loses its water. It then fuses at 2.5 to 3, coloring the flame reddish yellow, and acts like Anhydrite, giving a globule of CaS.

Gypsum is largely found in the clays of the secondary formation, which are found accompanying salt. In this association, the crystals are generally white, and often in flexible lamellæ. It is also found in very large beds in the tertiary formations. These crystals are whitish or sometimes colored by organic or bituminous matter, in lamellæ, which are not very flexible. In some exceptional cases, crystals are found which are very much bent even in their cleavage. Gypsum is also found in lamellar masses, made up of a large number of imperfect crystals, which often show hollows, as their union is not very perfect. It is also found fibrous, lining cavities. These fibers are generally parallel, and have a bright, silky lustre. It is found saccharoidal, with a grayish color, and is made up of little, imperfect crystals, not very coherent and often slightly transparent. This is the condition of the plaster of Paris in the environs of Paris. When found compact and with a conchoidal fracture, it constitutes the ordinary alabaster which is used for making ornamental objects. These are not of very great value, because, being very soft, they are easily deteriorated. In some places, as an accident, Gypsum is found in scales, which are very small, flat crystals, white and almost uniform in size. They are sometimes found in cavities in beds of Gypsum. They can hardly be mistaken for anything but Barite, but their association and a chemical test will distinguish them. Like Calcite, it is very often found in the condition of concretions, which are often very well crystallized. Crystals are often found on the sides of galleries of abandoned quarries; and concretions in steam boilers, when waters containing $\dot{C}a\,\ddot{S}$ in solution are used, or in the pipes through which these waters have flowed. In the exploitation of salt, especially at Bex, in Savoy, beautiful crystals are found, which have been deposited by the waters used to dissolve the salt.

Gypsum frequently resembles Heulandite, Stilbite, Talc and the Micas, from which its hardness and action with B. P. will readily distinguish it. The fibrous varieties resemble fibrous Quartz, Asbestus, some varieties of Calcite, and some of the fibrous Zeolites. The hardness, action of acids and of the B. P. distinguish it. When burnt and ground, it constitutes the Plaster of Paris of commerce. The white fibrous varieties are often made into personal ornaments. It is also used in agriculture, as a manure.

It is found in large beds in New York, Ohio, Illinois, Michigan, Virginia, Tennessee and Arkansas, and is usually associated with salt springs. Crystals are found at Lockport, N. Y., St. Mary's, Maryland, Poland and Canfield, Ohio. At the Mammoth Cave, in Kentucky, it occurs in the form of rosettes, stalactites, and many other fanciful shapes.

FORMULÆ OF THE CRYSTALS.

Pl. XIV.

Fig. 26. ∞P. $\infty \grave{P} \infty$. P. $-P$. *Fig.* 27. The preceding, with the prismatic faces smaller. *Fig.* 28. $\infty \grave{P} \infty$. ∞P. $-P$. *Fig.* 29. $\infty \grave{P} \infty$. ∞P. $\infty \grave{P} 2$. $-P$. $\frac{1}{8} \bar{P} \infty$. *Fig.* 30. $-P$. ∞P. $\infty \grave{P} \infty$. $\frac{1}{8} \bar{P} \infty$. *Fig.* 31. *Fig.* 29, with the pyramidal faces curved. *Fig.* 32. Lenticular crystal. *Fig.* 33. ∞P. P. $-P$. $\infty \grave{P} \infty$. with curved faces. *Fig.* 34. P. $-P$. with curved faces. *Fig.* 35. Twin crystal, with the combination *Fig.* 28. Composition-face parallel to $\infty \bar{P} \infty$. *Fig.* 36. Twin crystal, with the combination *Fig.* 26. *Fig.* 37. Twin by interpenetration. *Fig.* 38. Crystal, showing the twin plane $-\bar{P} \infty$. *Fig.* 39. Twin from the preceding. *Fig.* 40. Group of arrow-headed crystals.

Fluorite. Ca Fl. Isometric.

Syn.—Flussspath, Fluor Spar, Derbyshire Spar, Chaux fluatée.

It has a very easy octahedral cleavage. The most usual form of Fluorite is the cube, *Fig.* 1, *Pl.*XV. or the tetrahexahedron, *Fig.* 8 or combinations of both, *Fig.* 9. It is also found in rhombic dodecahedra and octahedra, *Fig.* 2, tetragonal-trisoctahedra and hexoctahedra, *Fig.* 11, with or without the cube, but these forms are rare. A complex crystal is shown in *Fig.* 12. Distorted form, *Fig.* 15. Twins, *Figs.* 13, 14. The fracture is generally lamellar, conchoidal and sometimes scaly in the compact varieties. Streak, white. Its lustre is vitreous and usually very bright. The colors of Fluorite are very variable. It is sometimes entirely colorless and transparent. It has all the shades of yellow, violet, green, red and blue. Crystals are sometimes found which show both the blue and green color. This is, however, not a true dichroism and is called Fluorism. These crystals are monorefringent and become phosphorescent under the influence of heat or light. Some substances are phosphorescent only while they are under the influence of the rays of light, but others remain so afterwards for a longer or shorter time. Fluorite is one of the latter. The same specimen often shows several distinct colorations. **H.**=4. **G.**=3.01–3.25. Composition, $\dot{C}a$=51.3, Fl=48.7.

Pyr. &c. B. P. It decrepitates, becomes milk white and fuses to an enamel, coloring the flame red ($\dot{C}a$). With Salt of Phosphorus, gives the reaction for Fl. When it contains foreign substances, it remains colored, except when the colorations are produced by organic matter. It is insoluble in HCl or $\dddot{N}$; $\dddot{S}$ dissolves it easily without effervescence, giving off H Fl. which attacks glass.

It is also found concretionary in layers, which usually show a very marked difference of color. The line of junction of these different layers is usually zigzag, caused by an attempt at crystallization. The color of these masses is often green, violet or rose. In this condition it is sometimes used to make ornamental objects. It was of this variety that the Murrhina vases, so much esteemed by the ancients, were made. In this condition it is very frequently associated with Quartz. It is also found saccharoidal or granular with a grayish color. It might in this condition be mistaken for Barite, but its density will distinguish it. It is found compact, as the gangue of some veins. These masses have a smooth, conchoidal fracture, sometimes scaly. Their colors are generally feeble and unequal in different parts of the mass. In this condition it might be confounded with Barite, Petrosilex and Agate. From the last two, it is distinguished because it is easily scratched by the knife; from Barite, by its density and a test with $\dddot{S}$. Where it is semi-crystalline, it sometimes resembles the emerald, but differs from it by its hardness. It may also resemble certain varieties of Wernerite, but is distinguished by the cleavage, which is distinct and vitreous in Fluorite, while it is resinous and scaly in Wernerite. It also resembles Apatite, but is insoluble in all but $\dddot{S}$, while Apatite is soluble in other acids without residue. Among the varieties, there is one which contains a small proportion of $\dddot{A}l$; it shows opaque and transparent parts, which are quite hard and strike fire with the steel. These contain $\dddot{S}i$ or $\dddot{A}l$, or sometimes both. Perhaps these two substances are in the state of Si Fl, or perhaps as a fluosilicate of lime. In many localities the crystals of

Fluorite which were formed have been afterwards destroyed, and have left cavities in the rock which have been afterwards filled with Quartz. By our ordinary methods we are forced to consider Fluorite insoluble, which shows that we do not understand the method by which this pseudomorph has taken place. Under the name of Derbyshire Spar, it is much used as an ornament. It is largely employed in making hydrofluoric acid, which is used for engraving glass or silicious minerals. It is largely used as a flux for reducing ores and hence its name. It is found in the U. S. in veins, sometimes associated with metals. At Trumbull, Ct., it is associated with Topaz; at Amity, N. Y., in narrow seams with Spinel and Tourmaline.

FORMULÆ OF THE CRYSTALS.

Pl. XV.

Fig. 1. ∞O∞. *Fig.* 2. O. *Fig.* 3. O. ∞O∞. *Fig.* 4. ∞O∞ O. *Fig.* 5. ∞O∞. 2O2. *Fig.* 6. O. ∞O. *Fig.* 7. ∞O∞. ∞O. *Fig.* 8. ∞O3. *Fig.* 9. ∞O∞. ∞O3. *Fig.* 10. ∞O∞. 4O2. *Fig.* 11. 4O2. *Fig.* 12. ∞O∞. ∞O. 2O2. 3O. 4O2. *Fig.* 13. Twin crystal. *Fig.* 14. Twin crystal. *Fig.* 15. 2O2. Distorted to the obliteration of some of the faces and forming a scalenohedron. *Figs.* 16, 17, 18, 19, 20. O. Forms obtained by cleavage.

Apatite. $\dot{C}a^3 \overset{\cdot\cdot\cdot\cdot\cdot}{\bar{P}} + \frac{1}{3} Ca (Cl, F)$. HEXAGONAL.

SYN.—Asparagus Stone, Chaux phosphatée, Phosphate of Lime.

It has an imperfect cleavage parallel to the base, and a less easy one parallel to the prism. Its fracture is unequal, conchoidal, rarely lamellar, but generally smooth and brilliant. Transparent or opaque. Crystals of Apatite have generally the forms of the hexagonal prism, which is often striated in the direction of the vertical axis. The colorless crystals from St. Gothard, frequently have the hemihedral forms with parallel faces, *Pl.* XV. *Fig.* 25. Sometimes the crystals are very much distorted, *Fig.* 1, *Pl.* XVI The crystals have often the appearance of having been subjected to the action of some solvent, which has rounded the faces and given them a vitreous appearance. This is sometimes carried so far that the faces of the crystal are entirely rounded, losing all trace of their form. Such crystals are usually disseminated in a gangue of Talc. The streak is white. The lustre of Apatite is generally vitreous on the plane faces, but when the faces are curved the lustre is rather resinous. The colors of Apatite are very various. In some places, as in the Alps, especially near St. Gothard, small transparent crystals are found, which are colorless and have a very brilliant lustre. They are oftenest found in cracks in the rocks, and the crystals have a large number of faces. Generally, however, Apatite is found in large crystals, which are yellow, green, blue or violet. The colors are not quite so brilliant as those of Fluorite. **H.**=4.5–5. **G.**=2.92–3.25. Composition, $\overset{\cdot\cdot\cdot\cdot\cdot}{\bar{P}}$ 40.92, $\dot{C}a$ 48.43=89.35; Cl 6.81, Ca 3.84 or $\overset{\cdot\cdot\cdot\cdot\cdot}{\bar{P}}$ 42.26, $\dot{C}a$ 50=92.26, $\dot{C}a$ $\overset{\cdot\cdot\cdot\cdot\cdot}{\bar{P}}$; F 3.77, Ca 3.97. Sometimes the $\overset{\cdot\cdot\cdot\cdot\cdot}{\bar{P}}$ is replaced by $\overset{\cdot\cdot\cdot}{\bar{A}}s$.

Pyr. &c. B. P. Fuses with difficulty on the edges at 4.5, coloring the flame red ($\dot{C}a$). It is soluble in HCl and $\overset{\cdot\cdot\cdot\cdot\cdot}{N}$, without residue. This reaction is not absolute, for when it contains Ca Fl, there is a slight residue. It is sometimes phosphorescent in the dark, especially in powder.

In Sweden, at Rossie, in the state of New York, and in Canada, very large crystals have been found which are almost always green. They have a calcareous gangue, and have a six-sided or basic termination. They are frequently contorted or bent. These crystals have very often a rounded appearance and a very resinous lustre. In certain rare cases, Apatite might be mistaken for emerald, but can be distinguished from it by its lustre, hardness and cleavage. Apatite is sometimes found in grains, with traces of cleavage. These grains are easily distinguished by their lustre. In Spain, an earthy variety of Apatite is found, which, without examination, might be mistaken for a silicate. Its fracture is semi-crystalline, or fibrous in some parts. It can be distinguished by the blowpipe and its solubility in acids. A compact variety, resembling an impure limestone, has been found near Charleston, S. C. It is used in making fertilizers.

FORMULÆ OF THE CRYSTALS.

Pl. XV.

Fig. 21. ∞P. P. Particularly on Moroxite. *Fig.* 22. ∞P. 0P. P. The most usual form. The prismatic faces often striated by ∞P2. *Fig.* 23. The preceding, with 2P2. *Fig.* 24. ∞P. 0P. $\frac{1}{2}$P. 2P2. *Fig.* 25. ∞P. 0P. P. 2P. 2P2. 3P$\frac{3}{2}$. P2. ∞P2. From St. Gothard and interesting, owing to the hemihedral occurrence of 3P$\frac{3}{2}$ and ∞P$\frac{3}{2}$. *Fig.* 26. ∞P. ∞P2. 0P. $\frac{1}{2}$P. P. 2P. P2. 2P2. 4P2. *Pl.* XVI. *Fig.* 1. Distorted crystal, with the following form. *Fig.* 2. ∞P. ∞P2. 0P. $\frac{1}{2}$P. P. 2P. 2P2. 3P$\frac{3}{2}$. 4P$\frac{4}{3}$. P2.

Pharmacolite. $(\frac{2}{3}\dot{C}a+\frac{1}{3}\dot{H})^3\ddot{\ddot{A}}s$. MONOCLINIC.

Crystallizes as an inclined rhombic prism of 111° 6′. It is usually found as the product of the alteration of metallic sulphides, especially of Nickel and Cobalt, in presence of lime. It is usually found in crystalline needles, or in cotton like efflorescences. Fracture, uneven. Lustre, vitreous. Cleavage, parallel to the clinopinacoid. Translucent, opaque. Color white, gray, or tinged red with cobalt. Streak, white. **H.**=2.-2.5. **G.**=2.64-2.73. Composition, $\ddot{\ddot{A}}s$=51.1, $\dot{C}a$=24.9, $\dot{H}$=24. Its formula varies somewhat, but it is always hydrated.

Pyr. &c. B. P. Yields $\dot{H}$ in a closed tube. In O. F., fuses with intumescence to a white enamel; on charcoal in R. F. gives an arsenical odor and fuses to a transparent bead. Insoluble in $\dot{H}$, but soluble in acids.

FORMULÆ OF THE CRYSTALS.

Pl. XVI.

Fig. 3. 0P. -P. ∞P. ∞$\grave{P}$∞. $\frac{1}{2}$$\dot{P}$∞.

Aragonite. $\dot{C}a\ddot{C}$. ORTHORHOMBIC.

SYN.—Arragon Spar, Flos Ferri, Eisenblüthe.

Crystals of Aragonite are rarely ever simple. When they are, they usually have the form *Figs.* 4–10, *Pl.* XVI. The exterior of these crystals shows a form which is generally more simple than their real structure. Their section frequently shows a hemitrope around the prism, *Fig.* 12. Instead of being like *Fig.* 12, sometimes the crystals are hemitrope several times and then have a section like *Fig.* 13. The complex crystals show a base which is grooved or, at least, striated, and this shows its twin formation. They also sometimes have a formation which is more or less complex, where the different crystals can be distinctly seen, as in *Fig.* 17, which is made up of several crystals. The angle of Aragonite is very near 120° and it thus takes those limit forms, which are so near to the hexagonal prism, as in Witherite and Strontianite.

Aragonite is found in great abundance and especially in iron mines, as at Bastenes, in France. At Aragon, in Spain, it occurs in crystals, apparently made up of the hexagonal prism and base. These crystals are really made up of four or six simple crystals and are very complex, as the striations on the surface show. They sometimes have a hollow centre or a cavity, from which other crystals start, *Fig.* 14. In other iron mines, especially at Framont, in the Vosges, and at Rossie, N. Y., a pseudo-hexagonal pyramid is found, which is made up of an acute rhombic octahedron and brachydome. The crystals of this last variety often pass into bacillary crystals, made up of almost parallel *baguettes*, which are sometimes arranged in diverging branches terminated on the surface by faces of crystals. These masses do not show any sign of cleavage, and usually have an unequal and conchoidal fracture. When the *baguettes* are very fine and arranged close together, the fracture is silky.

It has three cleavages; parallel to the brachypinacoid it is distinct, but it is imperfect, parallel to the prism and brachydome. Its fracture is scaly or subconchoidal. Its lustre is bright, vitreous or slightly resinous, especially in the concretionary varieties. Transparent or translucent. Color, white, gray, yellow, green and violet. Streak, white. **H.**=3.5–4. **G.**=2.927–2.947. Composition, $\dot{C}a$ 56, $\ddot{C}$ 44.

Pyr. &c. B. P. Heated in a tube, Aragonite decrepitates, and takes the form and density of ordinary Calcite, thus losing trom 3 to 4% of its own. Whitens, falls to pieces and then acts like Calcite. With acids, it effervesces slowly.

Aragonite is $\dot{C}a\ \ddot{C}$, crystallized under peculiar conditions. Calcite ordinarily crystallizes in the hexagonal system, while Aragonite is orthorhombic, the angle of the prism being 116° 10′. It is isomorphous with Witherite and Strontianite, with which it sometimes occurs. It not only differs from Calcite in its form, but also in its associations and its position. Thus, while hexagonal $\dot{C}a\ \ddot{C}$ may be reproduced under ordinary temperatures, when that temperature is raised to 60° or 70°, it crystallizes orthorhombic with the form of Aragonite. The conditions of crystallization may also be modified by causing it to take place in the presence of foreign substances. Thus the addition of salt allows of the formation of Aragonite at a lower temperature. This is a remarkable dimorphism produced by change of condition, and it is all the more remarkable, as it is the transition between the prismatic and rhombic carbonates. This same dimorphism was thought to be shown in Siderite, and Dufrénoy thought he discovered a carbonate called Junckerite in right rhombic prisms of 118°.

In large masses its colors are frequently disposed in concentric zones. A variety is found in small bent stalactites, which is called coral Aragonite. Its fracture is radiated and fibrous. It is also found compact when it resembles Calcite very much. It is, however, distinguished by its want of cleavage and its decrepitation with the blowpipe. It is distinguished by its density, from Witherite and Strontianite and, by the blowpipe, from Barite and Celestite. It is frequently found in lavas and other volcanic rocks, where it lines cavities. It might be confounded with some of the Zeolites, and especially with Natrolite, on account of its radiated form. But it can easily be distinguished with a glass, as the crystals of Natrolite are terminated by an octahedron, while those of Aragonite have either domes or very acute terminations. The distinction can be made by a blowpipe test, or with acids. It is also found pseudomorphic of certain fossil species. Thus the Inoceramus Cuvieri found in the chalk, is Aragonite, while the Echinoderms, asociated with it, are Calcite. Associated with certain metals, it takes different colors; with copper, it is green; with cobalt, it is rosy or violet.

FORMULÆ OF CRYSTALS.

Pl. XVI.

Fig. 4. $\infty P. \infty \breve{P} \infty. \breve{P} \infty.$ *Fig.* 5. The preceding, with P. *Fig.* 6. The preceding, with $2 \breve{P} 2$. *Fig.* 7. $\infty P. 2 \breve{P} \infty. 0P$; from Spain. *Fig.* 8. $\infty P. \infty \breve{P} \infty. 0P$. The base is striated parallel to the brachydiagonal. *Fig.* 9. $6 P \frac{4}{3}. 6 \breve{P} \infty$. This form is the basis of the pointed crystals. *Fig.* 10. $6 P \frac{4}{3}. \infty P. \infty \breve{P} \infty. 6 \breve{P} \infty. \breve{P} \infty.$ *Fig.* 11. Section of a twin crystal, showing the striations. *Fig.* 12. Section of a twin; composition face, ∞P. *Fig.* 13. Trilling, consisting of three individuals. *Fig.* 14. Crystal, consisting of six individuals. *Fig.* 15. Crystal, consisting of four individuals. The pointed crystals are often twinned in this way. *Fig.* 16. Trilling crystal. *Fig.* 17. Composite crystal, formed by interpenetration, as in the crystals from Spain. *Fig.* 18. Group of four crystals, $\infty P. 2 \breve{P} \infty$. *Fig.* 19. Twin crystal, $\infty P. \infty \breve{P} \infty. \breve{P} \infty$. *Fig.* 20. Twinning many times repeated, with successive reversed layers.

Calcite. $\dot{Ca} \ddot{C}$. HEXAGONAL.

SYN.—Calcareous Spar, Carbonate of Lime, Kalkspath, Kalkstein, Chaux carbonatée.

The crystalline forms of Calcite are very numerous, the largest of any known mineral. De Bournon, in 1808, described 691 forms of Calcite and 49 of Aragonite. We shall only speak of the simplest of the ordinary forms and of the most interesting of the composite ones. These give rise to nearly all the possible macles of the hexagonal system, most of which have been found in Calcite. When it is perfectly pure, it crystallizes in rhombohedra of 105° 5′; which angle however is not invariable, because Calcite is very often, we might say always, mixed with $\dot{Mg}$, $\dot{Fe}$ or $\dot{Mn}$. These carbonates are also rhombohedral, with angles only slightly different from those of Calcite, so that, while the crystalline form is not very much altered by this association, it is nevertheless somewhat modified. The angle of the rhombohedron is thus diminished or increased a little by the presence of these isomorphous carbonates, but it always oscillates around the angle 105°, and the variations are very small, not exceeding 2°.

Calcite has been found in 50 or 60 different rhombohedra. Of these, four are the most remarkable. The first is the primitive rhombohedron R, *Fig.* 22, *Pl.* XVI. which is rarely found as a crystal. The second, $-\frac{1}{2}$ R, is tangent to the first, and is remarkable for being quite flat, *Fig.* 21; it frequently occurs rounded. The third, -2 R, is the one to which the primitive rhombohedron is tangent; it is acute and is the one in which the Fontainebleau sandstone usually occurs, *Fig.* 23. It is referred to under Quartz. The fourth, 4 R, is very acute, *Fig.* 24. The fifth, $-\frac{8}{5}$ R, is the very rare one which is called the cubic rhombohedron, as it has an angle very near 90°, *Fig.* 27. When the rhombohedron 4 R, is truncated on the terminal angles, we have a form which very much resembles an octahedron *Fig.* 1, *Pl.* XVII. when truncated below the angle, it resembles an hexagonal prism.

Calcite is frequently found in hexagonal prisms, resulting from tangent modifications on the edges or angles of the rhombohedron. These two prisms may be readily distinguished by the different positions of the mirrors of cleavage; they frequently show the base and are often terminated, in which case it will be very easy to distinguish the rhombohedra and the prisms. This association of the prism and rhombohedron, frequently resembles a rhombic dodecahedron, but the angles differ, the vertical faces being inclined to each other 120°, while the three rhombohedral faces have the angle 105°. By a hemihedral modification of one or the other of these hexagonal prisms, a trigonal prism is produced. Calcite is also found as scalenohedra or as rhombohedra modified by the faces of a scalenohedron, either on the vertical or horizontal edges. The combinations of these forms with the hexagonal prism are also found, as well as with the hexagonal and dihexagonal pyramids. Besides these comparatively simple forms, there are other macled varieties formed by joining rhombohedra and hexagonal pyramids, or by twinning them on the vertical axis with the angle 60°. In the case of the scalenohedron, the hemitropy is formed around one of the faces of the primitive rhombohedron R. This hemitrope is remarkable for the great development of one the faces of the scalenohedron, *Figs.* 27, 28. The faces of the rhombohedron $-\frac{1}{2}$R are frequently rounded, and, as they are very flat, the crystals are often very much distorted. The carbonate of lime is then very impure and, the different carbonates not taking the same angle, the form is the resultant of them all, which will be a curve, and the effect is that the cleavage plane is also curved with it. Crystals of Calcite are frequently striated, the striæ usually resulting from a hemitropy, which is usually around a normal to one of the faces of the primitive rhombohedron. It has three easy cleavages parallel to the primitive rhombohedron. The fracture is conchoidal, but is rarely seen. The lustre is vitreous or earthy. Transparent, opaque. The colors are not generally very decided. It is usually white, sometimes yellowish, gray, red, green, blue, violet, yellow, brown and black. The streak is white. It is double refracting. **H.**=2.5–3.5. **G.**=2.508–2.729. Composition, $\dot{C}a$ 56, $\ddot{C}$ 44.

Pyr. &c. B. P. In a tube sometimes decrepitates. It is infusible, but gives a very luminous flame, coloring it red, (Ca). It is the same phenomenon on a small scale, that is produced with the Drummond Light. When heated, it loses $\ddot{C}$, and becomes $\dot{C}a$; when this is moistened on the finger, a sensation of heat is produced. It effervesces very rapidly with acids.

Calcite is one of the substances most extensively found in nature. It is the gangue of many veins either pure or mixed with other minerals, and alone composes the larger part of the sedimentary formations. It is found crystalline, imperfectly crystalline, lamellar, saccharoidal, bacillary, fibrous, concretionary, compact and earthy. It is also found in lamellar masses, in which the cleavage shows itself to a greater or less degree; these lamellæ become saccharoidal, and are often very small. A lamellar variety is found in some localities, which has a very peculiar appearance. It is in large, very thin lamellæ, placed one upon another, with very slight adhesion, so that they are easily separated and resemble a cleavage. It is not, however, for all these planes of separation are limited by a cleavage plane. The real cleavage thus exists across the lamellæ. Their edges usually show a modification which can be seen with a glass, and which would, if carried out, form the primitive rhombohedron. These lamellæ simply adhere to each other, but the whole mass is full of small empty spaces which separate them so that there is in this variety a very decided pearly lustre, which might cause it to be mistaken for Dolomite, which, however, never has this disposition of superposed lamellæ. Calcite is often found in bacillary masses formed by the union of imperfect crystals, which often come to the surface terminated with an acute rhombohedron, or sometimes with a very obtuse one; these varieties are sometimes diverging, and sometimes parallel. It is also found as stalactites made up of concentric zones. These masses are sometimes suspended from the upper part of the cavity, in which case they are called stalactites, but when they form on the floor of the cavity they are called stalagmites. The stalactites are usually conical and come down to a small point, and are frequently connected with the stalagmite below. The stalagmites are often flat, but are also made up of concentric zones. These masses are often colored, especially by iron, which gives a different color from one zone to another. Some of these masses are very much sought after in the arts, under the name of oriental alabaster.

The least foreign substance, which may be introduced into water charged with carbonate of lime will give rise to a concretion. A grain of sand which is thrown up and down by the force of a spring of this kind will form a pea-shaped concretion, especially if the water is made to bubble or move in any way, so as to lose a part of its $\dot{C}$ in excess. When the force of the spring is not strong enough to agitate these globules, after they have got to a certain size, they fall to the bottom and soon become soldered together by the calcareous deposit, which the water gives up, so as to form a mass which is called Pisolitic limestone. On breaking any one of the globules the concretionary structure can be recognized. Upon the sides of some mineral veins, Calcite is found in very fine fibers, having a fine silky lustre. It is found more frequently in saccharoidal masses, often colored by mixtures with foreign substances, and is then called limestone. One of the most remarkable of these varieties is the fetid limestone. This is colored black by organic matter, which is probably of animal origin, and when breathed upon or struck, emits a very disagreeable odor. When limestone it is found with a very fine granular fracture, almost compact, it is used for lithographic stones. It is also found with an entirely compact and earthy fracture, and it is in this state that it exists as the rock of by far the larger part of the geological epochs. These are either deposits left by the evaporated waters, or sometimes chemical precipitates. In this condition the carbonate of lime is very impure, and is mixed with foreign substances, especially silica, so that the fracture is frequently conchoidal. In large masses the rock appears like clay, but it does not adhere to the tongue, nor give out the argillaceous odor when breathed upon. If it is very argillaceous it is called marl, which has some of the properties of clays, but it can always be distinguished from clays proper by the action of acids. It is sometimes found exhibiting irregular cavities, somewhat like the silex called millstone, but is easily distinguished from it by acids, and by its hardness. The hardness, however, cannot always be depended upon for distinction, for it is clear that with sufficient silica limestone may be exceedingly hard. It is also found pulverulent, but much more rarely, and only between beds of limestone or in cracks, and never has any other than negative characters. It is soft to the touch and does not attract the tongue; the effervescence with acids will easily distinguish it from other pulverulent substances. It appears to have been deposited from calcareous waters which have lost their $\dot{C}$. It is also found in concretions where the caustic lime, in contact with water and $\ddot{C}$ of the air, has become carbonated on the surface. Waters which are charged with $\dot{C}a\ \ddot{C}$ easily form incrustations, and it is possible to direct these concretions, in such a way as to cover an object with carbonate of lime or to mould a delicate medallion with any degree of exactitude. These incrustations are generally made by placing the object in such a position, that the water drops upon it. In this way the water loses the excess of $\ddot{C}$, by virtue of which the carbonate of lime is dissolved, and the latter is then deposited. The most celebrated of these fountains are at Tivoli, Carlsbad and Auvergne. Calcite may form pseudomorphs, which when they replace organisms are commonly called petrifactions or fossils. These pseudomorphs are not formed after the manner of silica, as the carbonate of lime does not take the structure of the organisms, but is only a mould of the interior, and sometimes only of incrustations on the outside.

Under the name of marble, several varieties of Calcite are included, which are sought after in the arts, just as we have distinguished Jasper as Quartz under certain external conditions. A number of different qualities are requisite for marbles. Those which are colorless,or only slightly colored, should be translucent on the edges. The pure varieties which fulfil this condition, have a saccharoidal or at least a granular fracture, like the marbles of Carrara or Paros. Other marbles have a compact fracture, but they must be capable of receiving a bright polish. What is sought after particularly is the disposition of the coloring material; these colors may be uniform white, black, yellow, or red. What is most desired in such marbles, is that the color shall be uniform throughout. Marbles of variegated colors are also much sought after. Marbles colored in veins of black and white are called St. Anne. The marble of Numidia, so much sought after by the ancients,was yellow and red. The translucent onyx of Algeria is white, with orange yellow veins, but there are a very large number of other varieties which are very much sought for. Marbles are not always exclusively composed of carbonate of lime; thus the marble called verd-antique is filled with veins of Serpentine and Talc. There are other varieties of marble, which are called pudding-stone and breccia. These are pebbles of Calcite, held together by a cement of carbonate of lime which is of another color. When the fragments are rounded, they are called pudding-stone, when angular, breccia. There are also marbles, which contain the fossil remains of organisms, generally shells which are found in calcareous rocks of different color from themselves; some of these are black with white shells, or white with red shells. The one called black granite is black, but has white Encrinites scattered through it like crystals of Albite in Melaphyre. Among these shell-marbles, is one known by the name of Lumachelle, which is peculiar. The shell, although fossil, has not completely lost the pearly color and iridiscent hues, which give great brilliancy of effect to the stone. Another is the marble made up of the juxtaposition of kidney-shaped masses. The calcareous deposits which were formed have become cracked in every direction and these cracks have been afterwards filled with carbonate of lime of another color. When the section of one of these stones is properly taken, peculiar markings appear, which give the idea of ruins. These marbles are not very much sought after; they are found in Tuscany, and are known as ruin-marble. A remarkable variety of carbonate of lime is the one which still crystallizes as rhombohedra, but which contains from 10 to 12 % $\dot{P}b\ \ddot{C}$, which is prismatic in form. The rhombohedron will consequently have a different angle than when it is pure. It is found crystalline and lamellar and resembles Calcite.

FORMULÆ OF THE CRYSTALS.

Pl. XVI.

Fig. 21. $\frac{1}{2}$R. *Fig.* 22. R. Primitive and cleavage rhombohedron. *Fig.* 23. –2R. *Fig.* 24. 4R. *Fig.* 25. 13R. *Fig.* 26. 16R. 0R. *Fig.* 27. –$\frac{3}{2}$R. Angle=88° 18′ and resembles a cube very much. *Fig.* 28. ∞R. 0R. *Fig.* 29. ∞R. –$\frac{1}{2}$R. *Fig.* 30. –$\frac{1}{2}$R. ∞R. *Fig.* 31. R. ∞R. Resembles a dodecahedron.. *Fig.* 32. –2R. ∞P2. *Fig.* 33. –2R. R. *Fig.* 34. –2R. –$\frac{1}{2}$R. *Pl.* XVII. *Fig.* 1. 4R. 0R. Resembles an octahedron. *Fig.* 2. 16R. –$\frac{1}{2}$R. *Fig.* 3. –2R. –2R2. *Fig.* 4. 4R. R3. *Fig.* 5. ∞R. R2. –$\frac{1}{2}$R; R2 is generally striated parallel to the edge of R. *Fig.* 6. R2. $\frac{2}{5}$R2. R. *Fig.* 7. R3. The most frequent scalenohedron. *Fig.* 8. R3. ∞P2. *Fig.* 9. R3. ∞R. –2R. *Fig.* 10. R3. ∞R. –$\frac{4}{3}$R3. *Fig.* 11. R3. ∞R. $\frac{1}{4}$R3. *Fig.* 12. R3. R. *Fig.* 13. ∞R. R3. –$\frac{1}{2}$R. *Fig.* 14. R5. R3. 4R. R. *Fig.* 15. R3. $\frac{1}{4}$R3. –$\frac{5}{4}$R. R. –$\frac{1}{2}$R. In large complete crystals, from Derbyshire; with the base, as in the figure, from Ahrnthal, Tyrol. *Fig.* 16. –$\frac{3}{2}$R. 0R. R; $\frac{3}{2}$R=88° 18′, very much resembling a cube. *Fig.* 17. ∞R. –2R. 0R. *Fig.* 18. ∞P2. ∞R. 0R. 4R. –2R; ∞P2 is striated parallel to the edge of R. *Fig.* 19. Twin crystal; composition-face –$\frac{1}{2}$R. Composition often repeated. *Fig.* 20. Twin; composition-face ∞R. *Fig.* 21. Twin; composition-face R. *Fig.* 22. Twin; composition-face –$\frac{1}{2}$R. *Fig* 23. Twin, ∞R. –$\frac{1}{2}$R; composition-face 0R. *Fig.* 24. Twin, R3; composition-face 0R. *Fig.* 25. The same, with the modification $\frac{1}{4}$R3. *Fig.* 26. Twin; composition-face –2R. *Fig.* 27. Distorted scalenohedron showing the twin plane R. *Fig.* 28. Twin of same. *Fig.* 29. Twin; ∞R. R3. composition-face R. *Pl.* XVIII. *Fig.* 1. Twin of *Fig.* 13, *Pl.* XVII. composition-face R. *Fig.* 2. Twin, composition-face R, *Fig.* 3. Distorted form of *Fig.* 12, *Pl.* XVII. *Fig.* 4. Nail-head crystal from Przibram. *Fig.* 5. Prism terminated with R. and containing inside R3 tipped with a cube of Fluorite; Phenixville. *Fig.* 6. Prism rounded by indistinct scalenohedral planes and completed with –$\frac{1}{2}$R. ∞R; from Bristol, Ct. *Fig.* 7. The same in the more usual form. *Fig.* 8. R. 0R. The basal planes being sunken. *Fig.* 9. Top of a crystal, *Fig.* 29, *Pl.* XVI. showing the symmetrical arrangement of small crystals of Pyrite within the crystal. *Fig.* 10. The same as the preceding.

Dolomite. $\dot{C}a\,\ddot{C}+\dot{M}g\,\ddot{C}$. HEXAGONAL.

SYN.—Bitterspath, Perlspath, Pearl Spar, Braunspath, Brown Spar, Rhomb Spar, Tharandit, Chaux carbonatée magnesifère, Miemit.

It crystallizes as a rhombohedron, the angle of which, on account of its variation in composition, oscillates between 106° 10′ and 106° 20′. The form of the crystals is not very varied. It is generally the primitive rhombohedron, which is quite rare in Calcite. The faces of the crystals are often curved, which, however, does not prevent the cleavage from being perfectly plane, and they are often covered with a powder, which is not generally found on the plane faces. These faces do not reflect the light ray well, and the crystals consequently have a pearly lustre, especially on the cleavage faces.

It is rarely found in hexagonal prisms. The varieties which have this form are rarely ever pure, and generally show in their section hexagons of different colors. There is also found a rhombohedron, $-\frac{1}{2}$ R, which is always more or less distorted, and is not very easily recognizable. It is often rounded and bent into a saddle shape to such an extent, that even the cleavage is affected. These crystals are not generally covered with a powder, but on the contrary have almost always a bright and pearly lustre, which has given to the mineral the name of Pearl Spar. It is found in some localities as the rhombohedron −2 R, the faces of which are curved, and covered with a pulverulent coating; but around the terminal angle the faces are plane, brilliant, and are not coated. It has a rhombohedral cleavage, which is much less easy than that of Calcite, so that the fracture of the crystal is often conchoidal. Its lustre is vitreous or pearly; translucent, opaque. The colors of Dolomite are not very decided, usually grayish, brownish, white or black; sometimes it is rosy, owing to the presence of iron, or manganese. A very rare variety, called Miemit, has a decided green color, owing to the presence of iron. **H.**=3.5–4. **G.**=2.8–2.9. Composition, $Ca\ddot{C}$ 54.35, $\dot{M}g\ddot{C}$ 45.65.

Dolomite is a double carbonate of $\dot{C}a$ and $\dot{M}g$. Frequently the two carbonates are together atom for atom, but most of the Dolomites which are found differ from this composition. In the analyses, however, of a large number of Dolomites, it will be seen that the composition oscillates around the formula, but many of the specimens contain more $\dot{C}a$ or $\dot{M}g$ than this. It is frequently associated with other hexagonal carbonates, as Siderite and Magnesite.

Pyr. &c. B. P. Acts like Calcite, but with nitrate of cobalt the presence of $\dot{M}g$ can be ascertained. With acids, the Dolomite which is near the type composition does not effervesce unless heated. With the others which differ more or less from this type, there will be an effervescence when cold, which will become less active as the compound changes to the type formula, when it will cease but goes on when heated. With a little practice it will be easy to distinguish the effervescence produced by Dolomite from that of Calcite, as the former is always very much slower.

It is also found in bacillary masses which do not, however, resemble Calcite. It is usually a radiation of very small needles, making an angle of 60° to 120°. It occurs to a very large extent in saccharoidal masses, which in their exterior aspect, resemble Calcite somewhat, but their structure is not quite the same. Calcite is made up of interlaced lamellæ, resulting from the penetration of large crystals, while in Dolomite it is little crystals which are juxtaposed, and which feel under the finger like sand. The feeling is like that of passing the hand over a rough sandstone or over a rasp. It has been found compact and with a conchoidal fracture, but never earthy. It is frequently white, but it is never translucent on the edges. It is frequently colored black or gray by bituminous substances, but it never has the odor of the fetid limestone. Dolomite, in whatever state it occurs, might be easily mistaken for Calcite, but two very simple reactions will distinguish them. If treated with $\dddot{S}$ to complete solution, there will be crystallized needles of $\dot{M}g\ \dddot{S}$ if it is Dolomite, but a powder of $\dot{C}a\ \dddot{S}$ if it is Calcite. Or, after having driven off the $\ddot{C}$ by means of heat, it can be attacked by $\ddot{\bar{N}}$; if lime water is then added, a precipitate will be produced if it is Dolomite, but none in the case of Calcite. The cobalt solution may also be used. It is found at Lockport, N. Y., Smithfield, N. Y., and many other places in the U. S. It is sometimes used for making lime; some varieties are used as marble. It is also used in the manufacture of Epsom Salts.

FORMULÆ OF THE CRYSTALS.

Pl. XVIII.

Fig. 11. R. −2 R. 0 R. R 3. ∞ P 2. *Fig.* 12. Twin crystal; R. −2 R. 0 R. Composition face ∞ R. *Fig.* 13. Section of a crystal showing the different colored bands in the formation. *Fig.* 14. R; faces very much curved. *Fig.* 15. Saddle-shaped crystal, $-\frac{1}{2}$R.

Scheelite. $\dot{C}a\ \dddot{W}$. TETRAGONAL.

SYN.—Schwerstein, Scheelspath, Scheelin calcaire, Tungstate of Lime.

It is always found crystallized, but has many different crystalline types; they are generally octahedra, with a large number of modifications and sometimes two or even three octahedra of different orders. It is generally white or gray, with a very adamantine lustre. The crystals are frequently hemitrope. It has an octahedral cleavage, and a much more difficult one parallel to the base. Its fracture is uneven. The lustre is adamantine or vitreous, sometimes bright and sometimes dull. It is rarely transparent, but frequently translucent. Its colors do not show much variation; it is white, brown and yellowish; or greenish, when mixed with foreign substances. Streak, white. **H.**=4.5–5.8. **G.**=5.9–6.076. Composition, $\dot{C}a$ 19.4, $\dddot{W}$ 80.6.

Pyr. &c. B. P. It is fusible with difficulty to a transparent glass, and gives the reactions for $\dddot{W}$. In acids as in $\ddot{\bar{N}}$, it dissolves with difficulty, leaving a yellow residue which is $\dddot{W}$.

Its most distinctive character is its density, for, of all non-metallic substances, it is the heaviest. Sometimes the crystals are quite small and cover a gangue which is usually Quartz. The octahedra 2 P ∞ are usually very acute, and have curved faces. When the crystals are formed in this way they are usually complete and seem to have penetrated into the gangue. The gangue has pre-existed however, for the crystals often make a crust which covers the surface of a crystal of Quartz, allowing the form to be distinctly seen. The formation of this variety of Scheelite must have taken place under peculiar circumstances, which are totally unknown to us, carrying with it some corrosive action which has attacked the gangue. This fact is all the more remarkable, as this gangue is made up of bodies, which are usually considered as not attacked by most of the agents at our command, as Quartz and Mica. It was found at Lane's Mine, Munroe, and Huntington, Ct., associated with Wolfram, also in Nevada and North Carolina.

FORMULÆ OF THE CRYSTALS.

Pl. XIX.

Fig. 1. P. The primitive form. *Fig.* 2. P. $\frac{1}{5}$P. *Fig.* 3. 0 P. $\frac{1}{5}$P; often rounded. *Fig.* 4. P. 2 P ∞; occurs frequently. *Fig.* 5. P. 2 P ∞. 4 P 2; the last form is hemihedral. *Fig.* 6. 2 P ∞. P. 0 P. *Fig.* 7. 2 P ∞. P. $\frac{1}{2}$P. 0 P. *Fig.* 8. $\frac{1}{2}$P. P. 2 P ∞. $\frac{3}{5}$P ∞. 4 P 2, in which 4 P 2 occurs as a right octahedron of the third order. *Fig.* 9. P. 2 P ∞. 4 P 2. $\frac{3}{2}$P 3; the latter as a left octahedron of the third order. *Fig.* 10. Interpenetration twin crystal of the preceding combination.

MAGNESIUM.

Brucite. $\dot{M}g\,\dot{H}$. HEXAGONAL.

SYN.—Nemalite, Hydrate de magnésie, Texalith.

The crystals are generally tabular in form and are combinations of the rhombohedron and base. They cleave easily parallel to the base and are flexible but not elastic. Its lustre is pearly or silky; translucent or opaque. It is sectile. Its colors are white, gray, blue or green. Streak, white. **H.**=2.5. **G.**=2.35-2.46. Composition, $\dot{M}g$ 68.97, $\dot{H}$ 31.03.

Pyr. &c. B. P. Swells and becomes opaque, giving very bright lustre to the flame; it is however infusible. With cobalt solution, it gives the characteristic color of $\dot{M}g$. It sometimes becomes opaque, absorbing $\ddot{C}$ from the air. When pure, it is soluble in acids without effervescence or residue.

Resembles Talc, and feels soapy to the touch. The blowpipe will distinguish it, for Brucite swells, while Talc does not. It is also soluble in acids, while Talc is not. It is found as Nemalite at Hoboken, N. J. Large crystals are found at Woods Mine, Texas, Lancaster Co., Pa. It also occurs in crystals, lamellæ or fibers associated with Serpentine and Chromite.

FORMULÆ OF THE CRYSTALS.

Pl. XIX.

Fig. 11. R. 0R. 2 R. *Fig.* 12. R. 0R. -4 R. $-\frac{1}{2}$R.

Epsomite. $\dot{M}g\,\dddot{S}+7\,\dot{H}$. ORTHORHOMBIC.

SYN.—Epsom Salt, Bittersalz.

Crystallizes as a right rhombic prism of 90° 34′. Its forms are usually hemihedral, generally sphenoids. It has a cleavage parallel to the brachypinacoid. Its lustre is vitreous; transparent, translucent; color and streak, white. **H.**=2.25. **G.**=1.685-1.751.

Pyr. &c. B. P. Gives off water in a tube, and dissolves in its water of crystallization. On Ch., fuses and gives an infusible mass; very soluble in $\dot{H}$, and has a bitter taste.

It is found in certain localities, especially in Spain, in long parallel fibrous masses, which become detached in the air, and lose a portion of their water. It resembles Mirabilite, but is distinguished by the addition of lime water, or an alkaline carbonate. It is also found in fibers 50-60 c. m. in length, which form in cavities. Also in the limestone caves of the west, particularly in the Mammoth Cave, Ky.

FORMULÆ OF THE CRYSTALS.

Pl. XIX.

Fig. 13. $\infty P.\ P.$ *Fig.* 14. $\infty P. \frac{P}{2}$.

Boracite. $\dot{M}g^3 \ddot{\ddot{B}}^4 + \frac{1}{2} Mg\,Cl.$ Isometric.

Syn.—Magnésie boratée, Borazit.

The crystals of Boracite are remarkable for their hemihedry. It occurs sometimes as the cube and tetrahedron, *Fig.* 15, *Pl.* XIX. sometimes as the cube, tetrahedron, trigonal trisoctahedron, and rhombic dodecahedron. Sometimes it is found with the cube, tetrahedron and rhombic dodecadedron, but rarely with the cube and rhombic dodecahedron without the tetrahedron. This constant hemihedry is accompanied with peculiar physical properties, the mineral being pyroelectric in the highest degree. It is almost constantly associated with Gypsum and sometimes with Anhydrite. Its fracture is conchoidal, but the fracture of the crystal sometimes shows traces of cleavage parallel to the octahedron. Its lustre is vitreous, sometimes very decidedly so in the transparent varieties. It is generally transparent or translucent, but sometimes opaque. Boracite is generally colorless, though occasionally it is gray, bluish or greenish. The streak is white. **H.**=7, in crystals, but only 4, when massive. **G.**= 2.913–2.974. Composition, $\ddot{\ddot{B}}$ 62.6, $\dot{M}g$ 26.8, Mg Cl 10.6.

Pyr. &c. B. P. Gives off water in a closed tube and is fusible with intumescence to a globule, coloring the flame green. The bead on cooling becomes covered with crystalline needles; this is quite peculiar to Boracite, only one other mineral, Pyromorphite, doing the same; from which it can be easily distinguished by every other chemical and physical character. It is soluble in acids without effervescence, and this solution colors the alcohol flame green with $\ddot{\ddot{B}}$.

FORMULÆ OF THE CRYSTALS.

Pl. XIX.

Fig. 15. $\infty O \infty.\ \frac{O}{2}$. *Fig.* 16. $\infty O \infty.\ \infty O.\ \frac{O}{2}$. The most usual form. *Fig.* 17. $\infty O \infty.\ \infty O.\ \frac{O}{2}.\ -\frac{O}{2}$. *Fig.* 18. $\infty O.\ -\frac{O}{2}.\ \frac{O}{2}.\ \infty O \infty.\ -\frac{2O2}{2}$. *Fig.* 19. $\frac{O}{2}.\ \infty O \infty.\ \infty O$. *Fig.* 20. $\frac{O}{2}.-\frac{O}{2}.\ \infty O \infty.\ \infty O.\ \frac{5O\frac{5}{3}}{2}$. *Fig.* 21. $\infty O \infty.\ \infty O.\ \frac{O}{2}.\ -\frac{O}{2}.\ \frac{2O2}{2}$.

Magnesite. $\dot{M}g\ \ddot{C}.$ Hexagonal.

Syn.—Magnesie carbonatée, Breunnerite, Talkspath.

The primitive form is a rhombohedron of 107° 29′. The angle is variable, as there is usually some $\dot{F}e$ or $\dot{C}a$ present. It is generally crystallized, and the crystals are usually imbedded in some gangue, generally talcose schist or Chlorite. They are generally the primitive rhombohedron, with a basal termination and the faces are often rounded. It has an easy cleavage parallel to the faces of the rhombohedron. Fracture, conchoidal or flat; lustre, vitreous or silky; transparent, opaque. Color white, yellow or brown. **H.**=3.5–4.5. **G.**=3.–3.08. Composition, C̈ 54.4, Ṁg 47.6.

Pyr. &c. B. P. Heated in a closed tube, it gives off water, and acts like Dolomite. It is more easily soluble in acids than Dolomite, but it must be heated to effervesce. It is infusible, but glows intensely (Ṁg).

The color is white, if the carbonate is pure, but it is generally brownish or reddish, on account of the presence of iron or manganese. It is found in many localities in the U. S., and is much used for making Epsom Salts.

Spinel. Ṁg Äl. Isometric.

Syn.—Pleonaste, Ceylonite.

It occurs generally in octahedra, but more rarely in rhombic dodecahedra, which sometimes show modifications of the tetragonal trisoctahedra. The octahedra are very frequently flattened parallel to the faces of the octahedron and are very often hemitrope. The cleavage is parallel to the octahedron. Fracture, conchoidal; lustre vitreous or dull; transparent, opaque. It has a great variety of colors, which are generally shades of red, blue, green, brown, yellowish brown and black. The red Spinel is sometimes called the Balas ruby, and contains alumina, magnesia and chromium. The chromium is in the allotropic condition, which gives to its salts their red color. The crystals sometimes contain traces of manganese, which gives to them a somewhat violet color. Its streak is white. **H.**=8. **G.** =3.5–4.9. Composition, Äl 72, Ṁg. 28. The magnesia may be replaced by lime, iron, maganese or zinc, separately, or in combination. Alumina generally takes the place of a base, in Spinel however, it plays the part of an acid. Spinel is not really a mineral species, but is rather the name of a family of minerals, which are similar in composition and crystalline form, and probably in the conditions of their formation. They all group themselves around the oxygen ratio of 1 : 3. The R̶ is sometimes Äl and F̶e, and sometimes T̈i, which seems, with some Ṙ, to play the part of R̶ (Ṙ T̈i=R̶). It has been decided to include under the name of Spinel all those substances in which Äl predominates, the other R̶ being in variable proportious. Those which contain F̶e are Spinels, but are also minerals and ores of iron, as Magnetite, Chromite, Franklinite, &c. We shall consider them under iron.

Pyr. &c. B. P. Infusible, but changes color. Soluble in Borax and Salt of Phosphorus. Soluble with difficulty in concentrated S̈.

The white or violet blue variety is called Saphar, the predominating base of which is Ṁg. The blue color seems to be owing to iron, which is found both as Ḟe and F̶e. It is largely found in this country in a calcareous gangue. The black Spinels contain iron only, either as an acid or as a base. Their real color is green, more or less intense, which distinguishes them from Magnetite the color of which is black. These crystals sometimes attain an immense size, and their interior is often hollow or cavernous, which, however, does not alter the form of the crystal. They are also found disseminated in chloritic schist, in which case it is not very easy to distinguish them from Magnetite, which is found in the same association and with the same form. The color of the powder is the best test. When the Spinels contain zinc they are known under the name of Gahnite. Their usual association is chlorite slate, but, as the color of their powder does not differ from the others, a blowpipe test is required to distinguish them. The black is less hard than the blue variety, and the blue than the red. Magnetite, is still less hard, it being between Quartz and Fluorite. The black and sometimes the blue varieties of Spinel are found in granular masses, which are slightly lamellar. All of the Spinels are infusible, which will distinguish them from all granular masses, except the minerals of the Chrysolite group, from which the action of acids will give an easy distinction. There will be no deposit of gelatinous silica, and moreover the Spinels are not soluble in acids, except that of iron which is only dissolved with difficulty, when heated in HCl. and never in N̈. The color of these masses usually resembles the color of other granular masses. It is distinguished from Garnet by its fusibility, from Magnetite by not being attracted by the magnet, from Franklinite by giving no reaction for zinc. Very large black crystals have been found at Amity and Munroe, N. Y., and Franklin, N. J. Small gray crystals, at Hamburg and Vernon, N. J. Blue crystals at Antwerp, N. Y., and at many other localities in the U. S. The varieties used in the arts are usually brought to this country separated from their gangues. They come from the East, especially from Ceylon and Birmah. These Spinels are used by jewelers, and are called Balas ruby; they are much less esteemed than the oriental ruby.

FORMULÆ OF THE CRYSTALS.

Pl. XIX.

Fig. 22. O. *Fig.* 23. ∞O. *Fig.* 24. O. ∞O. *Figs.* 25 and 26. Distorted forms of O. *Pl.* XX. *Fig.* 1. O. ∞O. 3O3. *Fig.* 2. Twin crystal; composition-face O. *Fig.* 3. Twin crystal. *Fig.* 4. Distorted form. *Fig.* 5. ∞O. O. 2O2. Form of Pleonaste.

ALUMINIUM.

Corundum. Äl. Hexagonal.

Syn.—Corindon, Sapphire, Ruby, Oriental Amethyst, Smirgel, Emery.

It crystallizes in a rhombohedron of 86° 4′, which very much resembles a cube. It is isomorphous with Hematite. It is sometimes found in rhombohedra, but generally it is truncated at the base so as to resemble a regular octahedron, differing however in the angles.

$$1 \wedge 1 = 86^\circ\ 4'$$
$$0 \wedge 1 = 122^\circ\ 26'$$

This is usually seen in the stony varieties, where the rhombohedral cleavage is very marked. It is generally found in hexagonal pyramids which sometimes show the basal cleavage; the faces of these crystals are covered with unequal striations, which are parallel to the base of the hexagonal prism. The prism is sometimes found, especially in the earthy varieties, and frequently has the faces of a rhombohedron upon it. The base of this crystal is generally striated and shows a play of colors, and when polished usually shows a six-pointed star. It cleaves parallel to the base and the rhombohedon. Sometimes the rhombohedral, and sometimes the basal cleavage is the easiest. In the transparent varieties the basal cleavage is usually the most prominent. The fracture is unequal or conchoidal; sometimes lamellar. Lustre, vitreous, sometimes pearly on the base, and occasionally showing a bright opalescent star of six rays in the direction of the axes. Transparent, translucent to opaque. Color, red, blue, purple, yellow, brown, gray and white. Streak, colorless. **H.**=9. **G.**=3.909–4.16. Composition, Al 53.4, O 46.6.

Pyr. &c. B. P. It is infusible. After heating with nitrate of cobalt, it becomes blue ($\ddot{Al}$). It becomes electric by friction, and remains so for some time. Reduced to powder, it yields an argillaceous odor when breathed upon. It is not acted upon by acids.

Next to the diamond, it is the hardest of all known substances and is used for pointing the tools which are used by jewelers in cutting other stones. It has very many colors, sometimes it is limpid and colorless, and is in this state used to imitate the diamond. The clear colored varieties are called by the jewelers oriental stones, and have a very great value. According as they are red, blue, green, violet or yellow, they are the ruby, sapphire, emerald, amethyst or topaz. It is also found stony with the same colors. It occurs also in crystalline masses, made up of crystals, which are often very large and more or less irregular. Their color is generally blue, with a greenish or brownish tinge. It is also found in granular masses, made up of little grains which are crystalline, and which are bluish or reddish. When massive, it is known as emery. In this state, the $\ddot{Al}$ seems to be associated with sesquioxide of iron. These masses are usually dark brown, with a grayish powder, in which red and blue grains can be seen with a glass. In this state, the hardness is about the only characteristic. The density being 3.9–4, it can be distinguished from every other earthy substance which has this aspect. It cannot be mistaken for a metallic oxide. The different varieties of Corundum are much used in the arts. A perfect ruby of 3½ carats is as valuable as a diamond of the same size. Large crystals of sapphire, some of them semi-transparent, have been found at Newtown, N. J. Imperfect rubies have been found at Warwick, N. J., and bluish crystals in Delaware and Chester Counties, Pa.

FORMULÆ OF THE CRYSTALS.

Pl. XX.

Fig. 6. R. 0R. *Fig.* 7. $\frac{4}{3}$P2. *Fig.* 8. $\frac{4}{3}$P2. 0R. *Fig.* 9. 4P2. *Fig.* 10. $\frac{4}{3}$P2. R. ∞R. *Fig.* 11. 4P2. 0R. R. *Fig.* 12. 9P2. 4P2. $\frac{4}{3}$P2. 0R. R. *Fig.* 13. ∞P2. 0R. $\frac{4}{3}$P2. R. *Fig.* 14. ∞P2. $\frac{8}{3}$P2. $\frac{4}{3}$P2. −2R.

Diaspore. $\ddot{Al}\,\dot{H}$. ORTHORHOMBIC.

Its form is a right rhombic prism of 93° 42′. It is rarely found in distinct crystals, but rather in crystalline masses. It is isomorphous with Göthite. The crystals of Corundum which are found in Dolomite are frequently covered on their edges with dark crystalline filaments, which consist of Diaspore. It sometimes occurs crystallized, especially at Schemnitz, in Hungary, and at St. Gothard, Switzerland. These crystals are very small, and are in a talcose or aluminous gangue, and show remarkable dichroism. Its cleavage is easy, parallel to the brachypinacoid, so that the crystalline masses are always lamellar. Its lustre is vitreous on the faces, but pearly on the cleavage. Translucent or opaque. The colors of Diaspore are not very marked; they are grayish, greenish, yellowish, brownish and purplish. The yellow, red and brown colors are owing to iron, which may be hydrous or anhydrous, and which is interposed between the lamellæ. By digesting the mineral in HCl, this color may be dissolved out. Sometimes the crystals are trichroic, being violet-blue in one direction, purplish-blue in another and pale asparagus-green in a third. **H.**=6.5–7. **G.**=3.3–3.5. Composition, $\ddot{Al}$ 85.1, $\dot{H}$ 14.9.

Pyr. &c. B. P. It gives off water. It is infusible, but after heating, it becomes soluble in concentrated $\dddot{S}$. It is one of the few substances which has this peculiarity. It is not attacked by acids, but if the powder is heated in a tube it swells and gives off water, and when it no longer swells it can then be attacked.

It is quite fragile, and is usually found in little veins in emery. These are mostly crystalline masses, showing the cleavage parallel to $\infty \breve{P} \infty$, and a pearly lustre on this cleavage. The crystals are often bent in the direction of the vertical axis. Externally, Diaspore might be confounded with Cyanite, Pyroxene, Amphibole and Feldspar, but a blowpipe test easily distinguishes it. Pyroxene, Amphibole and Feldspar are fusible; Cyanite is infusible, but does not give off water. It has been found at Trumbull, Ct., associated with Topaz, and at the emery mine, at Chester, Mass.

FORMULÆ OF THE CRYSTALS.

Pl. XX.

Fig. 15. $\infty \breve{P} \infty$. ∞P. $\infty \breve{P} 3$. P. $2 \breve{P} 2$. The faces of P and $2 \breve{P} 2$ are usually curved. *Fig.* 16. Horizontal projection. $\infty \breve{P} \infty$. ∞ P. $\infty \breve{P} 3$. P. $2 \breve{P} 2$. $2 \breve{P} \infty$.

Aluminite. $\ddot{A}l\,\dddot{S}+9\,\dot{H}$.

SYN.—Websterite.

It never occurs crystallized. It has an earthy fracture and a dull, earthy lustre. Its color is white, and it is entirely opaque. **H.**=1–2. **G.**=1.66. Composition, $\ddot{A}l$ 29.8, $\dddot{S}$ 23.2, $\dot{H}$ 47.0.

Pyr. &c. B. P. It is infusible. In a tube, it gives off $\dot{H}$ and, at a higher temperature, $\dddot{S}$. It dissolves in acid without effervescence and without residue. It is insoluble in water, and has no argillaceous odor.

It is found in certain argillaceous beds, especially at New Haven, in England, in white, mamelonated masses, which are sometimes oolitic or even earthy, resembling chalk.

Alunogen. $\ddot{A}l\,\dddot{S}^3+18\,\dot{H}$. MONOCLINIC.

SYN.—Keramohalite, Haarsalz.

It is found in tabular crystals or in masses, in which some crystalline fibres are visible, but the crystals are never well defined. Lustre, vitreous or silky. Translucent. Color, white, yellowish or reddish. **H.**=1.5–2. **G.**=1.6–1.8. Composition, $\ddot{A}l$ 15.4, $\dddot{S}$ 36.0, $\dot{H}$ 48.6.

Pyr. &c. B. P. Heated in a tube, gives off $\dot{H}$ and, at a higher temperature, $\dddot{S}$. With nitrate of cobalt, becomes blue. It is soluble in $\dot{H}$.

These masses are usually white and silky, and have a peculiar, styptic taste. When fresh, it is white, but gradually becomes brown on the fracture, owing to the decomposition of some of the sulphate of iron, which it almost always contains. It occurs in mines and elsewhere, formed by the action of Pyrites on the rocks. It is also found in volcanic regions, covering the sides of cavities, and is usually perpendicular to them. It is found as an efflorescence in many places in the United States.

Alunite. $\dot{K}\dddot{S}+3\,\ddot{\bar{A}}l\,\dddot{S}+6\,\dot{\bar{H}}$. HEXAGONAL.

SYN.—Alaunstein, Alumstone.

It crystallizes as a rhombohedron of 89° 10′, which is consequently very near a cube. It has an easy basal cleavage, which is usually curved. The fracture is conchoidal or uneven; sometimes splintery and often earthy. Transparent or translucent. Color, white, grayish or reddish. Streak, white. **H.**=3.5–4. **G.**=2.58–2.752. The composition is not constant, but varies around the formula. Composition, $\ddot{\bar{A}}l$ 37.13, $\dddot{S}$ 38.53, $\dot{K}$ 11.34, $\dot{\bar{H}}$ 13.0.

Pyr. &c. B. P. Decrepitates, but is infusible, and in a closed tube it yields water and sometimes $\dot{N}H^4\dddot{S}$; and at a higher temperature $\dddot{S}$ and $\ddot{S}$. It then becomes partly soluble in water, giving alum and leaving $\ddot{\bar{A}}l$. It is insoluble in water, but soluble in acids, frequently giving gelatinous silica, which however does not appear to be in a state of combination. Sometimes a part of it is soluble in water and this is generally alum.

A substance which resembles Alunite in composition, but which differs from it in external characters, is called jaspery Alunite. It is a kind of a slate having variegated colors, which are sometimes in parallel bands and sometimes disposed in an irregular manner very much resembling Jasper. Alunite is usually found in concretionary, mamelonated and cavernous masses, of a yellowish or brownish color. Its cavities are usually covered with points of crystals, which cannot be seen without a glass. It is sometimes fibrous. It is the product of the reaction of HS and the vapor of water, which are given off from certain volcanoes, on the surrounding rocks. HS, by oxidation, yields $\dddot{S}$ which attacks the rock and the silica is set at liberty and gelatinizes, the $\dddot{S}$ combining with the $\ddot{\bar{A}}l$ and alkalies of the rock. This sometimes produces the massive, and sometimes the jaspery Alunite. When it is jaspery, it is anhydrous. It resembles certain decomposed varieties of Quartz, but is distinguished by its hardness; from Apatite, Fluorite, Calcite, Barite and Celestite, it is distinguished by the action of acids and the blowpipe. The density will also distinguish it from the last two. Its infusibility and complete solubility in acids without forming gelatinous silica are its most distinguishing characteristics. It is used for making alum. The rock, after roasting, is lixiviated and the alum collected by crystallization. Jaspery Alunite is so hard that it is used in Hungary for millstones.

FORMULÆ OF THE CRYSTALS.

Pl. XX.

Fig. 17. R. *Fig.* 18. R. 0R. *Fig.* 19. R. -2R. $\frac{6}{7}$R. $\frac{6}{5}$R. $\frac{1}{64}$R.

Kalinite. $\dot{K}\dddot{S}+\ddot{\bar{A}}l\,\dddot{S}^3+24\,\dot{\bar{H}}$. ISOMETRIC.

SYN.—Alum, Kalialaun.

It is usually found in fibrous or compact masses and crusts, and is generally a product of the decomposition of Pyrites in contact with clay. Its lustre is vitreous. Transparent, or translucent. Color white, usually stained. Streak, white. **H.**=2–2.5 **G.**=1.75. Composition, $\dot{K}\dddot{S}$ 18.4, $\ddot{\bar{A}}l\,\dddot{S}$ 36.2, $\dot{\bar{H}}$ 45.4.

Pyr. &c. B. P. Fuses in its water of crystallization, and forms a spongy mass. Shows the alumina reaction with $\dot{C}o\,\ddot{\ddot{N}}$. Soluble in 16 to 20 times its weight of cold water and about its own weight of boiling water.

There are a very large number of alums found in nature, the principal bases of which are, besides $\dot{K}$, $\dot{N}H^4$, $\dot{N}a$, $\dot{M}g$, $\bar{\dot{F}}e$, $\dot{F}e$ and $\dot{M}n$. Most of them are isometric in their forms, but they are generally found as fibrous or silky tufts. It is found abundantly in Maryland and Tennessee.

Cryolite. $3\,Na\,F + Al^2\,F^3$. TRICLINIC.

SYN.—Eisstein, Alumine fluatée alcaline.

It crystallizes as a doubly oblique rhombic prism of 88° 30′, and has a perfect basal cleavage. Its lustre is vitreous or slightly pearly, and is nearly the same on the three cleavages as on the crystal. Its fracture is lamellar or scaly. It is generally white, and has about the same kind of lustre as a stearine candle on the fracture. It is sometimes colored slightly red, or may be even brick red, when it is mixed with partially altered Siderite. Occasionally it is black. **H.**=2.5. **G.**=2.9-3. Composition, Al 13.0, $\dot{N}a$ 32.8, Fl 54.0.

Pyr. &c. B. P. Heated in an open tube, it gives up H Fl. Soluble in $\dddot{S}$, giving off H Fl. It is easily fusible even in the flame of a candle, without the aid of the blowpipe. If it is then thrown into water, there seems to be a commencement of decomposition, for an alkaline carbonate or lime water throws down $\ddot{A}l$. This mineral, which within a few years has acquired a certain importance from its commercial uses, has only been found in Greenland.

FORMULÆ OF THE CRYSTALS.

Pl. XX.

Fig. 20. ∞P. P. $2\breve{P}2$. $\breve{P}\infty$. 0P. Prism is marked with striæ parallel to the edges between ∞P and $\breve{P}\infty$ and ∞P and P, as shown by the dotted lines. *Fig.* 21. Twin crystal; composition-face ∞P. The form of these crystals is orthorhombic, but cryolite has been lately determined to be triclinic.

Turquois. $\ddot{A}l^2\,\bar{\dot{P}} + \dot{H}$.

SYN.—Türkis, Calaite.

It is never crystallized, but is found in amorphous masses, which sometimes have the appearance of being little fibers in a gangue. It is also found mamelonated. There is always a little lime, manganese and iron, mixed with the $\ddot{A}l$ and it is sometimes colored by an hydrate of copper. Its fracture is conchoidal and unequal. It is hardly ever transparent, even on the edges of thin scales. It is azure-blue, greenish-blue, or light green. Its color becomes altered in the air, if it passes frequently from a dry to a damp place, especially if the temperature is changed. The color is then, in the commercial phrase, *killed* and it becomes green. Its streak is white or greenish. **H.**=6. **G.**=2.6-3.83. Composition, $\bar{\dot{P}}$ 32.6, $\ddot{A}l$ 46.9, $\dot{H}$ 20.5.

Pyr. &c. B. P. It is infusible. In a tube, it decrepitates, yields $\dot{H}$ and turns brown or black. It colors the flame green. With borax, it affords the colors of iron and copper. In acids, it dissolves completely and gives to the solution the colors of copper.

Under the name of soft Turquois, a substance is used by the lapidaries, which has no real relation to the mineral; its true name is Odontolite.

It is a bone phosphate of lime; generally bones, or fibers of any bone, which have been subjected to infiltrations of salts of copper, and has about the same lustre as Turquois. It is found abundantly in the Urals and to some extent in France. The oriental Turquois is much esteemed as an ornament, but has no very great value. It comes from the east, generally from Nichabour in Persia. An impure variety is found in Saxony and also in Nevada.

Wavellite. $\dddot{A}l^3 \ddot{\ddot{P}}^2 + 12 \dot{H}$. ORTHORHOMBIC.

SYN.—Striegisan, Alumine phosphatée.

It crystallizes in a right rhombic prism of 126° 25′. Its cleavages are parallel to the prism. Its crystals are very rarely capable of being measured, and are generally only very fine needles. Its lustre is vitreous, pearly or resinous. Color white, gray, green, yellow, brown and black, with every gradation of intensity. It is often translucent. **H.**=3.25–4. **G.**=2.337. Composition, $\ddot{\ddot{P}}$ 34.4, $\dddot{A}l$ 37.3, $\dot{H}$ 28.3. It sometimes contains a little fluorine.

Pyr. &c. B. P. In a closed tube gives off $\dot{H}$ and sometimes reacts for Fl. Swells up, becomes of lighter color and splits into fine particles which color the flame green, but remains infusible. Gives off H Fl when treated with $\dddot{S}$. It is soluble in H Cl and caustic potash.

It generally covers the sides of cracks in schistose rocks. It is generally found in little spherical mamelons, which are made up of converging crystals. The ends of these fibers show microscopic crystals which have a velvety lustre and which is quite characteristic. The radiation of these mamelons is exceedingly regular. It is sometimes found in concentric zones, similar to a variety of Natrolite from Swabia and it sometimes resembles certain forms of Aragonite. On a whole specimen, it will be easy to make the distinction of the termination and velvety look of Wavellite, but, when the specimen has been broken, recourse must be had to acids. It is more difficult to distinguish it from Natrolite, but acids give the distinction, for the latter gives gelatinous silica and melts to a globule, while Wavellite is entirely soluble and swells without fusing. Wavellite is found at Bellow's Falls, N. H., at the Washington mine, N. C., in York Co.. Pa., and as stalactites at Steamboat, Chester Co., Pa.

FORMULÆ OF THE CRYSTALS.

Pl. XX.

Fig. 22. ∞P. $\bar{P}\infty$. $\infty\breve{P}\infty$.

Chrysoberyl. $\ddot{\bar{B}}e \dddot{A}l$. ORTHORHOMBIC.

SYN.—Cymophane, Alexandrite.

It crystallizes in a right rhombic prism of 129° 38′. It is one of those limit forms, which resemble the forms of the hexagonal system. The crystals frequently contain chromium and are usually small, though they sometimes attain considerable size. They show a remarkable dichroism, being green perpendicular to the base and blood-red parallel to it. It is also exceptionally red, probably owing to chromium in another allotropic condition It is always found crystallized and is isomorphous with Chrysolite, having forms very much resembling it. The combination *Fig.* 4, is rare. The crystals are usually macled 2 by 2 or 6 by 6, and the macropinacoid is then very much striated in these composite crystals. They do not show reentrant angles as the angle is so near 120°. The 2d form, *Fig.* 6, is not so rare. Crystals, which are found in the fracture of the rock, show very thin lamellæ. They are mostly associated with Calcite, Feldspar or Garnet, and sometimes with Pyroxene. The color of these varieties is usually yellow or greenish-yellow. The third form, *Figs.* 8, 9, 11, is peculiar to crystals of the chrome variety. Chrysoberyl only shows traces of cleavage, which are parallel to the faces of the brachydome. Its lustre is a peculiar vitreous one. It has pearly reflections and opaline tints, which are produced by interior fractures. Its fracture is conchoidal and vitreous. It is often transparent but sometimes only translucent. It is sometimes opaque. When it is transparent, it is always yellowish or greenish. Its colors are varied. It is sometimes pale yellow, sometimes yellow, at other times the color is dark green, nearly opaque. This color is owing to sesquioxide of chromium, which replaces a certain part of $\dddot{\bar{\text{A}}}\text{l}$ and is in the allotropic condition, in which it gives a green color. Streak, colorless. It is sometimes considered as an aluminate of glucina, the oxygen ratio of which is the same as that of Spinel, 1 : 3. The crystalline form, however, is not the same. We can hardly consider $\ddot{\bar{\text{B}}}$ as isomorphous with $\dot{\text{R}}$; it is rather with the series $\dddot{\bar{\text{R}}}$, as $\dddot{\bar{\text{A}}}\text{l}$, $\dddot{\bar{\text{F}}}\text{e}$, $\dddot{\bar{\text{C}}}\text{r}$. **H.**=8.5. **G.**=3.5–3.84. Composition, $\dddot{\bar{\text{B}}}\text{e}$ 19.8, $\dddot{\bar{\text{A}}}\text{l}$ 80.2.

Pyr. &c. B. P. It is infusible, and is not acted upon by acids.

It is found in Brazil and Ceylon as rolled pebbles. A blowpipe assay is necessary to distinguish them. It is found in the U. S. at Haddam, Ct., Greenfield, N. Y.; Orange and Summit, Vt., and Norway, Me.

FORMULÆ OF THE CRYSTALS.

Pl. XXI.

Fig. 1. $\infty\bar{P}\infty$. $\infty\breve{P}\infty$. $\breve{P}\infty$. The face $\infty\bar{P}\infty$ is vertically striated. *Fig.* 2. $\infty\breve{P}2$. $\infty\bar{P}\infty$. $\infty\breve{P}\infty$. P. $\breve{P}\infty$. *Fig.* 3. The preceding, with $2\breve{P}2$. *Fig.* 4. The combination *Fig.* 2, with $\infty\breve{P}\frac{3}{2}$. *Fig.* 5. $\infty\bar{P}\infty$. $\infty\breve{P}\infty$. $\breve{P}\infty$. P. $2\breve{P}2$. $\infty\breve{P}2$. $\infty\breve{P}3$. $6\breve{P}6$ (*e*). *Fig.* 6. Twin crystal; composition-face $3\breve{P}\infty$. *Fig.* 7. $\infty\breve{P}\infty$. $\infty\bar{P}\infty$. P. $2\breve{P}2$. *Fig.* 8. Twin crystal; composition-face $3\breve{P}\infty$, made up of 6 parts by the crossing of 3 crystals, united along the dotted lines. *Figs.* 9, 10, 11. Twin crystals; composition-face $3\breve{P}\infty$. *Figs.* 12, 13. Twins, made by the crossing of 3 pairs of twins, each section a pair twinned by $3\breve{P}\infty$ and united to the next pair by $\infty\breve{P}\infty$. *Fig.* 14. Twin crystal similar to *Fig.* 8.

IRON.

Iron. Fe. Isometric.

Syn.—Gediegen Eisen, Fer natif.

It has an octahedral cleavage, a very hackly fracture and a metallic lustre. Its color is iron-gray, and its streak shining. It is ductile and is attracted by the magnet. **H.**=4.5. **G.**=7.3–7.8. It contains variable quantities of other substances, principally nickel, associated with a small proportion of cobalt. The quantity of nickel may vary from 1 to 20 per cent.

Iron is supposed to have been found in nature, in a pure metallic state, in certain mines of Pyrites. Proust analyzed several specimens of this character and found them to be perfectly pure. Some specimens analyzed have since been proved to be artificial. Since this time, it has also been found in a mine in Dauphiné, Auvergne and Brazil; but such iron is exceedingly rare. It has been found in petrified wood, produced probably by the reduction of a salt of iron by the organic matter of the wood. It is found native, as grains disseminated through volcanic rocks, at the Giant's Causeway and in Auvergne. It is easy to prove its presence by dipping the rocks into a solution of $\dot{C}u\,\dddot{S}$, when the rock becomes coated with copper.

The usual way in which native iron is found, however, is in Meteorites. The authenticity of Aerolites was admitted only at the close of the last century, after the Aerolite of Tuscany in 1794 and those of L'Aigle in 1803 had been found. They are generally of two kinds. They may be entirely composed of iron, which is metallic. This iron is associated with chromium, nickel, and sometimes with cobalt, manganese and sulphur. When these masses are polished and treated with acid, they show the traces of crystallization. Sometimes they are associated with bituminous substances. These masses are ordinarily spongy, the cavities being always filled with Chrysolite, or substances analogous to it. The Gibbs meteorite in Yale College weighs 1635 lbs., the Tucson at the Smithsonian Institute, 1,400 lbs. Two in S. America are estimated as weighing, the one 32,000 lbs., and the other, 14,000. The Pallas Meteorite, which contains crystals of Chrysolite, found in Siberia, weighed 1,600 lbs. Other Meteorites, on the other hand, are of a stony character and contain the iron scattered through them in little bunches. The exterior of these masses is generally scorified and covered over with a coating, which seems to have been melted on the surface. The earthy part is partly soluble in acids, depositing gelatinous silica and it appears to recall the composition of Chrysolite or of Labradorite. The insoluble part seems to be allied to Pyroxene and to Orthoclase and Albite. The iron is always more or less altered and may be alloyed with other metals, which as before are chromium and nickel. They also contain phosphuret and sulphuret of iron. These Aerolites are sometimes very large; some have been found to be several thousands of pounds in weight. In Auvergne, masses of volcanic origin are found, which contain some iron partly oxidized. This iron has been produced by reduction of an oxide, probably in the same way as we produce it in the laboratories.

Magnetite. $\dot{F}e$, $\dddot{\bar{F}}e$. ISOMETRIC.

SYN.—Magnetic Iron Ore, Magneteisenstein, Magneteisenerz, Fer oxydulé.

The crystals of Magnetite are generally octahedral in their form; they are not often complete, but are generally attached to the mass which contains them which may be compact Magnetite, Slate, Chlorite or Serpentine. The octahedra are sometimes in twin crystals. The octahedron is often associated with the rhombic dodecahedron, which sometimes predominates. The faces of the rhombic dodecahedron are always striated parallel to the long diagonals of the rhomb. The crystals are sometimes very large and these striæ become so marked, that they appear like steps. The cleavage is parallel to the octahedron. Its fracture is conchoidal, generally unequal and sometimes quite lamellar, or granular. It is often compact in the uncrystallized varieties. On its natural faces, it has a semi-metallic lustre. It is generally opaque, but in dendrites is sometimes transparent. Its color is black, and the streak also black. **H.**=5.5–6.5. **G.**=4.9–5.2. Composition, O 27.6, Fe 72.4. Magnetite has the formula $\dot{\mathrm{F}}\mathrm{e}\ \bar{\mathrm{F}}\mathrm{e}$, which is sometimes written $\ddot{\mathrm{F}}\mathrm{e}^3$. It is a Spinel of iron, or a ferrite of protoxide of iron. Like all the Spinels, it crystallizes in the isometric system, in forms which are derived from the octahedron, to which form its cleavage is parallel.

Pyr. &c. B. P. It is fusible with difficulty. It is insoluble in $\ddot{\bar{\mathrm{N}}}$, but is with difficulty soluble in hot HCl. It is magnetic, and often has polar magnetism, especially when in large bodies.

Magnetite is often found in granular masses, made up of small crystals, which separate sometimes with the slightest effort of the finger. It sometimes happens that these grains are covered on the surface with a superficial coating of oxide, which makes them iridescent. Such ore is called *shot-ore* by the miners. These granular varieties by the action of the elements often become a fine black sand. Such sand is the only ore of iron in New Zealand, and it is found on the sea-shore, where the constant action of the water has washed out the impurities and made it quite pure. Granular Magnetite is often found disseminated in large or small quantities through certain rocks, such as Serpentine and Dolomite. It is sometimes in sufficient quantities to affect the magnet, and often deceives those who do not understand the extent of the deposit. In schistose rocks, Magnetite is often found in quite large masses which have the structure of the rock. Such masses are probably of metamorphic origin. Magnetite is sometimes found in entirely compact masses. These masses are sometimes very pure and make an excellent ore of iron for the manufacture of steel. Sometimes they are very impure and are mixed with Specular Iron, Pyrite, Menaccanite and Quartz. When Magnetite has polarity, it is the natural magnet. It is distinguished from all the minerals which it resembles by having a black streak, and by being magnetic. It is one of the most important iron ores. It is found largely in the states of N. Y. and N. J. and in other localities in the U. S.

FORMULÆ OF THE CRYSTALS.

Pl. XXI.

Fig. 15. O. *Fig.* 16. ∞O; the faces are striated parallel to the longer diagonal. *Fig.* 17. ∞O. O. ∞O∞. *Fig.* 18. Hemitrope, O. *Fig.* 19. ∞O. O; showing the striations; from Haddam, Ct. *Fig.* 20. ∞O. ∞O∞. O. 3O3 5O$\frac{5}{3}$; Achmatowsk.

Franklinite. $(\dot{\mathrm{F}}\mathrm{e}, \dot{\mathrm{M}}\mathrm{n}, \dot{\mathrm{Z}}\mathrm{n})(\bar{\mathrm{F}}\mathrm{e}, \bar{\mathrm{M}}\mathrm{n})$. ISOMETRIC.

It crystallizes generally as octahedra with small faces of the rhombic dodecahedron. It has an indistinct octahedral cleavage. Its fracture is conchoidal. Opaque. Color, black. Streak, dark reddish brown. It is very slightly magnetic. **H.**=5.5–6.5. **G.**=5.069. Composition, $\bar{\mathrm{F}}\mathrm{e}$ 66, $\bar{\mathrm{M}}\mathrm{n}$ 16, $\dot{\mathrm{Z}}\mathrm{n}$ 17. It is a Spinel of zinc and iron, a Magnetite, in which a part of the $\dot{\mathrm{F}}\mathrm{e}$ is replaced by $\dot{\mathrm{Z}}\mathrm{n}$ and a part of the $\bar{\mathrm{F}}\mathrm{e}$ by $\bar{\mathrm{M}}\mathrm{n}$. It has the same relation to Magnetite, that Gahnite has to Spinel.

Pyr. &c. B. P. Infusible. With borax, gives the reactions for zinc and iron. Soluble in HCl, giving up a small quantity of Cl.

To distinguish it from Magnetite, it is sometimes necessary to make a blowpipe test. In the reducing flame, it gives off zinc. It is used as an ore of iron and of zinc. It is only found in large quantities at Franklin, N. J. It is associated with the silicate and oxide of zinc, making the association of red and black colors, which is quite rare in mineralogy.

FORMULÆ OF THE CRYSTALS.

Pl. XXI.

Fig. 21. O. *Fig.* 22. O. ∞O∞. *Fig.* 23. O. ∞O; one of the most frequently occurring forms.

Hematite. F̶e. HEXAGONAL.

SYN.—Red Ochre, Micaceous Iron Ore, Iron Glance, Specular Iron, Red Hematite, Glanzeisenerz, Eisenglanz, Rother Glaskopf, Rotheisenstein, Speglande Eisenglimmer, Fer speculaire, Fer oligiste, Fer oxidé rouge, Hematite rouge.

Its form is the rhombohedron of 86° 10′. It is isomorphous with Corundum and has the same cleavages, parallel to the rhombohedron and to the basal pinacoid, only more difficult. The rhombohedral cleavages are only traces and are hard to determine. It is found as distinct crystals, or as thin scales. The best crystals are found at Framont, in the Vosges, and at Elba. These usually have the form given in *Pl.* XXI. *Fig.* 28; R is the primitive rhombohedron, $\frac{4}{3}$ P 2 a scalenohedron, $\frac{1}{4}$ R another rhombohedron, with faces which are always striated and sometimes curved. In the volcanic rocks, it is found in another form as thin hexagonal scales, which have also been artificially produced by causing the vapor of water to pass over perchloride of iron. It is perhaps formed in the same way in the volcanoes. These thin scales are called Specular Iron, on account of their bright, mirror-like surfaces. They are hexagonal planes, truncated by a very obtuse rhombohedron, the faces of which are usually striated. The faces of the hexagonal prism are also sometimes found, but they are always very small. When the hexagonal form is very much flattened by the reduction of two of its opposite sides, the form of the figure is frequently rhombic. These lamellæ are often piled the one on the other, but not parallel. There is usually a hollow in the centre and the edges are a little raised. The lamellæ themselves are sometimes curved. The thin scales are sometimes found as red, transparent and ochreous coatings on the outside of other minerals, but this is comparatively rare. Its fracture is conchoidal, rarely lamellar. Its lustre is metallic; sometimes splendent or earthy. The varieties that have a metallic lustre are called Specular Iron, but the earthy varieties are called Red Hematite. It is black and often iridescent on its natural faces, on account of a very thin pellicule of oxide on the surface. In thin scales, it is transparent and of a blood-red color. Streak, blood-red, or brownish-red. Sometimes slightly magnetic. **H.**=5.5–6.5. **G.**=4.5–5.3. Composition, O 30, Fe 70.

Pyr. &c. B. P. It is infusible, but when exposed for a long time to the reducing flame, it gives a magnetic globule. It dissolves with difficulty in hot hydrochloric acid, more especially if it contains titanium.

It is also found fibrous in little veins. These fibers are thin lamellæ, analogous to those described. In some schistose formations, it takes the character of the rock and is itself schistose. The variety called Red Hematite is a less perfect form of the red scales, of which we have just spoken, and the schistose varieties seem to be a passage to this form of the mineral. It is generally found in large mamelonated masses, frequently of sufficient size to justify mining to a large extent. These masses are usually made up of fibers which are soldered together, and which have a metallic lustre, usually grayish with red spots, but always having a red streak. The outside is sometimes black, caused generally by a superficial coating of manganese. Usually these masses are easily broken in the lines of radiation, forming a kind of pyramid by the fracture. Hematite constitutes the rouge which is used to burnish and polish metals. It is also found in masses, which are entirely compact and earthy and which are often very extensive. In this state it is generally impure being mixed with clay &c. A bacillary variety is also found, which breaks up into hexagonal prisms, which are sometimes very long. It appears to be of accidental formation, probably resulting from the decomposition of Siderite, which has afterwards cracked with an appearance of regularity, owing to the reduction in bulk. The angles of these hexagons are not constant. It is also found pisolitic and oolitic, but it does not appear to have been formed directly in that condition. It appears to be the result of the de-hydration and pseudomorphism of grains of Limonite, which are abundantly found in some localities, and of which we shall presently speak. In France, Mexico and elsewhere, these masses are sometimes large enough to be mined to advantage. It is also found in large pseudomorphic masses, resulting from a decomposition analogous to that to be described under Pyrite. It is one of the richest and most important of the ores of iron. The Specular varieties resemble some of the copper and silver ores; Red Hematite resembles Cuprite and Cinnabar, and the compact variety, sometimes Magnetite. From all these it can be distinguished by its red streak and behavior before the blowpipe. It is found largely at Lake Superior, Missouri, New York and elsewhere.

FORMULÆ OF THE CRYSTALS.

Pl. XXI.

Fig. 24. R. The primitive rhombohedron. *Fig.* 25. R. 0R; occurs tabular, when 0R predominates. *Fig.* 26. R. $-\frac{1}{2}$R. *Fig.* 27. R. $\frac{1}{4}$R. ∞P 2. *Fig.* 28. $\frac{4}{3}$P 2. R. $\frac{1}{4}$R; usual combination from Elba. *Fig.* 29. R. $\frac{1}{4}$R. $\frac{2}{5}$R 3. $\frac{4}{3}$P 2; from Elba. *Pl.* XXI. *Fig.* 1. $\frac{4}{3}$P 2. R. $\frac{1}{4}$R. $\frac{3}{5}$R. $\frac{2}{5}$R 3. *Fig.* 2. Twin crystal by interpenetration; combination, $\frac{4}{3}$P 2. R. 0R. *Fig.* 3. 0R. R. ∞P 2; thin tabular crystal from Vesuvius. *Fig.* 4. Twin crystal. The opposite halves of two individuals of the preceding form are united on a face of the prism ∞R. *Fig.* 5. Twin crystal by interpenetration. *Fig.* 6. $\frac{4}{3}$P 2. 0R; with the edge in front. *Fig.* 7. The same with the face in front. *Fig.* 8. Twin from the two preceding; composition-face 0R.

Goethite. $\dddot{Fe}\,\dot{H}$. ORTHORHOMBIC.

SYN.—Pyrrhosiderite, Lepidocrocite, Nadeleisenerz, Sammetblende.

The primitive form is a right rhombic prism of 94° 52′. The crystals are usually long needles on Limonite or Quartz, which they often penetrate. They are usually striated and formed by the juxtaposition of a number of crystals. It bears the same relation to Hematite that Diaspore has to Corundum. It is isomorphous with Diaspore and like it, crystallizes as a right rhombic prism with an easy cleavage, parallel to the brachypinacoid. Its lustre is adamantine, semi-metallic or earthy. Color, yellowish-white or brownish. When transparent, it is often blood-red. Its streak is brownish-yellow, or yellow. **H.**=5–5.5. **G.**=4–4.4. Composition, $\overline{\overline{\mathrm{F}}}\mathrm{e}$ 8.99, $\dot{\mathrm{H}}$ 10.1.

Pyr. &c. B. P. Heated in a tube, it gives off water and becomes red. Infusible but gives in the R. F. a magnetic globule. Gives reactions for iron, and generally for manganese. Easily soluble in HCl.

It is also found in thin transparent scales of an orange color on Quartz and Limonite. On Limonite they have a rosy color and usually a pearly lustre. It is found in great abundance on Hematite, at the Jackson iron mine in Lake Superior and on Limonite at Easton, Pa.

FORMULÆ OF THE CRYSTALS.

Pl. XXII.

Fig. 9. ∞P. $\infty \bar{P} 2$. $\infty \breve{P} \infty$. P. $\breve{P} \infty$. *Fig.* 10. $\infty \bar{P} \infty$. $\infty \bar{P} 2$. $\breve{P} \infty$. *Fig.* 11. $\infty \breve{P} \infty$. $4 \bar{P} \infty$. $\breve{P} \infty$. $\infty \bar{P} \infty$.

Limonite. $\overline{\overline{\mathrm{F}}}\mathrm{e}^2 \dot{\mathrm{H}}^3$.

SYN.—Bog Ore, Yellow Ochre, Brown Ochre, Brown Hematite, Raseneisenstein, Stilpnosiderite, Brauner Glaskopf, Brauneisenstein, Mine de fer limoneuse.

It is never crystallized, but is found as concretions having a fibrous and radiated structure, amorphous and earthy. Its fracture is fibrous and frequently shows a gradation of colors in bands. The colors are different shades of brown, and the streak is a yellowish brown. In the impure varieties, the fracture is entirely earthy. When concretionary, it takes the form of large or small stalactites, or mamelonated coatings. The surface is usually smooth and often lustrous. This is generally owing to a very thin coating of manganese. Its lustre is very variable. The fibrous varieties have generally a silky, the botryoidal varieties a metallic lustre. Very often it is dull and earthy. Color, generally different shades of brown, sometimes nearly black in the botryoidal varieties. When earthy, brownish or ochre-yellow. Streak, yellowish-brown. **H.**=5–5.5. **G.**=3.6–4. Composition, $\overline{\overline{\mathrm{F}}}\mathrm{e}$ 86.6, $\dot{\mathrm{H}}$ 14.4.

Pyr. &c. B. P. Gives off water and becomes red. Soluble in acids.

A variety called Aetites by the ancients was used as amulets and supposed to possess great virtue. They are all geodic varieties, which are generally impure and mixed with sand. They show a concentric structure. The interior is sometimes empty and sometimes contains a movable hard substance, of the same composition as the outside, which separated from the rest when the mass contracted in drying, and which frequently makes a noise when shaken. It is also found pisolitic, the grains held together by a cement of the same substance, but usually much poorer in iron. In France these masses constitute a very important ore of iron. They are entirely analogous, except in their composition and color, to the pisolites of Calcite, which have been described. The center of each globule is generally a small piece of some foreign substance, usually a grain of sand, Magnetite or Menaccanite, or even Chromite. Their formation appears to be owing to the decomposition of Siderite, held in solution in water. The oolitic varieties, except in color, are quite analogous to the oolites of Hematite. It often forms extensive beds, which are worked as an ore. The cement which holds them together is often clay. This is usually separated by washing, in order to enrich the ore. It is also found as irregular masses, caused probably by some metamorphic action, and, in a large number of cases, by the decomposition of Pyrite. These masses are usually very impure, and are mixed with earthy substances. They give rise to a number of pseudomorphs. The earthy varieties are called ochres, which are clays or marls impregnated with hydrated sesquioxide of iron. They are sometimes sufficiently rich to be mined as ores of iron. Their color is variable from brown, almost black, through all the shades of brown to pale yellow, known as Yellow Ochre. The colors give little or no idea of the yield of the ore in iron, for they vary not only with the quantity of iron contained, but also with its state of aggregation. It often happens, that a pale yellow ochre is richer than a brown one. This variation of color is not only found in natural, but also in artificial products, as in the different varieties of rouge, the color of which varies with the method of manufacture. An important variety of Limonite is called Bog Ore. This ore is very abundant, especially in Sweden, Russia, and some parts of this country. It is found in an entirely different association from the preceding varieties. It is not produced by the decomposition of Siderite, but seems to be owing to the decomposition of organic, generally vegetable matter in ferruginous waters. These masses are brown, more or less dark, with a smooth, even, and sometimes conchoidal fracture, which may even be resinous. It is always impure, containing besides the gangues, sulphur, phosphorus and some portion of organic matter, frequently fragments of plants which are not entirely decomposed, but which preserve their woody fiber, while passing into the state of peat. This production or Limonite by means of the decomposition of plants, is going on extensively now. The mines from which it is taken are not exhausted, because the ore is found continuously, so that it is easy to watch the progress of its formation. It is remarked, that when the roots of a tree are decomposed in the midst of a ferruginous sand, the sand around them is rendered colorless for a certain distance. The acid products of decomposition, such as ulmic, ceramic and aproceramic acids, when disengaged, dissolve out the oxide of iron from the neighboring sand, carrying it off with them. These salts on arriving at the air, are oxidized, the organic acids are burned, and a part of the $\ddot{C}$ resulting from their decomposition is fixed as $\dot{F}e\ \ddot{C}$. This carbonate once formed is dissolved by virtue of the excess of $\ddot{C}$ and, when it decomposes, Limonite is produced. It is always very impure, because, it contains sulphur and phosphorus besides the earthy material. Limonite is one of the most important of the ores of iron. It is mined extensively in the western part of New England, New York, Pennsylvania and other states. It is sometimes ground and used as a polishing material, furnishing different colors of brown and yellow, according to the method of its preparation.

Pyrrhotite. $Fe^7\ S^8$. HEXAGONAL.

SYN.—Pyrrhotine, Magnetic Pyrites, Magnetkies, Fer sulfuré magnetique, Magnetopyrite.

This Pyrites is magnetic and is the compound of iron and sulphur, which is found in Meteorites. It is also found in nature. Sometimes the formula is $Fe^8\ S^9$, and sometimes $Fe^{10}\ S^{11}$. It is rarely perfectly crystallized. The best formed crystals are found in Meteorites. They are also sometimes found disseminated in a gangue. It is generally, however, found in lamellar, crystalline masses, which can be easily recognized by their bronze color. It has an easy cleavage, parallel to the base, and traces of cleavage parallel to the faces of the prism. Its fracture is lamellar, sometimes conchoidal or even granular. Lustre, metallic. Color, dark yellow, bronze, and often a yellowish-red. Streak, dark grayish-black. The powder is always magnetic. **H.**=3.5–4.5. **G.**=4.4–4.68. Composition, Fe^7 S^8=Fe 60.5, S 39.5.

Pyr. &c. B. P. In an open tube, it gives off S̈. On Ch., it becomes either F̈e or Fe, according to the flame in which it is heated. In the R. F., it fuses to a magnetic mass. Generally gives the reactions for Ni and Co. It is attacked by the non-oxydizing acids. Sometimes it gives off ḢS and leaves sulphur as a residue.

Its most ordinary associations are Magnetite and Apatite, but it is frequently mixed with other sulphides, which may somewhat modify its color. It is often found as a pseudomorph after Arsenopyrite. Its inferior hardness and color distinguish it from Pyrite. From the ores of Co and Ni, it is distinguished by giving the iron reaction and a very magnetic globule. It is found in a large number of places, in New England, in New York State, New Jersey, Pennsylvania and Tennessee.

FORMULÆ OF THE CRYSTALS.

Pl. XXII.
Fig. 12. ∞P. P. 0P.

Pyrite. Fe S^2. ISOMETRIC.

SYN.—Iron Pyrites, Schwefelkies, Eisenkies, Fer sulfuré jaune.

Shows almost always the hemihedry of parallel faces, with traces of a cubical cleavage. It is frequently found crystallized with striated faces, but sometimes with polished ones. Its usual forms are the cube, octahedron, and a combination of the cube and hemi-tetrahexahedron, the faces of which are almost always striated parallel to their intersection with the cube. Faces of two hemi-tetrahexahedra at different angles are also found at times; in such a case, the faces of the cube have entirely disappeared, and there is often a trace of octahedral faces and when these faces are large, they form the solid of twenty triangular faces, the icosahedron. The faces of the hemi-tetrahexahedron in this case are always striated, while those of the octahedron are smooth. The diploid is also found in combination with the hemi-tetrahexahedron. Crystals of Pyrite are often very large and have sometimes very brilliant faces. The cleavage is more or less distinctly cubical or octahedral. It is very fragile, and its fracture is rough, rarely conchoidal. Its lustre is metallic, sometimes splendent. The color on its natural faces and on its fracture is brass-yellow, with a very decided metallic lustre, and is quite uniform. This color caused it to be much sought after at one time, as an object of ornament. It was then known to jewelers under the name of Marcasite. Its streak is greenish or brownish-black. **H.**=6–6.5. It strikes fire with the steel without giving out any odor. **G.**=4.83–5.2. Composition, Fe 46.7, S 53.3.

Pyr. &c. B. P. It fuses in the flame of a candle. Heated in a tube, sulphur sublimes. In the R. F., a residue is obtained which attracts the magnet. The non-oxydizing acids do not act upon it. It is attacked by N̈ with effervescence and gives off HS.

It is also found as concretions in balls covered with points of crystals. These are sometimes sufficiently brilliant, to give the ball a velvety appearance. The concretionary variety is also found as a coating on the various gangues. The fracture of these masses is generally compact, rarely fibrous, sometimes silky or velvety. It is also found as dendrites, which are ramified around straight or curved lines. It also occurs as amorphous masses with an unequal rongh fracture, but otherwise having the same aspect as the crystals. It frequently by pseudomorphism takes the shape of organisms, especially in the Lias. These forms seem to be owing to the reducing action, which organic matter in decomposition exercises on the sulphates. In these imitative forms Pyrite frequently preserves the structure of the organism. It appears generally to have formed after all the hollow parts have been filled up by some substance, which appears usually to be Calcite. When it is in the state of incrustations, it is never as perfect in preserving the structure as Quartz, and it is usually impossible in any case to study the details of the organism as it is in Quartz. It is susceptible of a remarkable decomposition, which has taken place under circumstances which are not perfectly understood, which is called Hepatic Pyrites, (from *hepar*, the greek word meaning liver.) It still has the form of Pyrite but its composition is generally that of Limonite. The sulphur has completely disappeared, we do not know how, without causing the crystal to be any the less perfect. When the alteration is not complete, Pyrite is found in the interior. This same decomposition has sometimes taken place in nature on a very large scale, producing large bodies of valuable iron ore, which are now worked. Pyrite is not altered in the air, and does not decompose in collections. Its name is from *pur*, fire, in allusion to its striking fire with a steel. It is distinguished from Chalcopyrite by its much greater hardness, and by its paler color. It is distinguished from Marcasite by being of a much lighter yellow. From Gold, by its not being sectile. It is extensively used in the manufacture of green vitriol and sulphuric acid. It is found very generally disseminated throughout the rocks of the U. S.

FORMULÆ OF THE CRYSTALS.

Pl. XXII.

Fig. 13. $\infty O \infty$. The striations result from an oscillatory combination between the cube and pentagonal dodecahedron. *Fig.* 14. $\frac{\infty O 2}{2}$. The most usual form. *Fig.* 15. $\frac{\infty O 2}{2}$. $\left[\frac{4 O 2}{2}\right]$. *Fig.* 16. $\left[\frac{4 O 2}{2}\right]$; this form occurs sometimes. *Fig.* 17. $\frac{\infty O 2}{2}$. O. $\infty O \infty$. *Fig.* 18. $\frac{\infty O 2}{2}$. O. $\left[\frac{3 O \frac{3}{2}}{2}\right]$. *Fig.* 19. O. $\frac{\infty O 2}{2}$; a very frequent combination. *Fig.* 20. The same as the preceding, but with the faces equally developed, so that it resembles an icosahedron. *Fig.* 21. $\frac{\infty O 2}{2}$. $\left[\frac{3 O \frac{3}{2}}{2}\right]$. *Fig.* 22. The same combination, but with the other dodecahedron. *Fig.* 23. $\left[\frac{3 O \frac{3}{2}}{2}\right]$. $\infty O \infty$. *Fig.* 24. $\infty O \infty$. O. 2 O 2. $\left[\frac{4 O 2}{2}\right]$. $\frac{\infty O 2}{2}$. *Fig.* 25. Twin crystal; composition-face $\infty O \infty$. *Pl.* XXIII. *Fig.* 1.

$\frac{\infty O 2}{2} \cdot - \frac{\infty O 2}{2}$; interpenetration twin. *Fig.* 2. The preceding, with the modification ∞ O ∞. *Fig.* 3. The cube resulting from *Fig.* 2; the striations show its derivation.

Marcasite. Fe S^2. ORTHORHOMBIC.

SYN.—Cockscomb, Spear and Cellular Pyrites, Leberkies, Kammkies, Speerkies, Fer sulfuré blanc.

It crystallizes in a right rhombic prism of 106° 5′, with a cleavage parallel to the prism, and only traces parallel to the brachypinacoid. The simple crystals are found as prisms, or as the union of two domes in such a way as to resemble the regular octahedron. They can only be distinguished by measurement of the angles. The crystals however are rarely single. The crystals, *Fig.* 6, which are quite short, are united by the faces of the prism and make complex forms. The form *Fig.* 10, where the composing crystals are complete, is rare. They are generally composed as in *Fig.* 7, and as in *Fig.* 8, with an indented contour, often also, as the angle of the crystals is 73° 58′ or almost that, of a regular pentagon. The crystals are found united by fives as in *Fig.* 9. This disposition is quite common, and is called Cockscomb Pyrites. Its fracture is rough and somewhat granular. It has a very decided metallic lustre, especially in the varieties which do not show a green tinge which appears to be owing to a commencement of decomposition. Its color is a much lighter yellow, and more greenish than Pyrite. Streak, grayish or brownish-black. **H.**=6–6.5. **G.**=4.678–4.847. Composition, Fe 46.7, S 53.3.

Pyr. &c. B. P. With the blowpipe and the different reagents, it acts exactly as Pyrites.

It has exactly the same composition as Pyrite, of which it is the dimorphic form. It is also found crystalline or concretionary. In some formations, especially in the Chalk, it occurs in balls, sometimes in cylindrical masses with a fibrous and radiated structure, the surfaces or which are covered with the points of crystals having the pseudo-octahedral form. Occasionally, though much more rarely than Pyrite, Marcasite takes the forms of organisms either by filling or by substituting molecule for molecule. It decomposes very easily in the air, and forms sulphate of iron. In order to preserve it in collections, it must generally be coated with varnish. Sometimes, though rarely, the product of decomposition is Hepatic Pyrites, as is the case with the balls that occur in the Chalk. A variety is sometimes found which does not decompose. It is used for the same purposes as Pyrite. It is largely found in the U. S.

FORMULÆ OF THE CRYSTALS.

Pl. XXIII.

Fig. 4. $\breve{P}\infty$. $\bar{P}\infty$. ∞P. P. 0P. *Fig.* 5. ∞P. 0P. $\frac{1}{8}\breve{P}\infty$. $\breve{P}\infty$. *Fig.* 6. P. $\breve{P}\infty$. *Fig.* 7. Twin crystal; composition-face ∞P. *Fig.* 8. Twin, consisting of four individuals. *Fig.* 9. Twin consisting of five individuals. *Fig.* 10. Group of four crystals; composition-face, ∞P. *Fig.* 11. Twin crystal.

Melanterite. $\dot{F}e\,\dddot{S}+7\,\dot{H}$. MONOCLINIC.

SYN.—Green vitriol, Copperas, Sulphate of Iron, Eisenvitriol, Fer sulfaté.

It crystallizes as an inclined rhombic prism of 82° 21, and the crystals are usually needle shaped. It has an easy cleavage parallel to the base, and a more difficult one parallel to the prism. Fracture, conchoidal. Lustre, vitreous. Tranparent, translucent. Color, different shades of green or white. Streak, colorless. **H.**=2. **G.**=1.832. Composition, $\dddot{S}$ 28·8, $\dot{F}e$ 25.9, $\dot{H}$ 45.3.

Pyr. &c. B. P. In closed tube gives $\dot{H}$,$\ddot{S}$ and $\dddot{S}$. On Ch., gives the reactions for S and Fe. Soluble in twice its weight of water. It is usually found as an efflorescence, being a product of the decomposition of Pyrite. It is white when it has not been exposed to the air, but soon becomes green, blue or brown by oxidation. It is used in dyeing and tanning, in making ink and Prussian blue. It is found abundantly in the U. S.

FORMULÆ OF THE CRYSTALS.

Pl. XXIII.

Fig. 12. ∞P. $0P$. This crystal resembles a rhombohedron. *Fig.* 13. ∞P. $0P$. $\bar{P}\infty$. *Fig.* 14. The preceding, with $\infty\grave{P}\infty$. *Fig.* 15. The combination *Fig.* 2, with $-P$. $\grave{P}\infty$. $\infty\grave{P}\infty$. *Fig.* 16. *Fig.* 15, with $-\bar{P}\infty$.

Copiapite. $\ddot{F}e^2\,\dddot{S}^5+18\,\dot{H}$. HEXAGONAL. ?

SYN.—Gelbeisenerz.

It is found as fibrous masses or crystalline aggregations, which are rhombic or hexagonal in form. It has a perfect basal cleavage. Lustre, pearly. Translucent. Color canary or sulphur-yellow. **H.**=1.5. **G.**= 2.14. Composition, $\dddot{S}$ 42.7, $\ddot{F}e$ 34.2, $\dot{H}$ 23.1.

Pyr. &c. B. P. In a tube yields $\dot{H}$ and $\dddot{S}$. On Ch., gives a magnetic mass, and the reaction for S. Insoluble in water.

Occurs in Saxony, in the Hartz, in France and Chili, as the result of the decomposition of Pyrite.

Vivianite. $\dot{F}e^3\,\dddot{P}+8\,\dot{H}$. MONOCLINIC.

SYN.—Blue Iron Earth, Eisenblau, Blaueisenerde, Fer phosphaté, Fer azuré, Mullicite.

There are several phosphates of iron found in nature, but this is the most important one. It crystallizes as an oblique rhombic prism of 111° 12′, with a very easy cleavage parallel to the clinopinacoid. The usual form is a combination of both pinacoids, the prism, pyramid and macrodome. Crystals are usually found as needles on metallic substances. Their faces are very generally curved. It has a vitreous lustre, except on the faces of the clinopinacoids which are pearly or metallic. Transparent, translucent, but becomes opaque on exposure. In fresh fractures, it is white and colorless, but becomes blue, owing to a partial peroxidation. When it has undergone the action of the air, it is sometimes deep blue, or bluish or dirty brown, sometimes even black. Streak, colorless. **H.**=1.5-2. **G.**= 2.58-2.68. Composition, $\dot{F}e$ 43, $\dddot{P}$ 28.3, $\dot{H}$ 28.7.

Pyr. &c. B. P. Decrepitates, fuses at 1.5 to a magnetic globule and colors the flame green ($\dddot{P}$). Gives off water. It is soluble in HCl without effervescence.

Its usual occurrence is in cavities of certain rocks, especially lavas or rocks in coal mines, which have been on fire. It is also found in the inside of buried bones and fossil shells. It is found earthy, in little bunches in clays. It is then generally of a more or less dark blue, but sometimes of a greenish or violet color, resembling certain salts of copper, but is easily distinguished by a chemical test. It is found at the Gap mine, Pa., and in large quantities at Mullica Hill, Gloucester Co., N. J., replacing Belemnites.

FORMULÆ OF THE CRYSTALS.

Pl. XXIII.

Fig. 17. $\infty\grave{P}\infty$. $\infty\bar{P}\infty$. $\bar{P}\infty$. P. ∞P. $\infty\grave{P}3$. *Fig.* 18. $\infty\grave{P}\infty$. $\infty\bar{P}\infty$. $\bar{P}\infty$. *Fig.* 19. $\infty\grave{P}\infty$. ∞P. P.

Leucopyrite. $\dot{F}e\,As^2$. ORTHORHOMBIC.

SYN.—Arseneisen, Arsenikkalkies.

The forms of the crystals are the same as those of Arsenopyrite, only differing in the value of their angles. The angle of the prism in Leucopyrite is 122°, but in Arsenopyrite it is 111°. Fracture, uneven. Lustre, metallic. Color, silver or steel-gray. Streak, grayish-black. **H.**=5–5.5. **G.**=6.8–8.71. Composition, As 72.8, Fe 27.2.

Pyr. &c. B. P. In a closed tube gives As, and in an open tube $\ddot{A}s$. On Ch., gives the odor of arsenic and in O. F. a coating of $\ddot{A}s$; in R. F. a magnetic globule. Gives the reactions for iron.

It very much resembles Arsenopyrite, and can generally only be distinguished from it by a chemical test, in which case only arsenic and no sulphur is found. Its density is high, about 7, while that of Arsenopyrite is 6,1.

FORMULÆ OF THE CRYSTALS.

Pl. XXIII.

Fig. 20. ∞P. $\bar{P}\infty$.

Arsenopyrite. Fe $(As, S)^2$. ORTHORHOMBIC.

SYN.—Mispickel, Arsenical pyrites, Arsenkies, Fer arsenical.

It crystallizes in the right rhombic prism of 111° 53′, with traces of cleavage parallel to the prism. The crystals are mostly prisms attached to a gangue. They may have a rounded termination, which is a union of domes, or pseudo-octahedron. The crystals are often macled, parallel, either to the prism or macrodome. They have frequently such a number of faces that they appear to be curved and striated. The brachydome is usually striated, which is very characteristic. Lustre, metallic and quite light in a fresh fracture, but it becomes dull on exposure to the air. Color silver or tin-white. Streak, grayish-black. **H.**=5.5–6. **G.**=6–6.4. Composition, As 46, S 19.6, Fe 34.4.

Pyr. &c. B. P. On Ch., it gives arsenical fumes. The heat from stroke of the steel is sufficient to produce this odor. Heated in a tube, it gives off S and As, which condense first as a red sublimate of sulphide of As and then as a mirror of As. Heated for a long time in R. F., it gives a magnetic scoria. Decomposed by $\ddot{N}$, giving a deposit of $\ddot{A}s$ and S.

Large masses of Arsenopyrite are semi-crystalline or indistinctly fibrous, they are rarely amorphous, compact and granular. It is distinguished from Marcasite by its color, its reaction for arsenic and its density, which is 6.12, while that of Marcasite is about 5. Resembles Cobaltite, but it can be distinguished by the crystalline form and a chemical test. It is found abundantly in the U. S.

FORMULÆ OF THE CRYSTALS.

Pl. XXIII.

Fig. 21. ∞P. ¼P̆∞. The characteristic striation of the brachydome is shown on one of the faces. *Fig.* 22. ∞P. ¼P̆∞. P̄∞. *Fig.* 23. P̄∞. P̆∞. ¼P̆∞. Resembles a square octahedron. *Fig.* 24. ∞P. ½P̄∞. P̆∞. *Fig.* 25. Twin crystal; composition-face ∞P. Horizontal projection. *Fig.* 26. Twin with the combination *Fig.* 21; composition-face P̄∞.

Scorodite. F̶e Ä̈s+4 Ḣ. ORTHORHOMBIC.

SYN.—Arseniksinter, Eisensinter.

It crystallizes as a rhombic prism of 98° 2′, generally in little crystals of a bluish-green color usually slightly altered. Cleavage imperfect, parallel to the brachypinacoid. Lustre, vitreous. Fracture, uneven. Translucent. Color, pale green or brown. Streak, white. **H.**=3.5–4. **G.**=3.1–3.3. Composition, Ä̈s 49.8, F̶e 34.7, Ḣ 15.5.

Pyr. &c. B. P. In a closed tube yields Ḣ and turns yellow. On Ch., fuses easily, colors the flame blue (As), and gives off arsenical fumes. Soluble in HCl.

Scorodite is found in Edenville, N. Y., and in Carabas Co., N. C.

FORMULÆ OF THE CRYSTALS.

Pl. XXIV.

Fig. 1. P. ∞P̄∞. ∞P̆∞. *Fig.* 2. P. ∞P̆∞. ∞P̆2. *Fig.* 3. P. 0P. ∞P̆∞. ∞P̆2. 2P̄∞. *Fig.* 4. The combination *Fig.* 1, with ∞P̆2 and 2P̄∞.

Pharmacosiderite. 3 F̶e Ä̈s+F̶e Ḣ³+12 Ḣ. ISOMETRIC.

SYN.—Cube Ore, Würfelerz, Fer arseniaté.

Its usual form is the cube, but tetrahedral faces have been found. The crystals are found in cavities. Its cleavages are very difficult, parallel to the cube. Its fracture is conchoidal and its lustre vitreous. Transparent. Its color is usually emerald-green. This green becomes brown in the flame of a candle. It is sometimes found brown or honey-yellow. Streak, green, brown and yellow. It is pyroelectric.

Pyr. &c. B. P. In O. F. melts rapidly to a metallic globule which is not magnetic. In the R. F., it becomes a magnetic scoria and arsenic is given off. Soluble in acids.

With the blowpipe and with acids, it is easily distinguished from the copper salts which it resembles.

FORMULÆ OF THE CRYSTALS.

Pl. XIX.

Fig. 15. ∞O∞. $\frac{O}{2}$.

Arseniosiderite. $\dot{C}a^6\,\dddot{A}s+4\,\overline{F}e^2\,\dddot{A}s+15\,\dot{H}$.

It is quite rare, and is found in fibrous concretions of a yellowish-brown or golden color. Lustre, silky. Streak, yellowish-brown. **H.**=1–2. **G.**=3.52–3.88. Composition, $\dddot{A}s$ 37.9, $\overline{F}e$ 42.1, $\dot{C}a$ 11.1, $\dot{H}$ 8.9.

Pyr. &c. B. P. The same as Scorodite.

Siderite. $\dot{F}e\,\ddot{C}$. HEXAGONAL.

SYN.—Chalybite, Spathic Iron, Spherosiderite, Spatheisenstein, Eisenspath, Fer spathique, Fer carbonaté, Siderose.

It crystallizes in rhombohedra of 107°, with a very easy cleavage parallel to the faces of the rhombohedron. It is generally found as the primitive rhombohedron with curved faces, the curve affecting even the cleavage. It is also found in lenticular and cockscomb crystals. Its fracture is lamellar, rarely conchoidal. Lustre, vitreous or pearly. When just taken from the mine and quite pure, it is sometimes entirely white, but it soon becomes altered in the air, and takes a grayish color, which sometimes becomes brown, brownish-red or green. Streak, white. **H.**=3.5–4.5. **G.**= 3.7–3.9. Composition, $\dot{F}e$ 62.1, $\ddot{C}$ 37.9.

Pyr. &c. B. P. On Ch., blackens and fuses at 4.5. Heated in a closed tube, it decrepitates, blackens and gives a magnetic globule. In O. F., the $\dot{F}e$ becomes $\overline{F}e$, in R. F. becomes magnetic. With acids, it effervesces slowly when cold.

It is found crystalline and compact. It was supposed for a long time, that an Aragonite of iron called Junckerite existed, which was isomorphous with the carbonates of baryta, strontia, lime and lead, but it has been proved not to be well founded. It is almost to be regretted that it is not so. It is sometimes entirely black, owing to the presence of manganese. This decomposition is remarkable, since carbonate of iron and manganese when perfectly pure do not easily become peroxidized. It is also found in beds of different colors, which have been formed by the decomposition of sulphates of iron in contact with organic matter, producing Pyrite and Siderite. It generally has a lustre analogous to that of Dolomite, but its density is nearly 4, while that of Dolomite is 2.8 to 3. When found in lamellar, saccharoidal masses or granular, it resembles Calcite, but the density distinguishes it. Exceptionally, it is found fibrous. This variety is distinguished under the name of Spherosiderite ; it is generally found in cavities in Basalt, forming little mamelons with a velvety surface, and a fibrous or scaly fracture. It might be mistaken for Sphalerite, but the action of acids will distinguish it. The lithoidal variety of Siderite is whitish or brownish in a fresh fracture, but is quickly altered on exposure to the air and becomes brown. Its fracture is compact and conchoidal, and is devoid of lustre. These masses have no very distinctive characteristics, and can only be distinguished by the action of acids. It is most extensively found in the coal formation, in which case it is black or brownish. When it is associated with bituminous matter, it is called Black band. One of the ordinary appearances of this variety is in the shape of Septaria. These are spheroidal, flattened and brownish masses, which have been cracked during their formation, and subsequently these cracks filled with Calcite and Quartz. It also forms pseudomorphs. It is distinguished from Dolomite and Calcite, which it resembles, by its higher specific gravity, and by giving a magnetic globule.

Siderite is one of the most important ores of iron. It is found in veins at New Milford, Conn., Plymouth, N. H., and Stirling, Mass., and other places. The argillaceous variety is very abundant in the coal formation.

FORMULÆ OF THE CRYSTALS.

Pl. XXIV.

Fig. 5. R. Faces usually curved. *Fig.* 6. R. 0R.

Menaccanite. (Ti, Fe, Mn, Mg)^2 O^3. HEXAGONAL.

SYN.—Washingtonite, Crichtonite, Ilmenite, Titanic Iron, Titaneisenstein, Titaneisen, Eisenrose, Iserin, Basanomelan.

It crystallizes as a rhombohedron of 85° 40′. It has a number of derived forms, many of which are hemihedral and twins. Lustre, metallic. Color, bluish-black. Streak, black or reddish-black. **H.**=5.6. **G.**= 4.5–5. The composition varies greatly. $\ddot{\mathrm{Ti}}$ 10–59, $\overline{\mathrm{F}}\mathrm{e}$ 1.2–82.47, $\dot{\mathrm{F}}\mathrm{e}$ 1.5–50.17.

Pyr. &c. B. P. It is infusible; with S.Ph. in the O. F. it is dissolved. In R. F., it gives a bead, which is colorless when hot, but dark violet on cooling. When pulverized and heated with HCl, it is slowly dissolved to a yellow solution, which, when filtered and boiled with tin, becomes blue.

It has the same general characteristics as Magnetite, except perhaps that it is a little bluer and does not attract the magnet. The crystals, which are quite rare, are sometimes very large, and generally have a reddish tinge on the surface, owing to a partial decomposition. It is frequently found as a sand. It is very rare that Magnetites do not contain some trace of Titanic Iron. It is also frequently found in the other ores of iron, as in Hematite, for in almost all of the furnaces where these ores are used, nitrocyanide of titanium is found. It has less lustre than Hematite, and has a black streak. It is of little value as an ore of iron, as it is too refractory to smelt in large quantities. It is found in large crystals in Orange Co., N. Y., also at Washington and elsewhere in Conn., and in Mass. and R. I.

FORMULÆ OF THE CRYSTALS.

Pl. XXIV.

Fig. 7. R. 0R. -2R. $\frac{\frac{4}{3}P2}{2}$. *Fig.* 8. 5R. 0R. *Fig.* 9. 5R. R. 0R.

Chromite. ($\dot{\mathrm{F}}\mathrm{e}$, $\dot{\mathrm{C}}\mathrm{r}$, $\dot{\mathrm{M}}\mathrm{g}$) ($\ddot{\mathrm{A}}\mathrm{l}$, $\overline{\mathrm{F}}\mathrm{e}$, $\ddot{\mathrm{C}}\mathrm{r}$). ISOMETRIC.

SYN.—Chromic Iron, Chromeisenstein, Fer chromaté.

It is one of the Spinels of iron, a sort of Magnetite, in which the $\overline{\mathrm{F}}\mathrm{e}$ is replaced by $\ddot{\mathrm{C}}\mathrm{r}$, and $\ddot{\mathrm{A}}\mathrm{l}$. It has the same forms as Magnetite, but is not so frequently crystallized nor so perfectly. Its usual form is the octahedron, more rarely the cube. Fracture, uneven. Lustre, semi-metallic. Opaque. Color, brownish-black. Streak, brown. Sometimes slightly magnetic. **H.**=5.5. **G.**=4.321–4.498. Composition, for the formula $\dot{\mathrm{F}}\mathrm{e}$ $\ddot{\mathrm{C}}\mathrm{r}$, $\dot{\mathrm{F}}\mathrm{e}$ 32, $\ddot{\mathrm{C}}\mathrm{r}$ 68.

Pyr. &c. B. P. In O. F. infusible; in R. F., slightly rounded on the edges and becomes magnetic. With fluxes, gives the reactions for chromium, which distinguish it from Magnetite. It is not attacked by acids.

It resembles Magnetite and cannot be distinguished from it with certainty, except by its chemical properties. It is perhaps a little blacker than ordinary Magnetite, but this is only an external character, which may not hold good. It is often found as a sand composed of octahedral crystals. It is used for the manufacture of pigments. It is found in Vermont, Massachusetts, Connecticut, Pennsylvania, Maryland and elsewhere.

Columbite. $(\dot{Fe}, \dot{Mn})\ (\ddot{Cb}, \ddot{Ta})$. ORTHORHOMBIC.

SYN.—Niobite.

Crystallizes as a right rhombic prism of 101° 26′. The forms which predominate are those of the square prism, the angles of which are frequently so highly modified, as to appear rounded. It has cleavages parallel to both macro and brachypinacoids. Fracture, uneven. Lustre, submetallic. Opaque. Streak, dark red to black. **H.**=6. **G.**=5.4–6.5. Composition, for $\dot{Fe}\ \ddot{Cb}$, $\dot{Fe}$ 21.17, $\ddot{Cb}$ 78.83. The columbic acid is always in excess.

Pyr. &c. B. P. With borax it is dissolved, giving the reactions for iron. If flamed in the R. F., gives a grayish-white bead, or, if there is an excess, the bead becomes opaque without flaming.

Its very dark color and a slight iridescence which is almost always found on the specimens, distinguishes it from other minerals. Its principal localities in the U. S. are Haddam and Middletown, Conn., where it occurs in crystals and in pieces of considerable size in the Feldspar quarries.

FORMULÆ OF THE CRYSTALS.

Pl. XXIV.

Fig. 10. $\infty\breve{P}\infty$. $\infty\breve{P}3$. $\infty\bar{P}\infty$. $0P$. $3\breve{P}3$. P. $\frac{1}{2}\breve{P}\infty$. *Fig.* 11. $\infty\breve{P}\infty$. $\infty\bar{P}\infty$. ∞P. $\infty\breve{P}3$. $0P$. P. $2\bar{P}\infty$. $\bar{P}\infty$. *Fig.* 12. $\infty\breve{P}\infty$. $\breve{P}\infty$. $\frac{1}{2}\breve{P}\infty$. P. $\infty\bar{P}\infty$. $2P$. $\infty\breve{P}3$. ∞P. *Fig.* 13. $\infty\breve{P}\infty$. $\frac{1}{2}\breve{P}\infty$. $\breve{P}\infty$. $2\breve{P}\infty$. ∞P. P. $3\breve{P}\frac{3}{2}$. $\infty\breve{P}3$. $2P$. $3\breve{P}3$. *Fig.* 14. $\infty\breve{P}\infty$. $\infty\bar{P}\infty$. $\infty\breve{P}2$. ∞P. $\breve{P}3$.

Wolframite. $2\ \dot{Fe}\ \dddot{W} + 3\ \dot{Mn}\ \dddot{W}$ or $4\ \dot{Fe}\ \dddot{W} + \dot{Mn}\ \dddot{W}$. ORTHORHOMBIC

SYN.—Wolfram.

Its primitive form is a right rhombic prism of 101° 5′, with a very easy cleavage parallel to the brachypinacoid, which gives it a lamellar structure, and another perpendicular to it, which is not so easy. The crystals usually show the prism and macrodome very prominently, but are rarely complete. They are usually striated, on account of the easy cleavage parallel to the brachypinacoid. They show the hemihedry with parallel faces. They are often macled and, as this macle is not parallel, the cleavage shows reentrant angles. Lustre, semi-metallic. Opaque. Color, dark grayish or brownish-black. Streak, dark reddish-brown or black. It is sometimes slightly magnetic. **H.**=5–.55. **G.**=7.1–7.55. Composition, for the first formula, $\dddot{W}$, 75.33, $\dot{Fe}$ 9.55, $\dot{Mn}$ 15.12; for the second, $\dddot{W}$ 76.2, $\dot{Fe}$ 5.6, $\dot{Mn}$ 17.94.

Pyr. &c. B. P. Fuses at 2.5–3 to a globule studded with crystals. With borax and S.Ph. gives the reaction of tungstic acid, red when hot, yellow when cold. It is attacked with difficulty by acids, and gives up tungstic acid when cold.

It is also found in lamellar masses made up of imperfect crystals. Its density fracture and streak distinguish it from Hematite and Magnetite, which it very often resembles. It has been found in Munroe, Trumbull, Conn., also in Maryland, N. Carolina, and Missouri.

FORMULÆ OF THE CRYSTALS.

Pl. XXIV.

Fig. 15. ∞P. $\infty \bar{P} \infty$. $\frac{1}{2} \bar{P} \infty$. $\breve{P} \infty$. *Fig.* 16. ∞P. $\frac{1}{2} \bar{P} \infty$. $\infty \bar{P} \infty$. $\infty \bar{P} 2$. $\breve{P} \infty$. P. $2 \breve{P} 2$. *Fig.* 17. Twin crystal; composition-face $\infty \bar{P} \infty$. *Fig.* 18. Twin crystal; composition-face $\frac{2}{3} \breve{P} \infty$.

MANGANESE.

Braunite. $\dot{M}n \ddot{M}n$, or $\ddot{M}n$. TETRAGONAL.

SYN.—Hartbraunstein.

It crystallizes as an octahedron, the angle of which is 119° 46′. It is not isomorphous with Hematite, which is hexagonal. Some of the varieties show an obscure octahedral cleavage, but crystals are comparatively rare. They are generally octahedra, which are less acute than those of Hausmannite; they frequently show the base, and have curved faces. Smaller crystals are sometimes found with the octahedron and di-octahedron. It has an uneven fracture and a metallic lustre, which is sometimes quite distinct. Opaque. It is black, with more or less of a bluish or brownish tint. Its streak is brown into black, without any tinge of red, which distinguishes it from Hausmannite, which is often found on the same specimen. **H.**=6–6.5. **G.**=4.75–4.82. Composition, Mn 86.95, O 9.85, Ba 2.25, $\dot{H}$ 0.95.

Pyr. &c. B. P. Infusible and does not give off $\dot{H}$. With fluxes, gives the reactions for Mn. The non-oxidizing acids hardly attack it and give off only a little Cl. With HCl, Cl is evolved. When it contains Rhodonite, gelatinous silica is deposited.

It is almost always found in granular masses, of a bluish-black color. The grains are very small and do not show cleavage. The bluish lustre will distinguish them from Magnetite and Hematite, but the color of their streak is a better distinction. Like Alabandite, Braunite is often found in little veins in the silicate or carbonate of manganese, as a product of their decomposition, in which case it is very impure and gives up gelatinous silica, or even effervesces slightly. It is also found associated with other minerals of manganese, as the violet variety of Epidote, the Titanite containing manganese, and Hausmannite.

Hausmannite. $\dot{M}n^3 \ddot{M}n$. TETRAGONAL.

SYN.—Glanzbraunstein.

The angle of the octahedron is 105° 25′. It is the red oxide of manganese of the laboratories. It has an easy cleavage parallel to the base and octahedral cleavages, which are more difficult. The simple crystals are octahedra, which are sometimes truncated. They are often grouped 2 by 2 or 4 by 4, the twin plane being parallel to the octahedron. Its fracture is uneven. Lustre, submetallic. Opaque. Its color is generally brown, almost black. Streak, chestnut-brown, almost red. **H.**=5–5.5. **G.**=4.722. Composition, Mn 72.1, O 27.9.

Pyr. &c. B. P. Alone, it is infusible. With the fluxes, gives the reactions for Mn. The non-oxidizing acids attack it, giving but little Cl. HCl, gives Cl.

Besides the crystals, crystalline masses are found which sometimes show the points of crystals; often in others there are no crystals but only cleavages. They are often deep black, which easily distinguishes them from the oxides of iron. If it is crystallized, the acute octahedra are easily distinguished. The best and most distinctive characteristic is that of its streak, which is dark reddish-brown with the red color most distinct. Sometimes these masses resemble Magnetite, but are distinguished because they are not magnetic, and Magnetite has a black streak; the red of Hematite is quite different. Wolframite, which sometimes resembles it, has a much higher density and a dark violet streak.

FORMULÆ OF THE CRYSTALS.

Pl. XXIV.

Fig. 19. P. *Fig.* 20. Twin crystal; composition-face $P\infty$. *Pl.* XXV. *Fig.* 1. Twinning repeated with four individuals.

Pyrolusite. $\ddot{\mathrm{M}}\mathrm{n}$. ORTHORHOMBIC.

SYN.—Weichmanganerz.

It crystallizes as a right rhombic prism of 93° 40′, with a cleavage parallel to the brachypinacoid. The crystals are always implanted in a gangue and only show their ends. It is sometimes found in radiated bacillary masses, showing in their fracture a lustre which is at the same time metallic and silky, with a gray iron color. Its fracture is irregular and unequal. Lustre, metallic. Opaque. Color, iron-black or dark steel-gray. Streak, black. **H.**=2–2.5. **G.**=4.82. Composition, Mn 63.3, O 36.7.

Pyr. &c. B. P. Is infusible, gives off no water. With fluxes, it gives the reactions for Mn.

It is very fragile, stains the fingers black and is easily broken up with the nail. It is also found concretionary, sometimes in tubercular masses. Sometimes in stalactites, when it is usually cavernous. It is made up of concentric beds of fibres fastened together at the surface, which is smooth, and the adhesion is consequently not very great. Such masses are very tender, their fracture is fibrous; sometimes it is compact when impure. It is also found amorphous, in large masses disseminated in earthy matter. It is then intensely black without lustre, showing sometimes in the fresh fractures a slight semi-metallic reflection, resembling several metallic sulphides. It is only distinguished by its streak and blowpipe reactions. It is found in all the geological formations, probably produced by the decomposition of the carbonate, as little masses and frequently as dendrites. It resembles Psilomelane, but is distinguished by its inferior hardness. From the ores of iron, it is distinguished by the blowpipe. It is used in glass works for making bleaching powders and also for the manufacture of oxygen. It is found in many of the iron mines of the U. S., forming a velvety coating usually on Limonite.

FORMULÆ OF THE CRYSTALS.

Pl. XXV.

Fig. 2. ∞P. $\infty \bar{P} \infty$. $\infty \breve{P} \infty$. $\breve{P} \infty$. $0 P$.

Manganite. $\ddot{\bar{\text{M}}}\text{n}\ \dot{\text{H}}$. ORTHORHOMBIC.

SYN.—Acerdèse.

It crystallizes as a right rhombic prism of 99° 40′, with an easy cleavage parallel to the brachypinacoid, and another more difficult, parallel to the prism. It is usually well crystallized. The crystals are usually formed by the faces of the prism, frequently terminated by a macrodome or an octahedron. Generally, however, the crystals have a very much larger number of faces, which produce very marked striations on the vertical faces, so that they are often fluted and rounded. The crystals which have the basal terminations are frequently joined together parallel to the prism. The common base is in that case striated and undulated, by the projection of the terminations of the different crystals. The crystals in such a case are very much striated and frequently very large. It also shows macles, formed by a hemitrope parallel to the brachydome. It is also found in bacillary masses, made up of crystals diverging from a common center. Its fracture is uneven. Lustre, semi-metallic. Opaque. Color, dark-brown or iron-black. Streak, reddish-brown to nearly black, darker than Limonite. Its color is intermediate between that of Braunite and Hausmannite, and might result from a mixture of two. It is well in the presence of such a powder to seek for water.

Pyr. &c. B. P. In a tube gives off $\dot{\text{H}}$, and is then infusible; this distinguishes it from the other oxides. With fluxes, it gives the reactions for Mn. In acids, even before calcination, it is dissolved and gives off Cl.

It is also found concretionary and stalactitic, resembling the forms of Pyrolusite, but usually harder. The distinction is easily made by the color of the streak and the presence of water. It is found also as very small crystals imbedded in compact Manganite, which has a granular fracture, which may also be unequal. The mineral is then in concentric layers. It is also found in amorphous masses, which may be distinguished from Pyrolusite by their hardness, their streak, and the presence of water. Psilomelane often contains a considerable amount of Manganite, which is the reason why it gives off so much water. Manganite is found very abundantly in nature, but it gives too little chlorine to be used as a substitute for Pyrolusite; it does not pay the expenses of working.

FORMULÆ OF THE CRYSTALS.

Pl. XXV.

Fig. 3. ∞P. $\infty \breve{P}2$. $\infty \bar{P}2$. $\bar{P}3$. $2P$. $2\breve{P}2$. $\frac{5}{3}\breve{P}2$; the last form is hemihedral, as a rhombic sphenoid. *Fig.* 4. Horizontal projection of the preceding combination. *Fig.* 5. Twin crystal; composition-face parallel to the vertical axis. *Fig* 6. Twin crystal; composition-face $\breve{P}\infty$.

Psilomelane. ($\dot{\text{B}}$a $\dot{\text{M}}$n) $\ddot{\text{M}}$n + $\ddot{\bar{\text{M}}}$n + n$\dot{\text{H}}$ $\ddot{\text{M}}$n.

SYN.—Hartmanganerz.

It is never crystalline, but massive and botryoidal; sometimes the free Pyrolusite which is mixed with it, shows traces of crystallization. This formula is the one around which the mineral varies. It appears to be a mixture in variable quantities of Pyrolusite, and Pyrolusite combined with bases, which are most generally $\dot{\text{B}}$a, sometimes $\dot{\text{C}}$a, rarely alkalies or $\ddot{\bar{\text{A}}}$l, with a variable proportion of water. Not having a very well defined composition, it has no very fixed characters. It is harder than Pyrolusite, and sometimes even strikes fire with the steel. This is owing to the presence of silica, which separates in a gelatinous state when it is treated with acids. Lustre, submetallic. Opaque. Color, iron-black. Streak, brownish-black and shining. **H.**=5.6. **G**=3.7-4.7. Composition, $\ddot{\bar{\mathrm{M}}}\mathrm{n}$ and $\dot{\mathrm{M}}\mathrm{n}$ 81.8, O 9.5, $\dot{\mathrm{K}}$ 4.5, $\dot{\mathrm{H}}$ 4.2.

Pyr. &c. B. P. In a closed tube yields $\dot{\mathrm{H}}$. Loses oxygen by ignition. Gives the reactions for Mn. Yields Cl with HCl.
It appears to be a product deposited by mineral waters. There is now at Luxeuil, in France, Thermal Springs which abandon a deposit, which is quite analogous, produced by the decomposition of $\dot{\mathrm{M}}\mathrm{n}$ $\ddot{\mathrm{C}}$ held in solution. This spring holds in solution also the earthy carbonates and silica. The deposit which forms is compact and is quite comparable to Psilomelane. It is also found in concretions resembling Pyrolusite, on breaking which, fibres are found, which are generally pure Pyrolusite. It is also found in compact masses, with a variable fracture, and with characters quite analogous to those of Pyrolusite, generally black with a black streak. Its hardness distinguishes it. In these compounds the $\ddot{\bar{\mathrm{M}}}\mathrm{n}$ seems to act like an acid and to saturate bases. It has the same uses as Pyrolusite.

Wad. $\dot{\mathrm{R}}\ \ddot{\bar{\mathrm{M}}}\mathrm{n}+\dot{\mathrm{H}}$. $\dot{\mathrm{R}}=\dot{\mathrm{K}}, \dot{\mathrm{B}}\mathrm{a}, \dot{\mathrm{C}}\mathrm{o}, \dot{\mathrm{M}}\mathrm{n}$.

SYN.—Bog Manganese, Asbolite, Asbolan, Kobaltmanganerz, Schwarzer Erdkobalt, Lampadite.

This mineral is found in nature in two different states. It is sometimes in large amorphous masses, brown nearly black, which soil the fingers with a chocolate-brown powder. It has such a low density, that it is frequently called Cork Manganese. It is easily cut with a knife, but can hardly be pulverized with a pestle, for it shows a certain degree of elasticity. It has no lustre. It frequently contains $\dot{\mathrm{C}}\mathrm{o}$, and is then called Earthy Cobalt. Sometimes it contains $\dot{\mathrm{C}}\mathrm{u}$. It is often very light and soils the fingers. Its color is a dull bluish or brownish-black, or reddish-brown. **H.**=0.5–6. **G.**=3–4.26. Composition, $\dot{\mathrm{M}}\mathrm{n}$ $\ddot{\bar{\mathrm{M}}}\mathrm{n}$ 79.12, O 8.82, $\dot{\mathrm{B}}\mathrm{a}$ 1.4, $\dot{\mathrm{H}}$10.66.

Pyr. &c. B. P. Acts like Psilomelane, the varieties containing $\dot{\mathrm{C}}\mathrm{o}$ or $\dot{\mathrm{C}}\mathrm{u}$ react for these metals. Treated with HCl, gives off Cl.

It may sometimes be found as a concretion. It seems to have been a muddy deposit, which has afterwards become dry, for the masses are usually very much cracked. When it is found in large masses it can be profitably used in commerce, for it is easily soluble and gives off a great deal of Cl. In some of the Limonite mines, it is found in a condition totally different and has a silvery look. It is then called Silvery Manganese. It is found on the surfaces of the ore, as a velvety covering, with almost a metallic lustre. With the glass, little soft elastic scales of very thin Wad are distinctly seen. In some localities the one passes into the other, and the entire mass is then compact, with traces of a silvery lustre. It is used for making Cl, but is too impure to make oxygen. It is sometimes used as a paint. It is found abundantly in N. Y. and N. H.

DISTINCTION BETWEEN THE OXIDES OF MANGANESE.—These oxides of Manganese are very difficult to distinguish with the blowpipe, as they all give the same violet bead with fluxes. Manganite is distinguished by giving off water, from Braunite, Hausmannite and Pyrolusite. Wad is distinguished especially by its lightness; for all the others, the best distinctions are taken from the color of their streaks.

HAUSMANNITE.—Acute octahedra with plane faces; traces of cleavage; streak brownish-red.

BRAUNITE.—Octahedra, curved faces without cleavage; granular with a bluish-black color; streak, brown.

PYROLUSITE.—Tender; stains paper black.

MANGANITE.—Black, with no bluish color; fracture granular; streak, brown. Hardness greater than the others; gives off water.

WAD.—Light; soils the fingers chocolate-brown and gives off water.

The only remaining oxide is PSILOMELANE which has no very distinct characters. It is generally necessary to make a chemical test for $\dot{B}a$, by treating with HCl and then with $\dddot{S}$. Its hardness is generally greater than that of the other oxides.

Alabandite. MnS. ISOMETRIC.

SYN.—Alabandine, Manganblende, Manganèse sulfuré.

It is sometimes found in nature in cubes and octahedra and in crystalline masses which are lamellar and have an easy cubical cleavage, but crystals are quite rare. Its fracture is uneven. In its fresh fractures, it has a metallic lustre with an iron-gray color. It becomes tarnished black or brown in the air. The streak is a characteristic green. **H.**=3.5–4. **G.** =3.95–4.04. Composition, Mn 63.3, S 36.7.

Pyr. &c. B. P. In an open tube, gives S; is fusible with difficulty on the edges. In O. F., it is hardly roasted, but after the roasting, during which some $\ddot{S}$ is given off, the red oxide of Mn gives Mn reactions. With soda and nitre it gives a green color; with borax, a violet glass. Soluble in dilute $\dddot{N}$, giving off HS.

It is a very rare mineral and is found in a semi-crystalline state, associated with other minerals of Mn, the silicate and carbonate. The Alabandite is deposited in gray veins or patches through the latter, which gives it a peculiar look. This is one of the empirical means of distinguishing it.

Triplite. $\dot{R}^3\dddot{P}+R^4F$. ORTHORHOMBIC.

SYN.—Eisenapatit, Eisenpecherz, Zwieselit, Manganèse phosphaté.

This is the usual phosphate of manganese. Cleavage unequal in three directions, perpendicular to each other. Fracture, small conchoidal. In the fresh fractures, it is of a violet color, which is often intense, but as soon as it is exposed to the air, it becomes black or dark-brown. It is always opaque. The fracture and the lustre are resinous, especially in the varieties which have been altered and which do not show much cleavage. Opaque. Streak, yellowish-gray or brown. Very fragile. **H.**=4–5.5. **G.**=3.44–3.8. Composition, $\dddot{P}$ 32.8, $\dot{F}e$ 31.9, $\dot{M}n$ 32.6, $\dot{C}a$ 3.2. The $\dot{R}$ of this formula is $\dot{F}e$ and $\dot{M}n$; the R is Ca, Mg, Fe.

Pyr. &c. B. P. Fuses easily at 1.5 to a black magnetic globule. Gives the reactions for chromium, iron and phosphoric acid. With $\dddot{S}$ gives HFl. Soluble in HCl.

The phosphates of manganese are the products of a more or less advanced decomposition. They are always impure and contain a more or less amount of iron. Its resinous lustre is characteristic, which, joined to the brownish-black color, allows it to be immediately distinguished.

Only one substance, Uraninite, has this color with the resinous lustre, but it is easily distinguished by its density, which is above 6.5, while that of Triplite is about 4.

Rhodochrosite. $\dot{M}n\ddot{C}$. HEXAGONAL.

SYN.—Dialogite, Himbeerspath, Rosenspath, Manganspath, Manganèse carbonaté.

It crystallizes in rhombohedra of 106° 51', frequently with curved faces. It has a rhombohedral cleavage, which is sometimes curved, but it is not easy. The crystals are usually very small, and the usual one shown is the rhombohedron. They are generally grouped and interpenetrated one with the other, making spherical mamelons, and forming a coating on a gangue, which is usually Barite or Quartz. On these coatings the points of crystals are sometimes seen, which have a velvety lustre. The fracture of these little mamelons is semi-lamellar or fibrous. The fracture is ordinarily uneven, sometimes fibrous and rarely compact. It is quite brittle. Lustre, vitreous or pearly on the fracture. Translucent. Opaque. Its colors are rose, which may be more or less dark, yellowish-gray, dark red or brown. Streak, white. **H.**=3.5–4.5. **G.**=3.4–3.7. Composition, $\ddot{C}$ 38.6, $\dot{M}n$ 61.4. It is rarely ever pure, and usually contains $\dot{C}a$ and $\dot{M}g$.

Pyr. &c. B. P. Decrepitates a little, becomes black and is infusible if it is pure. When it contains silica, it sometimes fuses slightly. It effervesces easily with acids, but not as easily as Calcite, as it must be heated a little.

Sometimes it occurs as an imperfect coating, which is pulverulent and earthy, slightly colored, and generally mixed with Calcite. The lamellar masses resemble Calcite, but the color and density distinguish it. It is often mixed with Siderite, Magnesite or Dolomite. It decomposes on exposure like the silicate, and is then covered with black spots of the different oxides, or of the sulphide, which is easy to determine by the color of the streak. It is found in Vermont, Massachusetts, New York and elsewhere.

COBALT.

Linnæite. $2\,Co\,S+Co\,S^3$. ISOMETRIC.

SYN.—Cobalt Pyrites, Siegenite, Kobaltkies, Kobaltnickelkies, Cobalt sulfuré.

It has an imperfect cubical cleavage. The ordinary form of the crystals is the octahedron, which, with the cube, always has plane faces. Its forms are usually holohedral. The fracture is unequal. Lustre, metallic. The color is gray, with a slight rosy tint, but becomes red by tarnish. Streak, blackish-gray. **H.**=5.5 **G.**=4.8–5. Composition, S 4.2, Co 58.

Pyr. &c. B. P. In O. F., it is roasted, but does not show any traces of As or Sb. In the R. F., it gives a magnetic residue, and with fluxes the reactions for Co. It is insoluble in the non-oxidizing acids, but in $\dddot{N}$ it is dissolved, giving a rose-colored liquid and deposit of sulphur.

It is sometimes accompanied by other minerals of cobalt, or associated with a gangue. It has been found at Mine la Motte, in Missouri, and at Mineral Hill, Maryland.

FORMULÆ OF THE CRYSTALS.

Pl. XXV.

Fig. 7. O. *Fig.* 8. O. $\infty O \infty$; the most frequent form.

Bieberite. $(\dot{C}o, \dot{M}g)\dddot{S} + 7\dot{H}$. MONOCLINIC.

SYN.—Cobalt Vitriol, Red Vitriol, Kobaltvitriol, Cobalt sulfaté.

It is usually found as a product of the alteration of Smaltite. It is a very incoherent mass with a styptic taste, and is generally found as a pulverulent coating. It usually contains $\dot{F}e\dddot{S}$. Lustre, vitreous. Translucent. Color, flesh and rose-red. **G.**=1.924. Composition, $\dddot{S}$ 28.4, $\dot{C}o$ 2.5.5, $\dot{H}$ 46.1.

Pyr. &c. B. P. Yields $\dot{H}$ and at a high heat, $\dddot{S}$. Gives the reactions for Co. It is rather a product of decomposition than a mineral.

Smaltite. (Co, Fe, Ni) As^2. ISOMETRIC.

SYN.—Smaltine, Chloanthite, Speiskobalt, Cobalt arsenical.

It occurs in crystals, the dominant form of which is generally the cube, but it always has octahedral faces which are convex, while those of Linnæite are plane. It is also found massive, with various imitative shapes. The cleavage is distinct, parallel to the octahedron and in traces, parallel to the cube. Lustre, metallic. Fracture, granular and uneven. Color, generally a silver or tin-white, sometimes iridescent or grayish from tarnish. Streak, grayish-black. **H.**=5.5–6. **G.**=6.4–7.2. Composition, As 72.1, Co 9.4, Ni 9.5, Fe 9.0.

Pyr. &c. B. P. On Ch., it gives As, and fuses to a globule. In a tube, it gives an arsenic mirror. With the fluxes, it affords the reactions for Co, Fe, and Ni. It is not attacked by the non-oxidizing acids.

The slightly crystalline masses usually have a reticulated structure. It is also found in amorphous masses of a steel-gray color. It is sometimes covered with a green coating, which results from the nickel that it contains. It is also found in masses, which have a semi-granular or semi-lamellar fracture. There is a great resemblance between these masses, Linnæite and Native Bismuth; but Linnæite is less lamellar and does not give off As; Bismuth is redder and cleaves in three directions and has a lamellar fracture. From Arsenopyrite and Leucopyrite, it can be easily distinguished by the crystalline form and blowpipe reactions.

It has been found in the U. S., at Chatham, Ct. It is used for making smalt; hence its name.

FORMULÆ OF THE CRYSTALS.

Pl. XXV.

Fig. 9. $\infty O \infty$. O; the most usual form. *Fig.* 10. $\infty O \infty$. ∞O.

Cobaltite. Co (S, As)^2 ISOMETRIC.

SYN.—Cobaltine, Kobaltglanz, Glanzkobalt, Cobalt gris.

The form of the crystals is the same as that of Pyrite, and it occurs in the same combinations. It shows the hemihedry of parallel faces in its crystals, which are well defined and which are its usual mode of occurrence. It has a cleavage, parallel to the cube, so that in masses it is always lamellar. The crystals are frequently octahedral, but they always have the faces of the hemi-tetrahexahedron. The faces of the cube are usually striated as in Pyrite, but those of the octahedron are smooth. The hemihedry distinguishes it from Linnæite, which has the same metallic lustre, but which is holohedral. It has the same formula as Arsenopyrite, but is not isomorphous with it, although it always contains some iron.

Lustre, metallic. Fracture, uneven. Color, silver-white, often a little rosy and also grayish, if much iron is present. Streak, grayish-black. **H.**=5.5. **G.**=6–63. Composition, As 45.2, Co 35. 5, S 19.3.

Pyr. &c. B. P. In a closed tube is unaltered; but in an open tube gives $\ddot{S}$ and a crystalline sublimate of $\ddot{A}s$. On Ch., affords fumes of S and As and fuses to a magnetic globule. With the fluxes gives the reactions for Co, Ni and Fe. It is soluble in warm $\dddot{N}$, giving a rose-colored liquid and depositing $\ddot{A}s$ and S. It is used in the manufacture of smalt and for painting porcelain.

FORMULÆ OF THE CRYSTALS.

Pl. XXV.

Fig. 11. $\frac{\infty O 2}{2}$. $\infty O \infty$; the most usual and characteristic form. *Fig.* 12. O. $\frac{\infty O 2}{2}$. *Fig.* 13. The same, with the two forms equally carried out; also a characteristic form. *Fig.* 14. $\frac{\infty O 2}{2}$. O. *Fig.* 15. $\infty O \infty$. O. $\frac{\infty O 2}{2}$. $\frac{\infty O 4}{2}$. $\frac{2 O \frac{4}{3}}{2}$.

Erythrite. $\dot{C}o^3 \ddot{\bar{A}}s + 8 \dot{H}$. MONOCLINIC.

SYN.—Erythrine, Cobalt Bloom, Kobaltblüthe, Cobalt arseniaté.

It has an easy cleavage, parallel to the clinopinacoid and is isomorphous with Vivianite. Lustre on the clinopinacoid, pearly, on the other faces adamantine, also dull or earthy. Transparent, translucent. It has a violet-rose color which is entirely characteristic, sometimes having the color of peach blossoms; sometimes greenish-gray. Streak, a little paler than the color. These small crystals are frequently grouped in such a way as to give the mineral a velvety appearance. The crystals are always very small. **H.**=1.5–2.5. **G.**=2.948. Composition, $\ddot{\bar{A}}s$ 38.43. $\dot{C}o$ 37.55, $\dot{H}$ 34.02.

Pyr. &c. B. P. In a closed tube it loses water, and from a rose color passes to a dark violet. In R. F., it gives off As. In acids, it dissolves without effervescence and without residue. Generally it is a rosy efflorescence on other minerals of cobalt. It often colors other minerals; it requires only 1 or 2 % to give a decided color.

Remingtonite. $\dot{C}o \ddot{C} + Aq$

This is the same carbonate which is obtained in the laboratories. It colors certain limestones, in which it enters to the extent of 4 to 5 %. This same color may be produced by Erythrite. The difference is difficult to distinguish by the eye, but can be distinguished by the blowpipe. It is soluble in HCl, and gives decided reactions for iron. Cobalt is distinguished by the borax bead. Generally the specimens are nothing but Aragonite, slightly colored.

It resembles Kermesite, but this volatilizes entirely before the blowpipe. It resembles some varieties of Cuprite, but is distinguished by the reactions with borax. It has been found at Finksburg, Md., at Chatham, Conn., and Mine la Motte, Missouri.

NICKEL.

Millerite. Ni S. HEXAGONAL.

SYN.—Capillary Pyrites, Haarkies, Schwefelnickel, Nickel sulfuré.

It is generally crystallized and perfectly pure. It has traces of a rhombohedral cleavage, which is generally almost impossible to determine, on account of the very small size of the crystals. Its lustre is metallic. Color, brass-yellow and often with an iridescent tarnish. Streak, bright. **H.**=3–3.5. **G.**=4.6 – 5.65. Composition, Ni 64.9, S 35.1.

Pyr. &c. B. P. In an open tube it gives sulphurous fumes; on Ch., after roasting in O. F., it gives in R. F. a magnetic globule. It is attacked by the non-oxidizing acids and gives the green color of nickel.

It can readily be distinguished from Pyrite by the disposition of its crystals, which are always capillary, sometimes extended on the gangue, and sometimes adhering by one extremity only and diverging. It is frequently called Capillary Pyrites. It is found with other metallic sulphides, Pyrite, Tetrahedrite and especially with Pyrrhotite, in which it appears to replace a portion of the Fe S, and is one of the principal ores of nickel. To be certain of the presence of nickel, the mineral must first be roasted, and then fused with borax or S.Ph. If there is no nickel, the result is a yellow glass not very highly colored; when there is, the glass is violet in the oxidizing flame, and gray in the reducing flame. It is found in capillary fibers at Antwerp, Jefferson County, N. Y., and in coatings of a radiated structure having a tufted appearance, at the Gap mine, Pa.

Niccolite. Ni As. HEXAGONAL.

SYN.—Copper Nickel, Kupfernickel, Rothnickelkies, Nickeline.

Crystals are exceedingly rare; it is usually found amorphous. On the fresh fracture the lustre is bright and metallic, but the surfaces become tarnished by exposure. They are often covered with a greenish coating of Annabergite, which can be used as an empirical character for distinguishing it. It is always opaque. Its color is a light copper-red, which is quite characteristic. The intensity of the color, however, is variable and is subject to tarnish; those specimens which contain antimony are much darker, while those containing arsenic are paler. Streak, pale brownish-black. **H.**=5–5.5. **G.**=7.33-7.671. Composition, Ni 44.1, As 55.9.

Ni As must be regarded as a general formula, for Niccolite is one of the hybrid species, in which arsenic and antimony replace each other, as the mineral may contain one atom of both together, or one atom of only one combined with one atom of nickel.

Pyr. &c. B. P. On Ch., it gives off a garlic odor with white vapors, if it contains arsenic; when antimony alone is present, which is very rare, there is only a coating of antimony without any odor. It is soluble in aqua regia.

The copper-colored mineral is often associated with a tin-white mineral which is Rammelsbergite, Ni As^2. This is one of its most characteristic associations. The color might cause it to be mistaken for metallic bismuth, which, however, is almost always crystalline, and has a density about 9. It is found at Chatham, Conn.

Ullmannite. $Ni\,(S, As, Sb)^2$. ISOMETRIC.

SYN.—Nickel Stibine, Nickelantimonkies, Nickelspiessglanzerz, Antimoine sulfuré nickelifère.

It is another of the hybrid species, in which Ni, Co, Mn, Fe, As, Sb and S may replace each other in every proportion. It has traces of a cubical cleavage. It is sometimes found in crystals; they are generally rounded cubes, on which octahedral faces are sometimes seen, which also are rounded. On the fresh fractures it has a metallic lustre and a tin-gray color, which however becomes dull. The natural surfaces are almost always altered and are covered with a greenish or reddish coating, which is an arsenite of nickel or cobalt.

Pyr. &c. B. P. Heated in a tube, it gives the orange-colored coating of As S and a black mirror of As or Sb. On Ch., it gives off white vapors of As and Sb and is then reduced to a magnetic cinder. Decomposed by $\ddot{N}$ and gives a green solution with separation of sulphur.

It is sometimes found in reticulated masses, which are frequently full of hollows. It is sometimes in masses of irregular structure, which resembles Arsenopyrite, from which it is distinguished by its density as well as by its fracture, which is irregular, while that of Arsenopyrite is lamellar. A chemical test, however, is always safest.

Annabergite. $\dot{N}i^3\,\dddot{A}s+8\,\dot{H}$. MONOCLINIC.

SYN.—Nickel Ochre, Nickeloker, Nickel arseniaté.

It is isomorphous with Erythrite. It is an accidental product and is found on other minerals of nickel, as a greenish efflorescence, with a color which is more or less intense. The pale varieties are usually mixed with free $\dddot{A}s$. When they are treated with boiling water, the intense green color of the mineral becomes evident. Streak, greenish-white. Composition, $\dot{N}i$ 37.2, $\dddot{A}s$ 38.6, $\dot{H}$ 24.2.

Pyr. &c. B. P. In a closed tube gives off water. Fuses easily and gives a metallic globule, which gives the reactions for Ni. It is soluble in acids.

It might be confounded with some of the ores of copper, but is distinguished by its chemical characters. This is soluble in acids with the characteristic colors of Ni. It has been found at Chatham, Ct.

Zaratite. $\dot{N}i\,\ddot{C}+2\,\dot{N}i\,\dot{H}+4\,\dot{H}$.

SYN.—Texasite, Emerald Nickel, Nickelsmaragd.

It is found as a crust, which is sometimes mamelonated. It has a vitreous lustre. Transparent to translucent. Color, emerald-green. Streak, paler than color. **H.**=3-3.25. **G.**=2.57-2.693. Composition, $\dot{N}i$ 59.4, $\ddot{C}$ 11.7, $\dot{H}$ 28.9.

Pyr. &c. B. P. In a closed tube yields $\dot{H}$ and $\ddot{C}$; infusible, gives the reactions for nickel. Soluble with effervescence in dilute HCl, when heated. It is quite rare, except at Texas, Pa., where it is found in Serpentine, associated with Magnesite and Dolomite.

ZINC.

Zincite. $\dot{Z}n$. HEXAGONAL.

SYN.—Red Zinc Ore, Rothzinkerz, Zinc oxydé.

It is the same oxide which is produced in the laboratories and in metallurgical operations. It is red, however, owing to the presence of a certain proportion of the oxide of manganese, which is always mixed with it, probably as M̈n. It is always found as crystalline masses, having cleavages parallel to the hexagonal prism, and another very easy one, parallel to the base, so that the masses are always lamellar, cracked and striated. In the fracture it has a bright, almost adamantine, lustre, and is translucent on the edges. Its color is characteristic, it is a dark orange-red, somewhat brown. Streak, orange-yellow. **H.**=4–4.5. **G.**=5.43–5.7. Composition, Zn 80.26, O 19.74.

Pyr. &c. B. P. Heated in a closed tube bleaches, but resumes its color on cooling. Infusible. In the R. F., gives metallic zinc, which volatilizes, oxidizes, and forms a white ring. Gives a green color with nitrate of cobalt. Shows the reactions for manganese. Soluble in acids.

It resembles somewhat red Stilbite, but is distinguished by its infusibility and its associations. It is used as an ore of zinc. It is found in masses in limestone at Franklin, N. J. It is also found associated with Franklinite at the same place.

Sphalerite. Zn S. ISOMETRIC.

SYN.—Blende, Black Jack, Zinkblende, Zinc sulfuré.

It is found crystallized, lamellar, compact and concretionary. The crystals are usually disseminated in a gangue, or united to form large crystalline masses which are always more or less lamellar. The crystals show the hemi-tetragonal trisoctahedron, or the tetrahedron and cube; it is usually found as a rhombic dodecahedron. The simple forms of Sphalerite are relatively rare. They are usually composite forms resulting from complicated laws of derivation, and, what renders them even more complex, they are often macled and hemitrope. One of the most frequent of these results from the hemitropy of the combination of the rhombic dodecahedron, and the tetragonal tris-octahedron producing a figure of 24 faces, 12 of which are triangular belonging to the tetragonal trisoctahedron, and 12 tetragonal, *Pl.* XXV. *Fig.* 24. It shows the hemihedry of inclined faces, giving generally tetrahedral forms. It has as easy cleavage parallel to rhombic dodecahedron, and a difficult one parallel to the octahedron. Fracture, conchoidal. Its lustre is very bright. In the clear varieties it is sometimes adamantine, and in the dark varieties it is almost metallic; sometimes it is resinous. Its colors are very variable, it is rarely colorless, but is generally honey-yellow, brown, black, red and green. When pure it is generally white or yellow. The variation in the color corresponds to a want of homogeneity in composition. The yellow variety sometimes contain cadmium, and all contain a very variable proportion of iron. Whatever may be its color, its streak is white or reddish-brown. **H.**=3.5–4. **G.**=3.9–4.2. Composition, Zn 67.0, S 33.0.

Pyr. &c. B. P. Infusible. In O. F. it gives off sulphurous vapors and often a cadmium coating. The roasting is long and difficult, and after it, in the R. F. it gives a coat of Zn which is yellow when hot, and white when cold. Soluble in HCl. With N̈ very little red vapor is given off, but much H S.

It is also found in lamellar, imperfectly crystallized masses, which may have every variety of color. These masses are sometimes lamellar and sometimes saccharoidal. The dark varieties may sometimes resemble Garnet, Vesuvianite and Cassiterite, but its density about 4, and the blowpipe reactions distinguish it. When it is concretionary it is in spheroidal masses radiated and with a fibrous fracture. These masses may be large or small and show rings of color. The fracture, where the masses are small, is usually compact. It might resemble in this state Barite, Celestite, Siderite or Apatite, but is easily distinguished by acids and the blowpipe.

It is found also with an entirely compact fracture resulting from a partial decomposition of the saccharoidal variety. These generally show some scales, and their fracture and the blowpipe distinguishes them. Some of the dark varieties resemble Cassiterite, and certain kinds of Garnet, but it is distinguished both by its cleavage and hardness. It is one of the most abundant ores of zinc. It is found in many places in the U. S.

FORMULÆ OF THE CRYSTALS.

Pl. XXV.

Fig. 16. $\frac{O}{2} \cdot - \frac{O}{2} \cdot$ *Fig.* 17. $\frac{O}{2} \cdot \quad \infty O.$ *Fig.* 18. $\infty O \infty. \infty O 2.$ *Fig.* 19. Two octahedra interpenetrating. *Fig.* 20. Twin crystal; composition-face O. *Fig.* 21. ∞O, showing the twin plane. *Fig.* 22. Hemitrope from the preceding. *Fig.* 23. $\infty O. \quad O. \quad \frac{3O3}{2} \cdot \quad \infty O \infty.$ *Fig.* 24. $\infty O. \quad \frac{3O3}{2}$; a characteristic form. *Fig.* 25. Hemitrope from the preceding. *Fig.* 26. ∞O, shortened in the direction of the diagonal of the faces. *Fig.* 27. The same, lengthened.

Goslarite. $\dot{Z}n \dddot{S} + 7 \dot{H}$. ORTHORHOMBIC.

SYN.—White Vitriol, Zinc Vitriol, White Copperas, Zinc sulfaté.

Crystallizes as a right rhombic prism of 90° 42′. It has a cleavage parallel to the macropinacoid. Lustre, vitreous. Transparent, translusent. Color, white, reddish or bluish. Streak, white. **H**=2–2.5. **G.**= 2.036. Composition, Zn 28.2, $\dddot{S}$ 27.9, $\dot{H}$ 43.9.

Pyr. &c. B. P. In a closed tube yields water. Gives the reaction for zinc and sulphur; easily soluble in water.

It is found as an efflorescence, or as an accidental product of decomposition of Sphalerite in certain mines.

Smithsonite. $\dot{Z}n \ddot{C}$. HEXAGONAL.

SYN.—Dry-bone, Zinkspath, Kapnit.

It is found in rhombohedra of 107° 40′, with a very easy rhombohedral cleavage, which gives it a lamellar structure. The crystals are sometimes the primitive rhombohedron and sometimes scalenohedra. In this last case, the crystals are usually very small and cover the amorphous variety. They are frequently covered with a coating which may be blackish or pure white. The rhombohedral crystals are usually much longer and have curved faces and a vitreous lustre, which is sometimes resinous, or even adamantine. Its fracture is uneven. Color, white, green, yellow or brown. Streak, white. **H.**=5. **G.**=4–4.5. Composition, $\dot{Z}n$ 64.8, $\ddot{C}$ 35.2.

Pyr. &c. B. P. In a closed tube loses $\ddot{C}$. Infusible. On Ch., with soda gives vapors which are yellow while hot, white when cold. It is soluble in acids with effervescence, which however is slow; the acid must be heated to produce it.

It is variously colored by foreign substances. In copper localities it is green. The blonde varieties are usually colored with iron or manganese. The stalactites, which are white, usually contain lime or magnesia. The concretionary masses have usually a stalactitic structure, occurring in little mamelons, sometimes covered with the points of crystals. Their lustre is generally feeble, but is sometimes vitreous. The fracture of these masses is sometimes lamellar or scaly, very rarely fibrous. They are sometimes compact and have the appearance of calcareous concretions. It is found amorphous and earthy, having the appearance of a coarse limestone, but its density and the blowpipe distinguish it. It is also found as a white arborescent, compact mass. It often resembles Siderite and the compounds of baryta and strontia, but is distinguished by the blowpipe. It is also distinguished from Sphalerite, Willemite and Calamine by its action with acids. It is much harder than any of the carbonates, with which it might be confounded by its form and color.

It is found in large quantities in Missouri, Iowa, Wisconsin and Tennessee.

Hydrozincite. $\dot{Zn}\,\ddot{C} + 2\,\dot{Zn}\,\dot{H}$.

SYN.—Zinc Bloom, Earthy Calamine, Zinkblüthe, Zinconise, Marionite.

It is never crystallized. It is white and has a dull, earthy lustre. Sometimes when it has just been taken from the mine, it is translucent, but it quickly becomes opaque. It usually forms a series of white coatings, with vacant spaces between each bed, which is irregular and bent. It is sometimes found in balls, having the same disposition in concentric layers. Its streak is shining. **H.**=2 — 2.5. **G.**=3.58–3.8. Composition, $\dot{Zn}$ 75.3, $\ddot{C}$ 13.6, $\dot{H}$ 11.1.

Pyr. &c. B. P. Heated in a tube, it gives off water and becomes yellow, but turns white again on cooling. It is reduced with soda on charcoal, the zinc is volatilized, burned and deposited as oxide. Acids dissolve it easily with effervescence.

This mineral was formerly considered as an accidental product in other ores of zinc. Within a few years, however, it has been found in very large masses in Spain, and is a real mineral of zinc. It has been found in Pennsylvania, Wisconsin and Arkansas.

TIN.

Cassiterite. $\ddot{Sn}$. TETRAGONAL.

SYN.—Tin Stone, Stream Tin, Zinnstein, Zinnerz, Étain oxidé.

Simple crystals are usually rare, but are found with the forms, *Pl*, XXVI. *Fig.* 1–5. Generally the crystals are macled around a plane parallel to the octahedron of the second order, P ∞. This may take place on the prism or on the octahedron. These macles show re-entrant angles and are often called *tin beaks* (becs d'étain.) Sometimes these macles occur several times, and give rise to geniculated crystals, *Figs.* 9–11. The prisms are striated and cylindrical, while the faces of octahedra are smooth and polished. Each locality has forms peculiar to it, so that generally the crystals from the same place can be recognized by their form. Cassiterite is isomorphous with Rutile. It is trimorphous. If $\ddot{Sn}$ is prepared in the laboratory by the action of water on the chloride or fluoride of tin, different forms are obtained according to the salt which is used. Both of these forms differ from the Cassiterite of nature. It has cleavages which are very distinct, parallel to both prisms. The fracture is ordinarily unequal, sometimes conchoidal, rarely lamellar. Lustre, adamantine. Transparent, opaque. The colors of Cassiterite are very variable. It is found exceptionally colorless and transparent, in a few localities. Generally its color is every gradation, intermediate between blonde and black, owing to mixtures with iron. Sometimes, it is red, gray, white or yellow. The color is usually not equally diffused in the mass or the crystal, but is generally in bands. Streak, white, grayish or brownish. **H.**=6–7. **G.**=6.4–7.1. Composition, Sn 78.67, O 21.23.

Pyr. &c. B. P. Infusible. In R. F., it is reduced with difficulty, but soda facilitates the reduction. With borax, it melts easily and the Sn becomes the base of an enamel. The glass is opaline and white, sometimes somewhat yellow when hot, on account of the presence of a little iron. It is only slightly acted upon by acids.

It is also found concretionary in zones of different colors, but it is easily distinguished by its density, which is between 6 and 7; and its hardness distinguishes it from Sphalerite and Barite, which it frequently resembles. It is found in rolled pebbles with different bands of colors, and is then called Wood Tin. It is distinguished by its hardness and its density. It may become substituted by pseudomorphism for certain crystals, especially for those of Orthoclase, which are hemitrope and half interpenetrated and which are found in the Trachytes. It is, however, only a kind of moulding, the cause of which is unknown. The Feldspar appears first to become clay, and the clay to be replaced by tin, without any perceptible change in the form of the crystal. It is recognized by its density, which also distinguishes it from Vesuvianite, some varieties of Garnet and Tourmaline, which it frequently resembles in form, lustre and color. From Sphalerite, it is distinguished by its blowpipe reactions and its hardness. It has been found in the U. S. at Lyme and Jackson, N. H.

FORMULÆ OF THE CRYSTALS.

Pl. XXVI.

Fig. 1. ∞P. P; P sometimes predominates. *Fig.* 2. ∞P. P. ∞P∞; also pyramidal as in *Fig.* 15. *Fig.* 3. The preceding, with P∞. *Fig.* 4. ∞P. ∞P$\frac{3}{2}$. 3P$\frac{3}{2}$. P. P∞. *Fig.* 5. 3P$\frac{3}{2}$. P. ∞P. *Fig.* 6. *Fig.* 1; showing the twin plane P∞. *Fig.* 7. Hemitrope from the preceding. *Fig.* 8. The same, but with the plane in a different position. *Fig.* 9. Hemitrope of *Fig.* 3. *Fig.* 10. Hemitrope of *Fig.* 2. *Fig.* 11. Trilling, formed by the repetition of the hemitropy. *Fig.* 12. Twin of four individuals. *Fig.* 13. Hemitrope of a pyramidal crystal. *Fig.* 14. The same. *Fig.* 15. Hemitrope of a pyramidal crystal like *Fig.* 2. *Fig* 16. Hemitrope of nine individuals.

Stannite. 2 (Ɵu, Fe, Zn,) S+Sn S^2.

SYN.—Tin Pyrites, Bell Metal Ore, Zinnkies, Étain sulfuré.

The composition of this substance in not very well defined; it appears to be a sulphostannide of iron and copper with a little zinc. It has been found very rarely with traces of crystallization and of cleavage. Its lustre is metallic. Fracture, uneven. Opaque. Its color is steel-gray, iron-black or dark green, rather brownish from tarnish. Its streak is black. **H.**=4. **G.**=4.3–4.522. Composition, Sn 27.2, Cu 29.3, Fe 6.5, Zn 7.5, S 29.6.

It is a very rare mineral, and has no resemblance to any other, except to the different Pyrites, from which it is distinguished by the action of nitric acid.

TITANIUM.

Rutile. Ti. TETRAGONAL.

SYN.—Nigrine, Ilmenorutile, Titane oxydé.

Rutile is isomorphous with Cassiterite; it is crystallized in the same system and has the same relation to its modifications, and the same cleavages. It generally has a lamellar appearance. The simple crystals are sometimes octahedra, with plane faces and sometimes prisms which are striated and dull, also frequently macled. The crystals are like those of Cassiterite, showing geniculated forms, which are frequently turned in two or three directions, *Figs.* 2–7. The color varies from reddish-yellow to brown, always preserving a reddish tint; sometimes bluish, violet or green. In thin plates by reflection it is brownish blood-red. Its fracture is conchoidal across the crystal, and lamellar parallel to it. Lustre, metallic, adamantine. Translucent, opaque. **H.**=6–6.5. **G.**=4.18–4.25. Composition, Ti 61, O 39.

Pyr. &c. B. P. It is infusible and gives with fluxes the reactions for Ti. With borax and S.Ph., it dissolves, giving in the O. F. a glass colorless when hot, and amethyst-violet when cold. It is not acted on by ordinary acids, but after fusion with alkaline carbonates, the acid solution boiled with tin foil gives a violet color.

Besides its crystallized varieties, it is often found as a mass in a gangue and then somewhat resembles Vesuvianite. When the density cannot be taken into account, the color, the nature of the gangue and, if need be, a chemical examination will distinguish it. It is found in bacillary and acicular crystals; these may be fine needles enveloped in a gangue, which is often perfectly clear crystallized Quartz, which proves that the Quartz was formed after the Rutile and with sufficient slowness not to disarrange the general arrangement of the crystals. These acicular crystals are often very fine. Sometimes these crystals are reticulated on the surface of a gangue. In some localities, especially at Mt. Blanc in Dolomite, they affect a peculiar disposition. It is a kind of tissue of a yellowish changeable color made up of acicular crystals, which often show geniculations like the large crystals and are sometimes woven together.

Under the name of Nigrine a black Rutile is described, which contains 14% of iron; it is never crystallized, but shows traces of cleavage. Its streak is yellowish-brown. Its infusibility distinguishes it from Tourmaline, Vesuvianite and Pyroxene. From Cassiterite, by giving no metal with soda. It is found in very large crystals in Pennsylvania and Georgia.

Titanic acid is found in nature in three different states and with three different forms. It is therefore trimorphous. In one of these forms it is isomorphous with Cassiterite, which under certain circumstances appears also to be trimorphous. Titanic acid is always found in one of these forms; two of which, Rutile and Octahedrite, are tetragonal, but with different dimensions. The third, Brookite, is orthorhombic.

FORMULÆ OF THE CRYSTALS.

Pl. XXVII.

Fig. 1. ∞P. ∞P∞. ∞P3. P. P∞. P3. 3P$\frac{3}{2}$. *Fig.* 2. Hemitrope; composition-face P ∞. Combination, ∞ P. 3. P. *Fig.* 3. Hemitrope, with two geniculations. *Fig.* 4. Twin crystal. *Fig.* 5. Twin, resembling an hexagonal prism and formed by the meeting of the extremities of the geniculated crystals. *Fig.* 6. Hemitrope; composition-face P ∞. *Fig.* 7. Hemitrope.

Octahedrite. T̈i. TETRAGONAL.

SYN.—Anatase, Dauphinite, Oisanite.

The forms and dimensions of the crystals are different from those of Rutile. It has an easy cleavage parallel to the base, and others more difficult, parallel to the octahedron. It is always crystallized. In crystals it shows two different types. From Dauphiny it is found in acute octahedra, which sometimes show a base. These crystals are found on Dioryte, and are accompanied by Albite and Asbestus. From Brazil, however, the base predominates; the crystals are very much flattened, and show the octahedron and dioctahedron. Its colors are very variable. In Brazil, in certain diamond sands, it is found almost colorless or pale honey-yellow; sometimes it is reddish-brown, but generally it is a characteristic dark brown, which is quite distinctive. It is greenish-yellow by transmitted light. In thin scales it is transparent. It is the most refringent of bodies, after the sulphate of mercury, so that its lustre is very bright. The lustre is often hidden in rolled crystals. Its fracture is lamellar or conchoidal across the crystals. Lustre, metallic or adamantine. Transparent, opaque. Streak, colorless. Its chemical characters are that of Rutile. Its density which is less, becomes the same as Rutile, on being heated. **H.**=4–5.5. **G.**=3.82–3.95, or 4.11–4.16 after heating. Composition, Ti, 61, O 39.

Pyr. &c. B. P. Same as Rutile. It has been found at Smithfield, R. I.

FORMULÆ OF THE CRYSTALS.

Pl. XXVII.

Fig. 8. P. *Fig.* 9. P. 0P. *Fig.* 10. P. $\frac{1}{5}$P. *Fig.* 11. P. $\frac{5}{19}$P5. *Fig.* 12. P. $\frac{1}{8}$P. 2P∞. *Fig.* 13. P. 0P. P∞. ∞P∞. *Fig.* 14. 0P. P. P∞. 2P∞.

Brookite. T̈i. ORTHORHOMBIC.

SYN.—Arkansite.

It is much rarer than Rutile. It is ordinarily found in thin plates, which are to be referred to the right rhombic prism of 99° 50′. The crystals are complex. The faces of the macropinacoid are striated and dull, the others are smooth and lustrous. There is an indistinct cleavage parallel to the base and prism. Transparent, opaque. Lustre, metallic, adamantine. Color, brown, yellowish or reddish. Streak, colorless, grayish or yellowish. **H.**=5.5–6. **G.**=4.12–4.23. Composition, Ti 61.0, O 39.0.

Pyr. &c. B. P. Same as Rutile. The hardness is less than Rutile, as is also the density, which however becomes the same on heating. It is found at Magnet Cove, Arkansas, and at Ellenville, Orange Co., N. Y.

FORMULÆ OF THE CRYSTALS.

Pl. XXVII.

Fig. 15. ∞P. $\infty \bar{P} \infty$. $\breve{P} 2$; the usual form of Arkansite. *Fig.* 16. The same, with $\infty \bar{P} \infty$ predominating. *Fig.* 17. ∞P. $\infty \bar{P} \infty$. $\infty \breve{P} \infty$. $\infty \bar{P} 2$. P. $\frac{1}{2}P$. 2 P. $\breve{P} 2$. $2 \breve{P} 2$. $\frac{1}{2} \bar{P} \infty$. $\frac{1}{4} \bar{P} \infty$. $2 \breve{P} \infty$. 0P; from the Urals. *Fig.* 18. ∞P. $\infty \bar{P} \infty$. $\infty \breve{P} \infty$. $\infty \bar{P} 2$. P. $\frac{1}{2}P$. $\breve{P} 2$. $2 \breve{P} 2$. $\frac{1}{2} \bar{P} \infty$. $\frac{1}{4} \bar{P} \infty$. $2 \breve{P} \infty$. 0P; from Ellenville, N. Y.

LEAD.

Lead. Pb. Isometric.

Syn.—Gediegen Blei, Plomb natif.

It has been found in several localities, but is however an accidental product, which is exceedingly rare. It has been found as globules in Galenite, as filaments or sheets in lava or basalt, or formed by reducing agencies on either the sulphide, sulphate or carbonate of lead. When it is found on the outcrops, as in S. A., it is probably owing to fires having been kindled upon the vein. It is crystalline, and shows its system to be isometric. Its lustre is metallic. Color, lead-gray; malleable and ductile. **H.**=1.5. **G.**=11.445.

Pyr. &c. B. P. Fuses easily in the O. F., covering the Ch. with a yellow coating; in the R. F. volatilizes.

Minium. $\ddot{P}b + 2 \dot{P}b$.

Syn.—Mennig, Plomb oxidé rouge.

In many localities oxides of lead are found as products of the decomposition of Galenite or Cerussite. It is always found as a pulverulent coating, on the surface of other ores. Lustre, greasy or dull. Opaque. Color, very variable; it is yellow when its composition is near Massicot and redder when it approaches Minium; generally it is orange-yellow, which is more or less dark. Streak, orange-yellow. **H.**=2.3. G.=4.6. Compostion, Pb 90.66, O 9.34.

Pyr. &c. B. P. In the R. F. easily reduced to metallic lead. Its most distinguishing character is its color and association with the other ores of lead.

Galenite. Pb S. Isometric.

Syn.—Galena, Bleiglanz, Bleischweif, Plomb sulfuré.

It is by far the most important of the ores of lead, the other minerals being frequently nothing but the products of its alteration. It has a perfect cleavage in three directions at right angles to each other, all equally easy. It is usually found crystalline or crystallized; the crystals are often very large, the dominant forms being the cube, rhombic dodecahedron and octahedron. The general form is the cube and octahedron. The crystals are sometimes macled and are usually the cube, macled by hemitropy around a normal to the face of the octahedron. More frequently it is found as crystalline masses made up by the union of several crystals. The hemitropy of the cube is frequently found in these masses and is recognised by the cleavage, which shows two systems of cubes. It sometimes has a peculiar look, which is owing to a slight inclination of two systems of crystals, showing a slight depression between the two. It is also found in comby masses, formed by a large number of little crystals. Its fracture is ordinarily lamellar or saccharoidal, passing sometimes to a granular texture. In some particular varieties it is entirely compact, with a smooth, conchoidal fracture, showing concentric zones like some specimens of Silex. It has a grayish-blue color. In its fresh fracture it has a metallic lustre, which is quite bright, but becomes dull on exposure. Streak, lead-gray. **H.**=2.5–2.75. **G.**=7.25–7.77. Composition, Pb 86.6, S 13.4.

Pyr. &c. B. P. In an open tube gives off $\dot{S}$. On Ch., decrepitates, and then in O. F. is roasted, giving off a sulphurous odor and lead fumes, which coat the coal at a short distance from the assay with a yellow ring. After being roasted, it gives metallic lead which is malleable. It is soluble in acids, giving off HS.

It usually occurs in lamellar or saccharoidal masses, in which way it occurs in nearly all the lead mines. It is usually found disseminated in Calcite, Quartz, Fluorite or Barite. The lamellæ of these masses are usually small, sometimes plane, sometimes curved, and it can usually be seen that they are all arranged in the same general direction. Sometimes these masses appear to be stratified. When the mineral is quite compact its color is much bluer. It is frequently found pseudomorphic, replacing crystals of other substances. The most usual pseudomorph is after Pyromorphite. Galenite is the simple sulphide of lead, but is rarely ever pure. It alloys itself by mixture with a certain proportion of sulphide of silver, which is isomorphous with it. There is probably no specimen of Galenite, which does not contain some silver. In some ores the proportion is very slight, while in others it is sufficiently large to make it an ore of silver. It is possible to a certain extent, to judge from external characters whether the ore is rich in silver or not, but these characters are only true within certain limits, and, if reliance is placed on them exclusively, very grave errors will be committed. In every case, recourse must be had to an assay for silver, in order to obtain a clear, or even approximate idea of the yield of an ore. The Galenites with large lamellæ, and especially those which are well crystallized, are generally poor in silver. Those which show an arborescent structure are often argentiferous. Those with small lamellæ, especially when their faces are curved, and have a granular fracture, and which are nearly compact and of a dark blue color, are generally rich in silver. The ones which are entirely compact, although they differ but little from those described, are rarely rich, but usually very poor. In the stratified Galenites, the glass usually shows parts which are lamellar and bluish, having the aspect of ordinary Galenite, but near to which there is a cement which is entirely compact and of a much whiter color. These parts are not Galenite, but a mineral which has about the same relation to lead, which Tetrahedrite bears to copper. It is composed principally of As, Sb, Pb, and other metals, and is always more or less argentiferous and is sometimes very rich.

On account of its lamellar structure, Galenite might be sometimes mistaken for Stibnite, but is distinguished by being harder and less brittle, and by its cleavage in 3 directions, while Stibnite has only one; the blowpipe also distinguishes it. It sometimes resembles Tetrahedrite, but is distinguished by its cleavage and by the absence of copper. At first sight, Molybdenite resembles Galenite, but is much softer, and lamellar in one direction only. Its lamellæ are flexible. Galenite is the most important ore of lead. It is sometimes used for glazing earthenware. It is found abundantly in the U. S., in Missouri, Illinois, Wisconsin, and also to some extent in N. Y. State, but the mines have not generally proved profitable. It has been found in many other states, particularly in N. C. But little of the Galenite of the eastern part of the U. S. is argentiferous. Some of the veins of the other states contain enough silver to pay for extraction.

FORMULÆ OF THE CRYSTALS.

Pl. XXVII.

Fig. 19. ∞O∞. O; the most usual form. *Fig.* 20. The same, with the faces equally developed *Pl.* XXVIII. *Fig* 1. ∞O∞. ∞O. *Fig.* 2. ∞O∞. ∞O. O. *Fig.* 3. The preceding, with the faces of the octahedron larger; a characteristic form. *Fig.* 4. ∞O∞. ∞O. O. $\frac{3}{2}$O$\frac{3}{2}$. *Fig.* 5. ∞O∞. 2O. *Fig.* 6. O. ∞O∞. ∞O. 2O; also characteristic. *Fig.* 7. Twin crystal; composition-face O, and flattened parallel to the same.

Bournonite. 3 (Cu, Pb) S+Sb^2 S^3. ORTHORHOMBIC.

SYN.—Wheel Ore, Rädelerz, Schwarzspiessglaserz, Antimoine sulfuré plumbo-cuprifère.

The primitive form is a right rhombic prism of 93° 40′. It has an imperfect cleavage parallel to the brachypinacoid. This is probably not a cleavage, but a separation of the different crystals. The simple crystals have two types, according as the dominant form is the prism or made up of the pinacoids, but such crystals of Bournonite are rare, as it is almost always found as macled crystals. Sometimes the macle is formed of two crystals only, but there are generally a large number of simple crystals arranged in a perfectly characteristic manner, somewhat resembling a cog wheel or pinion. In Brazil it is found in large crystals which have no lustre, but the surface of which is covered with points. It is also found in compact masses which appear to be parts of large crystals, that have been prevented from being developed. These can be easily distinguished by their fracture and peculiar lustre. They are covered in some specimens with blue or green spots, resulting from a superficial decomposition of the copper. It has a conchoidal fracture, which is usually smooth. Its lustre is metallic, and is very bright on the natural faces when they are smooth. Upon the fracture the lustre is somewhat resinous, without ceasing to be metallic; this, with the nature of the fracture, characterizes the mineral. Its color and streak are gray, varying from iron to steel-gray. It is very brittle. **H.**=2.5–3. **G.**=5.7–5.9. Composition, Pb 42.4, Cu 12.9, Sb 25, S 19.7.

Pyr. &c. B. P. In a closed tube, decrepitates and gives a dark red sublimate. On Ch., it is fusible and gives off Sb, which is condensed in a white ring. In the O. F., there is formed a yellow ring of Pb, and beyond that the antimony coat appears. In the R. F., a globule of copper is obtained, which is brittle. There is usually an odor of arsenic. Decomposed by nitric acid.

It resembles Tetrahedrite, but can be distinguished from it by the presence of lead.

FORMULÆ OF THE CRYSTALS.

Pl. XXVIII.

Fig. 8. 0P. $\infty\bar{P}\infty$. P. $\breve{P}\infty$. *Fig.* 9. 0P. ∞P. $\infty\bar{P}\infty$. $\infty\breve{P}\infty$. $\bar{P}\infty$. $\breve{P}\infty$. *Fig.* 10. $\infty\breve{P}\infty$. 0P. $\bar{P}\infty$. $\infty\bar{P}\infty$. ∞P; lengthened in the direction of the brachydiagonal. *Fig.* 11. $\infty\bar{P}\infty$. $\infty\breve{P}\infty$. 0P. P. $\frac{1}{2}$P. ∞P. $\infty\bar{P}2$. $\infty\breve{P}2$. $\bar{P}\infty$. $\frac{1}{2}\bar{P}\infty$. $\breve{P}\infty$. *Fig.* 12. Twin crystal; composition-face ∞P. *Fig* 13. Wheel-shaped crystal, formed by repeated twinning.

Anglesite. $\dot{P}b\ \dddot{S}$. ORTHORHOMBIC.

SYN.—Bleivitriol, Vitriolbleierz, Plomb sulfaté.

It is isomorphous with Barite and Celestite. The primitive form is a right rhombic prism of 103° 43′, while that of the other minerals is 104° and 106′. It has a very easy cleavage parallel to the base, and two others parallel to the faces of the prism. Its crystalline forms are nearly the same as those of Barite. It occurs also in a combination of the rhombic octahedron and the prism, in which the octahedron always predominates. It is also found macled. The fracture is lamellar or conchoidal. Lustre, adamantine, vitreous and resinous. Transparent, translucent, opaque. It is generally white, but sometimes black on the surface, or tinged with yellow, green or blue. **H.**=2.75–3, **G.**=6.12–6.39. Composition, Pb 73 6, $\dddot{S}$ 26.4.

Pyr. &c. B. P. Decrepitates and fuses in the flame of a candle. In the O. F., decrepitates and melts to a transparent bead, which becomes opaque on cooling. In the R. F., intumesces and then gives the reactions for lead, a metallic globule and yellow ring. Concentrated $\ddot{N}$ dissolves it completely. Soluble in 22,816 parts of water.

It is also found in lamellar masses, which can be recognized by their density, which is about 6, and their adamantine lustre; also in earthy masses, which are often either concretionary, or made up of beds fitting into each other. Such masses must be distinguished by their blowpipe characters. Its exterior characters ally it with Cerussite, but it is easily distinguished by the action of acids. It can be distinguished from all other minerals by its reactions for Pb and $\dddot{S}$. It is found associated with Galenite. The most celebrated locality in the U. S. was formerly Phenixville, Pa.

FORMULÆ OF THE CRYSTALS.

Pl. XXVIII.

Fig. 14. $\infty\breve{P}2$. $\bar{P}\infty$. *Fig.* 15. $\infty\breve{P}2$. $\bar{P}\infty$. $\infty\breve{P}\infty$. *Fig.* 16. The preceding, with P and $\breve{P}\infty$. *Fig.* 17. P; from Phenixville. *Fig.* 18. P. ∞P. *Fig.* 19. ∞P. $\frac{1}{2}\bar{P}\infty$. $\breve{P}\infty$. P. $\breve{P}2$. $\infty\bar{P}\infty$. $\infty\breve{P}\infty$.

Clausthalite. Pb Se. ISOMETRIC.

SYN.—Selenblei, Plomb séleniuré.

It is a selenium Galenite. Nothing in its external characteristics distinguishes it from the latter, with which it is associated in a very small number of metallic deposits. It has the same cleavage and is found in lamellar masses, which have either large lamellæ, or are saccharoidal or even granular; detached crystals have not yet been found. Cleavage, cubic. Fracture, granular or shining. Lustre, metallic. Opaque. Color lead-gray, distinctly bluish. Streak, darker than color. **H.**=2.5–3. **G.**=7.6–8.8. Composition, Pb 72.4, Se 27.6.

Pyr. &c. B. P. In a closed tube, it decrepitates. In an open tube, gives the odor of selenium. On Ch., fuses readily and gives a strong odor of selenium, coating the coal near the assay gray, with a reddish border (Se), and beyond this yellow (Pb). If pure, it is extremely volatile.

The distinctive character is its blue color, which approaches to black; but this character is a very delicate one and is difficult to appreciate. In reality the external aspect cannot determine it and it is indispensable, when the presence of this mineral is suspected, to test it. This test is a simple treatment before the blowpipe, which in the presence of selenium affords a strong and unpleasant odor.

Pyromorphite. $3\,\dot{P}b^3\,\dddot{P}+Pb\,Cl$. Hexagonal.

Syn.—Phosphate of Lead, Grünbleierz, Plomb phosphaté.

Its usual form is the hexagonal prism, with traces of a cleavage parallel to the prism and pyramid. It is generally crystallized as an hexagonal prism, which may have a variable number of faces on the edge of the base. These crystals are often distorted by the rounding of their faces, or by having the base hollow as in the crystals from Phenixville, Pa. The gray or colorless varieties, which are nearly pure, are often found in large striated and rounded crystals. The yellow varieties, which contain $\dddot{A}s$, are generally quite regular. They are, however, somewhat rounded at times. The green varieties are usually in small distinct crystals, but are sometimes barrel-shaped. This rounding takes place on the vertical faces only and does not affect the base. In the orange-red varieties, the rounding of the vertical faces is even more distinct, so that the crystals are sometimes almost spheroidal. This, with their color, almost distinguishes them. The fracture is conchoidal and smooth. Lustre, resinous. Translucent, opaque. The colors are very variable, green, yellow, brown or white, and are dependent upon the composition. The varieties which are nearly pure phosphate of lead, are generally colorless or grayish, with a slight yellowish tinge. When they have a color which is more or less green, it is usually owing to the presence of copper or chromium. The varieties which contain considerable $\dddot{A}s$ are yellow. All of these varieties may occur together, so that the color is intermediate between gray, green and yellow, and is sometimes very undecided. In some localities it is found with a dark orange-red color, which is probably owing to the presence of chromic or vanadic acid. Whatever may be the color, it has a peculiar lustre which is partly adamantine, and partly resinous. Streak, white, sometimes yellowish. **H.**=3.5-4.5. **G.**=6.5–7.1. Composition, $\dot{P}b$ 74.1, $\dddot{P}$ 15.7, Cl 2.6, Pb 7.6. $\dddot{P}$ and $\dddot{A}s$ being isomorphous, small quantities of the latter are often found in the crystals. Some varieties contain also copper, vanadium and chromium, without its being possible to make a separate species of them. Lime occurs also in small proportions.

Pyr. &c. B. P. In a closed tube gives a white sublimate of chloride of lead and As if the latter is present. Fuses at 1.5. In the R. F., it gives a globule of lead which is brittle; in the O. F., melts and gives a globule, which in cooling assumes polyhedric forms. This a characteristic peculiarity. Soluble in acids with difficulty, but without residue.

Besides these forms, it is often found in bacillary fibers which are often interlaced. These crystals are often dark and look very much like moss, this is especially true of the green varieties. It is also found in mamelonated masses with a velvety lustre, owing to the little crystals which appear on the surface. It is also found in concretionary masses of different colors. These masses are characterized by their lustre, which is feeble but at the same time adamantine and resinous, the latter predominating. It is also found in amorphous masses, which sometimes appear to be intermediate between compact and concretionary. They are crystalline masses made up of large crystals, which have been prevented from being developed. This is shown by the acicular crystals which are seen at a few points at the same time. These masses are very much like Anglesite and Cerussite, but their lustre is much less resinous, and their chemical characters are different, both with the action of acids and before the blowpipe. It sometimes occurs in red gummy-looking drops, when it contains vanadic acid. It is a valuble ore of lead.

It was formerly found in great abundance in the Wheatley Mine, Pa., and in the Davidsōn Mine, N. C.

FORMULÆ OF THE CRYSTALS.

Pl. XXVIII.

Fig. 20. ∞P. 0P; most usual form. *Fig.* 21. 0P. ∞P. ∞P2. *Fig.* 22. ∞P. 0P. P. *Fig.* 23. ∞P. 0P. P. ∞P2.

Mimetite. $3\,\dot{P}b^3\,\ddot{\bar{A}}s + Pb\,Cl$. HEXAGONAL.

SYN.—Campylite, Mimetine, Mimetisite, Plomb arseniaté.

It is generally found crystallized as an hexagonal prism, with a number of pyramids. Cleavage imperfect, parallel to the pyramid. Lustre, resinous. Translucent, opaque. Color, pale yellow, orange-yellow, brown or white. Streak, white. **H.**=3.5. **G.**=7–7.25. Composition, $\dot{P}b\,\ddot{\bar{A}}s$ 90.60, Pb Cl 9.34.

Pyr. &c. B. P. In a closed tube, gives vapors of chloride of lead and of arsenic. Fuses at 1, and in the R. F. gives a globule of lead, the usual lead coating and often a white one of chloride of lead. Soluble in $\dddot{\bar{N}}$.

It resembles Pyromorphite in form and association, and it is sometimes difficult to distinguish between the two species, especially when Pyromorphite contains arsenic. It is often botryoidal and compact. It is a rare mineral and was formerly found at Phenixville, Pa., associated with Pyromorphite. When found in large quantities, it is a valuable ore of lead.

FORMULÆ OF THE CRYSTALS.

Pl. XXVIII.

Fig. 22. ∞P. 0P. P; usual form. *Fig.* 24. ∞P. 0P. 2P. *Pl.* XXIX. *Fig.* 1. ∞P. 0P. P. 2P. *Fig.* 2. ∞P. P. *Fig.* 3. ∞P. P. 2P. *Fig.* 4. P. 0P.

Cerussite. Ṗb C̈. ORTHORHOMBIC.

Syn.—Carbonate of Lead, Weissbleierz, Plomb carbonaté, Plomb blanche, Céruse.

It is isomorphous with Witherite, Strontianite and Aragonite, with which it is sometimes associated. The angle of the prism is 117° 13′. There is a trace of cleavage parallel to the prism, which gives the mineral at times a lamellar appearance. The simple crystals have the same form as those of Aragonite. There is a form which is very much flattened, in the direction of the brachypinacoid, *Pl.* XXIX. *Fig.* 7. This form is peculiar to certain crystalline masses, made up of acicular crystals woven together, sometimes parallel to each other, sometimes divergent or even interlaced without any apparent law. These masses are generally milk-white and without lustre, or with a silky lustre, the crystals themselves being often made up of little capillary crystals. Besides these simple forms, there are macles, which are generally hemitropes around the face of the prism. If this macle is repeated several times, it gives the star-shaped twin, *Fig.* 13, but it is sometimes much more complex. As the angle is almost 120°, the interior opening being very small and generally full of massive Cerussite, there results a regular star with six branches. It is very brittle. Fracture, conchoidal. Lustre, adamantine. Transparent, translucent, opaque. Its colors are white, grayish-white and yellowish-white, blue or green. It is sometimes black on the surface, but this is superficial, being caused by sulphurous vapor, and does not interfere with the adamantine lustre. **H.**=3–3.5. **G.**=6.465–6.48. Composition, Pb 83.5, C̈ 16.5.

Pyr. &c. B. P. In a closed tube decrepitates, and seems to pass into a state of dimorphism similar to that which takes place with Aragonite. After decrepitation, it becomes reduced in the R. F. to a metallic globule. Soluble with effervescence in dilute N̈.

It is frequently found in saccharoidal masses, which are easily distinguished by their adamantine lustre and their density, which is about 6. Amorphous and earthy masses are also found, which are only characterized by their great density. They are colored differently, according to their association. It resembles Anglesite, but is distinguished by the action of acids. It is a valuable ore of lead. It was formerly found in large crystals at the Davidson Mine, N. C., and at Phenixville, Pa., and is generally found on the outcrops of veins.

FORMULÆ OF THE CRYSTALS.

Pl. XXIX.

Fig. 5. P. 2 P̆ ∞; resembles an hexagonal pyramid. *Fig.* 6. P. ∞ P̆ ∞. 2 P̆ ∞. ∞ P; the faces 2 P̆ ∞ and ∞ P̆ ∞ usually have horizontal striations. *Fig.* 7. ∞ P̆ ∞. P. ∞ P. ∞ P̆ 3; tabular crystal. *Fig.* 8. ∞ P̆ ∞. 4 P̆ ∞. 2 P̆ ∞. P. ∞ P. *Fig.* 9. P̆ ∞. ∞ P̆ ∞. P. ∞ P. ∞ P̆ 3. ∞ P ∞. *Fig.* 10. ∞ P. ∞ P̆ ∞. 0 P. P. 2 P̆ ∞. 3 P̆ ∞. 4 P̆ ∞. *Fig.* 11. ∞ P̆ ∞. ½ P̆ ∞. 2 P̆ ∞. 4 P̆ ∞. P. ∞ P. ∞ P ∞. ∞ P̆ 3. ½ P ∞. *Fig.* 12. Twin crystal, with the combination ½ P̆ ∞. 2 P̆ ∞. ∞ P̆ ∞. P. ∞ P: composition-face ∞ P. *Fig.* 13. Trilling, with the combination ∞ P̆ ∞. ∞ P. P. ½ P̆ ∞. *Fig.* 14. Twin crystal, with the form of *Fig.* 7. *Fig.* 15. Horizontal projection of a twin crystal. *Fig.* 16. Trilling, with the form of *Fig.* 7.

Crocoite. Ṗb C̈r. Monoclinic.

Syn.—Crocoisite, Rothbleierz, Plomb chromaté.

It crystallizes as an oblique rhombic prism of 93° 42′. It has several cleavages, one parallel to the prism distinct, the other parallel to the base and clinopinacoid less so; but these cleavages are difficult and as the substance is very fragile it is not easy to determine them. It is always in more or less perfect crystals, which are generally attached to a gangue and form a crust upon it. Its fracture is conchoidal. Lustre, adamantine, vitreous. Translucent. It is very brittle. Its color is dark red somewhat like that of Realgar, which, however, contains more yellow. Streak, orange-yellow. **H.**=2.5–3. **G.**=5.9–6.1. Composition, Ṗb 68.9, C̈r 31.1.

Pyr. &c. B. P. In a closed tube decrepitates and blackens, but recovers its color on cooling. Fuses at 1.5, spreads over the charcoal and gives in the R. F. a metallic globule, a chrome scoria and lead fumes. It is easily dissolved in acids, and in HCl gives a green solution and crystallized needles of chloride of lead.

FORMULÆ OF THE CRYSTALS.

Pl. XXIX.

Fig. 17. ∞P. -P. P; both hemipyramids are equally developed. *Fig.* 18. ∞P. -P. *Fig.* 19. The preceding, with ∞P̀2. *Fig.* 20. ∞P. 4P̄∞; Beresowsk. *Fig.* 21. ∞P. -P. 4P̄∞. *Fig.* 22. ∞P. ∞P̄2. ∞P̄∞. -P. P̄∞. 3P̄∞. 2P̄2. 0P. *Fig.* 23. ∞P. -P. P̄∞. 2P̄2. 0P. 2P̀∞. P̀∞. ½P̀∞. *Fig.* 24. ∞P̀2. -P. 3P̄∞. 4P̄∞. ∞P.

Stolzite. Ṗb Ẅ. Tetragonal.

Syn.—Tungstate of Lead, Scheelbleispath, Wolframbleierz, Scheelitine.

It is isomorphous with Wulfenite and is found under the same conditions. Its crystals are never flattened, but show the octahedron distinctly. It has an imperfect cleavage parallel to the base. Lustre, resinous. Translucent, opaque. Color, brown, red, green or yellowish-gray. Streak, colorless. **H.**=2.75–3. **G.**=7.87–8.13. Composition, Ṗb 49, Ẅ 51.

Pyr. &c. B. P. It decrepitates and fuses at 2 to a crystalline bead. In the R. F. on Ch., it gives a metallic globule. Affords the reactions for tungstic acid and lead. Decomposed by N̈, giving a yellow residue of tungstic acid.

It is of a much browner color than Scheelite, but generally the two species must be distinguished by trial. It has been found in this country at Southhampton, Mass.

FORMULÆ OF THE CRYSTALS.

Pl. XXIX.

Fig. 25. P. ∞P. *Fig.* 26. P. 2P. ∞P. P∞.

Wulfenite. Ṗb M̈o. Tetragonal.

Syn.—Molybdate of Lead, Gelbbleierz, Plomb molybdaté, Melinose.

It has an easy cleavage parallel to the octahedron, and a less easy one parallel to the base. The crystals are usually very short resembling square tablets, and have faces of an octahedron of the second order, which are rarely carried to a point. They are usually set on their edge on a gangue, the base having generally less lustre than the other faces. Sometimes the crystals are thicker, opaque and without lustre. They have always about the same form, and rest on the gangue in the same way. Exceptionally it is found with an octahedral form; they are usually small, and of a dark orange color. This variety always contains chromic acid. Its fracture is subconchoidal and without lustre, even when the natural faces are brilliant. Its lustre on the natural faces is resinous or adamantine, sometimes prominent and sometimes feeble, as it is often covered with a yellow powder. Translucent, opaque. Its color is yellow, with different shades; sometimes it is almost canary-yellow, sometimes orange-yellow, which is probably owing to the presence of a little chromic acid. When it contains tungstic acid it is brownish-yellow; it is also gray or white. Streak, white. **H.**=2.75–3. **G.**=6.05–7.01. Composition, Pb 61.5, Mo 38.5.

Pyr. &c. B. P. It decrepitates and becomes dark-brownish yellow, and fuses at 2. On Ch., melts, and gives a metallic globule and scoria. Affords the reactions for lead and molybdic acid. Decomposed on evaporation in HCl.

It is also found in crystalline masses, made up of lamellæ which are interlaced; these masses are cavernous, being made up of crystals which have left considerable spaces between them, thus diminishing the density of the mass, which should be about 7. The masses become at times entirely compact, and show no traces of crystalline structure. It has been found in this country at Southampton, Mass., and at Phenixville, Pa.

FORMULÆ OF THE CRYSTALS.

Pl. XXX.

Fig. 1. P. ∞P. 0P. *Fig.* 2. ∞P. 0P. $\frac{1}{3}$P. *Fig.* 3. 0P. ∞P 2. $\frac{1}{3}$P. ∞P. *Fig.* 4. $\frac{2}{3}$P∞. $\frac{1}{3}$P. *Fig.* 5. 0P. ∞P. ∞P 2. *Fig.* 6. 0P. P. $\frac{1}{3}$P. *Fig.* 7. P∞. $\frac{2}{3}$P∞. P. $\frac{1}{3}$P. *Fig.* 8. 0P. $\frac{1}{3}$P. $\frac{1}{2}$P∞. *Fig.* 9. ∞P∞. 0P. $\frac{1}{16}$P; Phenixville.

All the compounds of lead containing oxygen, appear to have been formed by the action of outside agents upon Galenite. They are usually found in the vein within the influence of surface waters, which at lower depths contains sulphides only. Thus, Stolzite and Wulfenite are found at Bleiberg in Carinthia, on the outcrop of a vein of Galenite, which is crossed by a vein containing Molybdenite and Wolframite. The occurrence of Anglesite, Cerussite, Pyromorphite and Mimetite is similar, although these minerals are more common. It is quite common to find pseudomorphs of all the oxidized minerals into sulphides or *vice versa*. All of the oxides of the metals, are usually found within surface influences, the sulphides below being unaltered.

BISMUTH.

Bismuth. Bi. HEXAGONAL.

SYN.—Gediegen Wismut, Bismuth natif.

It crystallizes as a rhombohedron of 87° 40′. It is isomorphous with Antimony, having the same cleavage parallel to the face of the rhombohedron and the base, which gives it the appearance of having a lamellar fracture. Lustre, metallic. Opaque. Color, silver-white, with a reddish tinge. It is subject to tarnish. Streak, same as color. Sectile. When cold it is brittle, but slightly malleable when heated. **H.**=2–2.5. **G.**=9.727. Composition, when pure, Bi.

Pyr. &c. B. P. It melts in the flame of a candle. On Ch., fuses and is entirely volatilized, leaving a yellow coating. It is not attacked by HCl. It is soluble in $\dot{\ddot{N}}$, from which it is precipitated by diluting the solution with water.

It is sometimes found in large masses, but is usually disseminated in a gangue. It has a definite position, all of the cleavages have the same direction, and most of the specimens look alike. The direction of the lamellæ can be seen on almost every specimen. It usually occurs in reticulated masses scattered through a gangue. It has a rosy tinge which makes it resemble Linnæite and Cobaltite, but it is distinguished by its lamellar fracture and by its being completely volatile. It has been found in small quantities at Lane's mine, Connecticut, and in S. Carolina.

Bismuthinite. $Bi^2 S^3$. ORTHORHOMBIC.

SYN.—Bismuth Glance, Wismutglanz, Bismuthine, Bismuth sulfuré,

It crystallizes as a right rhombic prism of 91° 30′. The crystals are always acicular. It has easy cleavages parallel to the brachypinacoid and base, and a less easy one parallel to the macropinacoid. Lustre, metallic. Opaque. Color, lead-gray or tin-white, with a yellowish or iridescent tarnish. Streak, same as color. **H.**=2. **G.**=6.4–7.2. Composition, Bi 81.25, S 18.75.

Pyr. &c. B. P. In an open tube, gives sulphurous fumes and a bismuth sublimate. On Ch., fuses at 1 into drops, brown while hot and yellow on cooling, coating the charcoal. Dissolves in $\dot{\ddot{N}}$ and gives a precipitate on dilution.

It is sometimes found massive, with a foliated or reticulated structure. It is also found associated with Native Bismuth and sometimes in large masses like Stibnite, with which it is isomorphous, having the same easy cleavage parallel to the brachypinacoid. It has the same general characters also, except that the crystals are usually much smaller. It was probably much less soluble in the solutions from which it crystallized than Stibnite. Its color too is much darker, and it is sometimes variegated. It cannot generally be distinguished from Stibnite except by the blowpipe or by dissolving it in $\dot{\ddot{N}}$, which does not act on Stibnite. It has been found in Haddam, Ct., and in N. Carolina.

Aikinite. 3 (Ꞓu, Pb) S+$Bi^2 S^3$. ORTHORHOMBIC.

SYN.—Acicular Bismuth, Aciculite, Nadelerz, Bismuth sulfuré plumbo-cuprifère.

It crystallizes as a right rhombic prism of 110°. The crystals are usually long and acicular, which gave it the name of Needle Ore, and are generally imbedded in a gangue. Fracture, uneven. Lustre, metallic. Opaque. Color, lead-gray slightly red. **H.**=2–2.5. **G.**=6.1–6.8. Composition, Bi 36.2, Pb 36.1, Cu 11, S 16.7.

Pyr. &c. B. P. In an open tube, gives sulphurous fumes and a bismuth sublimate. On Ch., fuses and gives the bismuth and lead coatings. With fluxes, the reactions for copper. Decomposed by N̈.

It has been found in the U. S. in North Carolina and Georgia, and is often associated with Gold.

Tetradymite. $Bi^2 Te^3$. HEXAGONAL.

SYN.—Telluric Bismuth, Tellurwismut, Bismuth telluré.

It is found in small crystals which are often tabular. They generally show faces of a rhombohedron of 81° 2′ and frequently the base, and are usually macled. Cleavage easy, parallel to the base. Lustre, metallic and very bright. Color, pale steel-gray. The laminæ are flexible and soil paper. **H.**=1.5–2. **G.**=7.2–7.9. Composition variable, as it often contains sulphur and selenium; for the formula $Bi^2 Te^3$, Bi 51.9, Te 41.1.

Pyr. &c. B. P. In an open tube, gives a white sublimate of tellurous acid. On Ch., fuses and gives a white sublimate. In the R. F., tinges the flame bluish-green.

It has been found in the U. S. in the gold mines of Virginia, N. Carolina and Georgia.

FORMULÆ OF THE CRYSTALS.

Pl. XXX.

Fig. 10. Twin, consisting of four individuals; composition-face R; 3 R. 0R.

ARSENIC.

Arsenic. As. HEXAGONAL.

SYN.—Gediegen Arsen, Arsenic natif.

It crystallizes as a rhombehedron of 85° 41′. It has an imperfect cleavage, parallel to the base. The crystals are rare and are very small rhombohedra. The fracture is granular, and sometimes saccharoidal and even conchoidal. Lustre, almost metallic. On a fresh fracture, which is often very brilliant, it is steel-gray or tin-white. It afterwards tarnishes and becomes dark gray or black. Streak, same as color. **H.**=3.5. **G.**=5.93. Composition, when pure, As.

Pyr. &c. B. P. Gives metallic arsenic in a closed, and Äs in an open tube. In the R. F., it volatilizes without residue and without melting, coloring the flame blue. It is not attacked by HCl, but is soluble in N̈.

It can often be easily detached in plates, with curved surfaces which fit into each other. The mamelons are generally large, and are covered with points of crystals. It is also found in bacillary masses as little irregular prisms, generally in a gangue of Quartz or Barite. It is also found as concretionary masses in little mamelons, which have been deposited on the salient points of different minerals, often on Rhodochrosite. This mamelonated character distinguishes it from Nagyagite or Hauerite, with which it is often associated. It has been found in the U. S. at Haverhill and Jackson, N. H., and at Greenwood, Me.

Arsenolite. Äs. ISOMETRIC.

SYN.—Arsenous acid. Arsenige Säure, Arsenikblüthe, Arsenit, Arsenic oxidé.

It is found rarely in little octahedral crystals, or as silky tufts on the sides of a fissure. It is generally found as a white, earthy powder, associated with Arsenic. It is probably dimorphous with Senarmontite. It is sometimes found botryoidal, stalactitic and earthy. Lustre, vitreous or silky. Transparent, opaque. Color, white, with a yellowish or reddish tinge, owing to impurities. Streak, same as color. Taste, astringent. **H.**=1.5. **G.**=3.698. Composition, As 75.76, O 24.24.

Pyr. &c. B. P. Sublimes in a closed tube. On Ch., is entirely volatile, giving white fumes and a garlic odor. Slightly soluble in hot water.

It has no other characters, than its association and chemical reactions.

Realgar. As S. MONOCLINIC.

SYN.—Rothe Arsenblende, Roth Rauschgelb, Arsenic sulfuré rouge.

It crystallizes as an inclined rhombic prism of 74° 26′. It is always crystallized or crystalline. The crystals usually have a very large nnmber of faces, but the primitive form is comparatively rare. The usual faces are shown in *Pl.* XXX. *Figs.* 11–14. It has an easy cleavage, parallel to the base and clinopinacoid, and a less easy one parallel to the orthopinacoid. It fracture is smooth and conchoidal, with a resinous lustre. The lustre is bright on the natural faces which have not been altered. Transparent, translucent, opaque. The color is a characteristic red. When it has been exposed to the light for a long time, it becomes orange-yellow. Streak, red when not decomposed, but generally orange-yellow. **H.**=1.5–2. **G.**=3.4–3.6. Composition, As 70.1, S 29.9.

Pyr. &c. B. P. In a closed tube, it fuses and volatilizes without decomposition. The vapor is condensed in an orange-colored sublimate, which is also fusible and volatile. In an open tube, it burns with an odor of sulphur and arsenic. It is soluble in caustic alkalies.

In collections it must be kept from the light and as much as possible from the air, otherwise it becomes decomposed and falls to powder. Realgar can usually be distinguished by its color, streak and hardness, and its association with Native Arsenic, and if necessary by its chemical characters. The only substances which resemble it are Crocoite and Cinnabar. The color of the streak of Crocoite is yellow and its association is different. The blowpipe characters, streak, and chemical reactions distinguish it from Cinnabar.

FORMULÆ OF THE CRYSTALS.

Pl. XXX.

Fig. 11. ∞P. $\infty \bar{P} 2$. $0 P$. $\grave{P} \infty$, *Fig.* 12. The preceding, with P and $\infty \grave{P} \infty$. *Fig.* 13. ∞P. $\infty \bar{P} 2$. $\infty \bar{P} 4$. $\infty \bar{P} \infty$. $\infty \grave{P} 2$. $\infty \grave{P} \infty$. P. $0P$. $\frac{1}{2} \grave{P} \infty$. $\grave{P} \infty$. $2 \grave{P} \infty$. $\bar{P} \infty$. $2 \bar{P} \infty$. $\bar{P} 2$. $2 \bar{P} 2$. $4 \bar{P} 2$. *Fig.* 14. ∞P. $0 P$. $\frac{1}{2} P$. $\infty \bar{P} 2$. $\infty \bar{P} \infty$. $\infty \grave{P} \infty$. $\grave{P} \infty$, $2 \bar{P} \infty$.

Orpiment. $As^2 S^3$. ORTHORHOMBIC.

SYN.—Auripigment, Gelbe Arsenblende, Rauschgelb, Operment, Arsenic sulfuré jaune.

It crystallizes as a right rhombic prism of 100° 40′. The crystals usually show the brachydome, the rhombic prism, macro- and brachyprisms, macro- and brachypinacoids and an octahedron. It has a crystalline structure and a very easy cleavage parallel to the macropinacoid. On the cleavage faces its lustre is pearly, somewhat silky, because the cleavage faces are frequently striated; elsewhere it is resinous. Translucent. Its color is a decided lemon-yellow; sometimes slightly orange colored, owing to a slight admixture of Realgar. Its streak is yellow and a little paler than the color. By cleavage, very thin lamellæ can be detached, which are flexible but not elastic. **H.**=1.5–2. **G.**=3.48. Composition, As 61, S 39.

Pyr. &c. B. P. In a closed tube it fuses and volatilizes, giving a dark yellow sublimate; acts otherwise like Realgar. Dissolves in aqua regia and caustic alkalies.

It has been found at Edenville, N. Y., on Arsenopyrite.

FORMULÆ OF THE CRYSTALS.

Pl. XXX.
Fig. 15. ∞P. $\infty \bar{P} 2$. $\infty \bar{P} \infty$. $\infty \check{P} 2$. $\infty \check{P} \infty$. $\check{P} \infty$. $2 \check{P} 2$.

ANTIMONY.

Antimony. Sb. Hexagonal.

Syn.—Gediegen Antimon, Antimoine natif.

It crystallizes in rhombohedra of 87° 35′, with easy cleavages parallel to the base, so that in mass it is always lamellar, and the faces of the lamellæ are more or less large. Its lustre is bright and metallic when pure. The varieties containing arsenic have a very brilliant lustre on the fresh fracture, but this soon becomes gray and dark from a superficial oxidation, which is more easily effected on arsenic than antimony. Color and streak, tin-white. Very brittle. **H.**=3.35. **G.**=6.646–6.72. Composition, when pure, Sb. It generally contains one or two per cent. of impurity.

Pyr. &c. B. P. On Ch., it is fusible and gives off vapors, which condense at some distance from the assay in a white ring. It is easy to distinguish this from arsenic, which is much more volatile. The white coating in the R. F. tinges the flame bluish-green.

It is generally found in lamellar masses, having almost a compact fracture. The varieties, which have very large lamellæ, are almost pure antimony. The others usually contain arsenic, the proportion of which can be distinguished to a certain degree by the size of the lamellæ. The proportion of arsenic is sometimes sufficiently large, to give the specimen the shelly character so common in Native Arsenic, the presence of which is easily verified by the blowpipe. It is sometimes found in large masses which are somewhat coated with a yellow oxide. It has been found in the U. S. at Warren, N. J.

Senarmontite. $\dddot{S}b$. Isometric.

Syn.—Antimoine oxydé octaédrique.

It crystallizes in regular octahedra, with a trace of octahedral cleavage. The fracture is almost lamellar, with a bright lustre which is always adamantine or resinous on the fresh fracture. It is generally semi-transparent or semi-translucent. When pure, it is colorless. Streak, white. **H.**=2–2.5. **G.**=5.22–5.3. Composition, Sb 83.56, O 16.44.

Pyr. &c. B. P. In a closed tube, fuses and partly sublimes. On Ch., it becomes reduced in the R. F. and is volatilized, coloring the flame greenish-blue. It is insoluble in $\dot{\bar{N}}$.

It is generally found in crystallized masses, with the crystals interlaced. If the crystals are very small, it becomes granular. It is generally white or without color, but sometimes grayish, when it has become mixed with clay. With the crystallized varieties are associated others that are compact and earthy and which have almost a conchoidal fracture, in which no crystalline character can be seen. Sometimes it resembles clay. It must then be determined by its density or blowpipe reactions. This mineral has only been found in Algeria, and is very rare in collections.

FORMULÆ OF THE CRYSTALS.

Pl. XXX.
Fig. 16. O; characteristic form.

Oxide of antimony is found crystallized in two systems, the isometric and the orthorhombic. It is therefore dimorphous. To the isometric oxide the name Senarmontite has been given, and to the orthorhombic, Valentinite. A third oxide has been found, but not crystallized, so that it is possible that it may be trimorphous.

Valentinite. $\dddot{Sb}$. ORTHORHOMBIC.

SYN.—Weiss-Spiessglaserz, Antimonblüthe, Antimoine oxydé.

It crystallizes in a right rhombic prism of 136° 58′, with an easy cleavage parallel to the prism. It is frequently found as little crystals on the surface of Stibnite. The usual form is a combination of the prism with the brachypinacoid and domes. Its fracture is lamellar in the direction of the brachypinacoid, which has a pearly lustre. The other faces are either adamantine, or dark and striated. Color, white, peach-red, gray or brown. Streak, white. **H.**=2.5–3. **G.**=5.566. Composition, Sb 83.56, O 16.44.

Pyr. &c. B. P. In the R. F., it is reduced and gives the white sublimate characteristic of antimony. Heated in a tube, it melts without volatilizing, which distinguishes it from tellurium. It is insoluble in $\ddot{\dot{N}}$.

In Algeria it has been found in large masses on the outcrop of veins of Stibnite. These masses are made up of spheres, formed by little acicular crystals, diverging from a centre and fastened together. They have a very bright lustre, which is at the same time silky and adamantine. The open spaces between the spheres are filled up by a yellow pulverulent mass, which is an intermediate oxide. The density of this mass is almost 5, which prevents it being confounded with other earthy minerals. In detached crystals it resembles Stibnite or Gypsum, but it can be distinguished by its chemical characters. The adamantine lustre and density make it resemble the compounds of lead, but there is no compound of lead, which is found in this state. The intermediate oxide which is associated with these masses, seems to be the result of decomposition of the ores of antimony by the surface waters, and is often found as crusts on their surface.

FORMULÆ OF THE CRYSTALS.

Pl. XXX.
Fig. 17. $\infty\breve{P}\infty$. ∞P. $\breve{P}\infty$. $2\breve{P}2$.

Stibnite. $Sb^2 S^3$. ORTHORHOMBIC.

SYN.—Gray Antimony, Antimonite, Grauspiessglaserz, Antimonglanz, Antimoine sulfuré.

It crystallizes as a right rhombic prism of 90° 54′, with a cleavage parallel to the brachypinacoid so easy, that the lamellæ can frequently be detached with the nail. It is found in crystals, more or less perfect, which show the two general forms, *Pl.* XXX, *Figs.* 18 and 20. The second form, *Fig.* 20, is less common. These crystals are generally striated and cylindrical. They are usually grouped together and divergent, or arranged in groups, which are sometimes several decimeters in length. The crystals are generally disengaged towards their extremity, so that the terminations can be seen. The fracture is generally sub-conchoidal. The cleavage faces are scratched with the nail, but the other faces are harder. The color is lead-gray, with a metallic lustre that is often very bright. It frequently becomes tarnished when exposed to the air, and is sometimes iridescent, and sometimes almost a dead black. Streak, same as color. The lamellæ are flexible but not elastic, and make a dark streak on paper. **H.**=2. G.=4.516–4.612. Composition, Sb 71.8, S 28.2,

Pyr. &c. B. P. It is fusible without the aid of the blowpipe. In the O. F. it becomes roasted, giving off sulphurous fumes. On Ch., in the R. F., it gives the antimony coat and colors the flame greenish-blue.

The crystalline varieties are sometimes bacillary and divergent at the same time. They then show a fibrous and lamellar structure and are often iridescent, which appears to be owing to a commencement of oxidation. The crystals are generally associated with an earthy or stony gangue, or with a mass of Stibnite, which is not well crystallized. The fracture of these masses is always lamellar, but only in one direction, which distinguishes it from Galenite, which has three cleavages. With Stibnite other substances are associated which have the same general aspect, in capillary crystals grouped together and divergent, which, in some localities, become woven together like Native Antimony. They are minerals of antimony having a definite composition, generally sulpho-antimoniurets of lead. Before the blowpipe, these minerals give the reactions for lead, as well as for antimony. There are also antimonio-sulphurets of iron, which are found in amorphous and granular masses.

In the U. S., Stibnite occurs in Maine, New Hampshire and Maryland. Argentiferous varieties have been found in Nevada.

FORMULÆ OF THE CRYSTALS.

Pl. XXX.

Fig. 18. $\infty P.\ P.\ \infty \breve{P} \infty$; the most usual form. *Fig.* 19. The preceding, with $2\breve{P}2$. *Fig.* 20. $\infty P.\ \frac{1}{3}P.\ \infty\breve{P}\infty$. *Fig.* 21. $\infty P\ \infty\breve{P}\infty.\ P.\ \frac{1}{3}P.\ 2\breve{P}2.\ \frac{4}{3}\breve{P}2.\ \frac{1}{2}\breve{P}\infty$.

Kermesite. $\dddot{Sb}+2\ Sb\ S^3$. MONOCLINIC.

SYN.—Pyrostibite, Antimonblende, Rothspiessglaserz, Antimoine oxidé sulfuré.

It crystallizes as an inclined rhombic prism. It is usually found as capillary crystals, having a basal cleavage. Lustre, adamantine or metallic. Translucent. Color, cherry-red. Streak, brownish-red. **H.**=1–1.5. **G.**=4.5–4.6. Composition, Sb 75.3, S 19.8, O 4.9.

Pyr. &c. B. P. In a closed tube, blackens, fuses and gives a white sublimate. If heated strongly, a black or dark red sublimate is formed, otherwise it acts like Stibnite.

Its formation appears to be owing to the decomposition of Stibnite, either by the vapor of water or damp air. It occurs in fine fibres which are usually capillary and divergent, or in silky tufts. Sometimes they lie flat upon the gangue and are irregularly woven together. Its principal character is its cherry-red color, which is rather brownish and perfectly distinct from that of Erythrite.

To Kermesite belongs a mineral called Silver Tinder. It has a variable composition and an argentine lustre. It is never crystallized and is quite light and occurs as a more or less earthy mass, associated with a gangue.

URANIUM.

Uraninite. $\dot{U}\,\ddot{\bar{U}}$. ISOMETRIC.

SYN.—Pitchblende, Uranpecherz, Urane oxydulé.

It has been found as an octahedron associated with a cube or rhombic dodecahedron, but usually it only shows traces of crystallization. Its fracture is unequal and sometimes conchoidal. Lustre, submetallic, greasy or dull. Opaque. Color, intense black, grayish, greenish or brownish. Streak, a lighter brown, grayish, olive-green a little shining. **H.**=5.5. **G.**=6.4–8. Composition, $\dot{U}$ 32.1, $\ddot{\bar{U}}$ 67.9. The degree of oxidation of the uranium appears to be variable, and is not very well determined. The mineral appears to be generally a mixture of several metallic oxides.

Pyr. &c. B. P. It is infusible. It colors the flame green, which has a different shade from that of boracic acid or copper. With fluxes, it gives a greenish-yellow bead in the O. F., which becomes green in the R. F. It is not acted on by the non-oxidizing acids. Aqua regia and $\dddot{N}$ dissolve it to a yellow solution, from which ammonia precipitates the uranate of ammonia, which is flaky and yellowish.

Its fracture and lustre make it resemble Triplite, but it is distinguished by its density, which is near 7, and which is its distinguishing feature. It is frequently covered with a pulverulent yellowish coating, which is uranic acid or some other intermediate oxide, or a subsulphate or phosphate of uranium.

Autunite. $\ddot{\bar{U}}^2\,\dddot{P}+\dot{C}a\,\bar{H}+7\,\bar{H}$. ORTHORHOMBIC.

SYN.—Lime-Uranite, Uranit, Kalkuranit.

It crystallizes as a right rhombic prism, the angle of which is very near a right angle. It has a very easy cleavage parallel to the base, which gives it a very lamellar structure, both on its natural and cleavage faces. It is usually found as a crystalline incrustation, which has a fan-shaped appearance, as they are formed of a large number of crystals placed together. It is ordinarily impossible to distinguish any form. Its lustre on the base is pearly, elsewhere almost adamantine. Translucent. It has an orange-yellow color, showing greenish reflections however, owing to traces of copper. Streak, yellowish. **H.**=2–2.5. **G**=3-05–3.19. Composition, $\ddot{\bar{U}}$ 62.7, $\dot{C}a$ 6.1, $\dddot{P}$ 15.7, $\bar{H}$ 15.5.

Pyr. &c. B. P. Melts in the R. F. to a black globule, first swelling and losing its water. It is soluble in $\dddot{N}$, giving a yellow solution.

In the U. S., it is found at Acworth, N. H., and at Philadelphia, Pa.

Torbernite. $\dddot{\bar{U}}^2 \,\ddot{\ddot{\bar{P}}}+\dot{C}u\,\dot{H}+7\,\dot{H}$. TETRAGONAL.

SYN.—Copper-Uranite, Kupferuranit, Chalcolith, Uranglimmer, Urane oxidé.

Its general characters are the same as those of Autunite; it has the same easy cleavage, and is usually well crystallized. The crystals are often tabular and sometimes show traces of octahedra. The crystals are sometimes made up of crystalline plates, piled the one on the other. The base has a pearly lustre and is pale green. The vertical faces are vitreous and darker in color. They form coatings on the gangue, reposing usually on the base and are quite analagous to Mica, except in their color. Lustre, pearly on the base, elsewhere adamantine. Transparent, translucent. Color, various shades of green. Streak, paler than color. Sectile. **H.**=2–2.5. **G.**=3.4–3.6. Composition, $\dddot{\bar{U}}$ 61.29, $\dot{C}u$ 8.44, $\ddot{\ddot{\bar{P}}}$ 15.57, $\dot{H}$ 15.05.

Pyr. &c. B. P. In a closed tube, yields water. Fuses at 2.5 to a blackish mass, coloring the flame green. Gives the reactions for uranium and copper.

FORMULÆ OF THE CRYSTALS.

Pl. XXX.

Fig. 22. P. 0 P. *Fig.* 23. ∞ P. P. 0 P. *Fig.* 24. P. 0 P. ∞ P ∞. *Fig.* 25. P. 0 P. P ∞. *Fig.* 26. 0 P. $\frac{2}{3}$ P. 2 P. *Fig.* 27. ∞ P. ∞ P ∞. 2 P. $\frac{2}{3}$ P. 0 P.

MOLYBDENUM.

Molybdite. $\dddot{Mo}$. ORTHORHOMBIC.

SYN.—Molybdine, Molybdänoker.

It crystallizes as a right rhombic prism of 136° 48′. It is isomorphous with Valentinite. It is found in capillary crystals tufted or radiated, or as a coating on Molybdenite, resulting from its decomposition. Lustre of the crystals, silky, of the incrustation, earthy. Color, straw-yellow. **H.**=1–2. **G.**=4.49–4.5. Composition, Mo 65.71, O 34.29.

Pyr. &c. B. P. On Ch., fuses and coats the coal with yellow crystals near the assay, which are white on the outer edge of the coating. The coating in the R. F., becomes blue, and then red after long heating. With borax in the O. F., it gives a bead, yellow while hot, and colorless on cooling; in the R. F., the bead becomes black.

It is found on New York Island, associated with Molybdenite, also in Pennsylvania and Georgia.

Molybdenite. $Mo\,S^2$. MONOCLINIC? HEXAGONAL?

SYN.—Molybdänglanz, Wasserblei.

It is found in tabular hexagonal prisms, with a cleavage parallel to the base. It is usually found in thin scales of a grayish-blue color, which are flexible but not elastic, and which soil the fingers with a bluish-gray stain. Lustre, metallic. Opaque. Color, lead-gray. Streak, same as color. **H.**=1–1.5. **G.**=4.44–4.8. Composition, Mo. 59, S 41.

Pyr. &c. B. P. On Ch., after heating some time it burns, giving off a sulphurous odor, and leaving a yellowish infusible residue, which is molybdic acid. Decomposed by $\ddot{\dddot{N}}$.

It resembles Graphite somewhat, but it is blue and acts differently. On glazed porcelain, it gives a greenish streak, while Graphite gives a black one.

COPPER.

Copper. Cu. ISOMETRIC.

SYN.—Gediegen Kupfer, Cuivre natif.

It crystallizes generally in cubes, octahedra, rhombic dodecahedra or tetrahexahedra. It is frequently found regularly crystallized in large arborescent masses, ramified and dendritic, which are formed of cubes and octahedra more or less lengthened out and joined together. It is also found in thin sheets, which cover or penetrate different substances. Its fracture is hackly. Lustre, metallic. Color, copper-red. Streak, metallic, shining; ductile and malleable. **H.**=2.5. **G.**=8.838. Composition, copper with a little silver, bismuth, etc.

Pyr. &c. B. P. Fuses easily. Soluble in acids.

Its density is less than that of melted copper, which is 8.921 when simply melted, but when hammered, it is 8.952. Its sonority is also different, which would seem to show that it has been formed in the wet way, either by chemical or galvanic precipitation. What appears to corroborate this opinion is that it is often found moulded on crystals and other substances, taking pseudomorphic forms. Thus, Calcite and other minerals which would not have resisted the high temperature of the fusion of copper are often found coated with it. It almost always contains a certain proportion of silver, which cannot always be seen, but which appears to show a simultaneous precipitation of the two metals.

It is always easy to recognize it by its metallic lustre and red color, which however is sometimes masked by superficial action, but it is easily seen by breaking or scratching with a knife. It is also found in very large amorphous masses in Lake Superior. The largest ever found weighed 420 tons and contained 90% of copper.

FORMULÆ OF THE CRYSTALS.

Pl. XXXI.

Fig. 1. $\infty O 2$; a frequent form. *Fig.* 2. Hemitrope of the preceding; composition face O. *Fig.* 3. The same, flattened; resembles an hexagonal pyramid.

Cuprite. $\dot{C}u$. ISOMETRIC.

SYN.—Octahedral Copper Ore, Rothkupfererz, Ziegelerz, Cuivre oxidulé, Chalcotrichite, Kupferblüthe.

Its crystals are generally the octahedron which is either simple or modified, sometimes the rhombic dodecahedron, but rarely the cube. Sometimes the cube is lengthened to capillary forms. It has an octahedral cleavage which makes it often look lamellar on its fracture. Its fracture is conchoidal or uneven, sometimes lamellar. On the natural faces its lustre is quite bright and is semi-adamantine or semi-metallic. In some varieties it is earthy. Its crystals are sometimes transparent or translucent in thin pieces. Its color is dark blood-red; sometimes almost black. Its streak is dark cochineal-red. **H.**=3.5–4. **G.**=5.85–6.15. Composition, Cu 88.8, O 11.2.

Pyr. &c. B. P. In the O. F., it is infusible and gives a black scoria. In the R. F., it gives a button of metallic copper, which is malleable and ductile. Soluble in HCl and $\dddot{N}$.

It is often found as crystalline masses, made up of crystals disseminated in a gangue or interlaced among themselves. These are generally small and indistinct, and give to the whole mass a ramified or scoriaceous look. At Chessy, near Lyons, France, Cuprite was formerly found in large crystals. They are transformed on their outside into Malachite, without having the regularity of their faces disturbed. This is sometimes superficial, and sometimes only a little core of Cuprite is left, and in others the pseudomorphism is complete. The faces of some of the crystals are simply concave, while others are completely hollow. This is generally true of the octahedron, the edges of which, however, are very prominent, the faces being very much sunken. It also occurs in semi-lamellar masses of a lighter red than the crystals, having almost a granular or compact fracture. These masses have a density above 5, and are almost always accompanied by some stain of copper. It might be mistaken for Hematite or Magnetite, but the streak and blowpipe characters are different. When found in capillary crystals, which are generally elongated cubes, it is called Chalcotrichite. This is frequently associated with Limonite. It has a peculiar lustre, and its capillary condition and reactions with the blowpipe, make it easy to distinguish. This variety is quite rare.

When found in large quantities, it is a valuable ore of Copper. In the U. S. it has been found at Somerville, N. J., and Cornwall, Pa., and Lake Superior.

FORMULÆ OF THE CRYSTALS.

Pl. XXXI.

Fig. 4. O; the most frequent form. *Fig.* 5. ∞O; also characteristic. *Fig.* 6. ∞O. O. *Fig.* 7. Octahedron, with cavernous faces. *Fig.* 8. O. 2O2. ∞O2. 3O$\frac{3}{2}$. 2O. ∞O. ∞O∞.

Chalcocite. €u S. ORTHORHOMBIC.

SYN.—Vitreous Copper, Copper Glance, Chalcosine, Kupferglanz, Cuivre sulfuré.

It crystallizes as a right rhombic prism of 119° 35'. By fusion and decantation, it can be artificially obtained in octahedra. This dimorphism is only a question of temperature. The angle of the prism gives rise to limit forms. There is a trace of cleavage parallel to the prism. Crystals are not abundant, but they show appearances of limit prisms, which resemble hexagonal types. The base of the prism shows slight striations, which are parallel to brachypinacoid. These striations serve to distinguish the simple crystals from those which are formed by the union of six crystals placed together upon the face of the prism. The base of such crystals shows radiated striæ. It has another peculiar macle, formed by four crystals joined in a cross around the faces of the brachydome. Fracture, conchoidal. Lustre, metallic. Color and streak, dark blue, almost black. It is ductile, and can be easily cut with a knife into curved shavings. **H.**=2.5 to 3. **G.**=5.5 to 5.8. Composition, Cu 79.8, S 20.2.

Pyr. &c. B. P. It melts in the flame of a candle, giving off sulphurous acid. Gives the reactions for copper. It is soluble in hot N̈, but is not acted upon by cold.

Besides the crystallized varieties it is found in semi-lamellar, semi-compact masses, with almost a conchoidal fracture of a bluish color, which in old fractures is very dark. It occurs sometimes in schistose masses. It has also been found in ramified or stalactitic forms, which appear to be pseudomorphs after Native Copper. It is usually ductile, but has been found in masses which do not possess this property, although having the same physical appearance. It is then probably alloyed with a considerable proportion of Argentite. It resembles Argentite, but the latter is more sectile, has a different lustre and gives different reactions with the blowpipe. The mines at Bristol, Conn., furnished some of the most beautiful specimens, that have ever been found. It has been found in New York, New Jersey, Virginia and other states.

FORMULÆ OF THE CRYSTALS.

Pl. XXXI.

Fig. 9. 0 P. ∞ P. ∞ P̌ ∞. *Fig.* 10. 0 P. $\frac{1}{3}$ P. $\frac{2}{3}$ P̌ ∞. *Fig.* 11. 0 P. ∞ P. ∞ P̌ ∞. $\frac{1}{3}$ P. $\frac{2}{3}$ P̌ ∞. *Fig.* 12. Group of twin crystals; the lower half with the composition-face ∞ P, the upper half twinned with a face of $\frac{4}{3}$ P̌ ∞. *Fig.* 13. Cruciform twin, having 0 P and ∞ P of one crystal parallel respectively to ∞ P̌ ∞ and 0 P of the other.

Bornite. (Ꞓu, Fe) S. ISOMETRIC.

SYN.—Erubescite, Purple Copper Ore, Peacock Ore, Buntkupferkies, Cuivre pyriteux hepatique, Philipsite.

It is found in cubes, octahedra and rhombic dodecahedra, which are either simple or macled on one of the faces of the octahedron, but crystals are very rare. It has traces of an octahedral cleavage. Its fracture is generally unequal and conchoidal. Lustre, metallic. Its color is reddish-brown or a dark violet-blue, with a great variation in colors, owing to tarnish. Streak, blackish bronze-yellow. **H.**=3. **G.**=4.4–5.5. Composition, for ($\frac{4}{5}$ Ꞓu+$\frac{1}{5}$ Fe) S, is Cu 70.13, Fe 7.76, S 22.11.

Pyr. &c. B. P. In a closed tube, gives a faint sublimate of sulphur. In the O. F. it is roasted with a sulphurous odor and gives in the R. F. a half-melted globule, which is attracted by the magnet. Soluble in N̈.

It is generally found compact, and shows such a brilliant play of colors, that it is not possible to confound it with anything else. It is found at Bristol, Conn., in crystals. It occurs massive in Mass., New Jersey, Penn. and elsewhere.

FORMULÆ OF THE CRYSTALS.

Pl. XXXI.

Fig. 14. ∞ O ∞. O. *Fig.* 15. ∞ O. 2 O 2.

Chalcopyrite. Ꞓu S+Fe S+Fe S^2. TETRAGONAL.

SYN.—Copper Pyrites, Kupferkies, Pyrite cuivreuse, Cuivre pyriteux.

The prism is very near the cube. It is found in crystals or in concretionary or amorphous masses. The crystals are sometimes octahedra but more frequently sphenoids, which, on account of the small difference in the edges, resemble regular tetrahedra. But of the three edges which join at the same solid angle, two only are of the same kind, so that the modifications that can be produced upon them are not in pairs of 3, but of 2 and 1 which shows it to be tetragonal. It has traces of cleavage parallel to the vertical faces of the prism. Fracture, conchoidal or uneven. It has a metallic lustre and a brass-yellow color, which is deeper than that of Pyrite. It is subject to tarnish and is often iridescent. **H.**=3.5–4. **G.**=4.1–4.3. Composition, Cu 34.6, Fe 30.5, S 34.9.

Pyr. &c. B. P. In a closed tube, decrepitates and gives off S.

On Ch., melts, giving off sulphurous acid, and yields a magnetic globule. Soluble in N̈.

It often shows on amorphous masses a play of colors, which is owing to a superficial alteration. This iridescence is sometimes found on the fracture when the fissures have been acted on by the same agency that affected the surface. These colors are very bright, but are generally lighter than those of Bornite. The concretionary varieties are generally little mamelons separated from each other, which are quite smooth on their surface and stalactitic in character. The amorphous masses, which are the most common, are recognized by the yellow color, which is often variegated, the unequal fracture and hardness. It does not strike fire with the steel, which character distinguishes it from Pyrite, with which it is often associated.

It is a valuable ore of Copper and is found in many localities in the U. S.

FORMULÆ OF THE CRYSTALS.

Pl. XXXI.

Fig. 16. P; the two sphenoids equally developed. *Fig.* 17. $\frac{P}{2}$. $-\frac{P}{2}$. *Fig.* 18. P. 0 P. 2 P ∞. P ∞. *Fig.* 19. The same, but with one sphenoid further developed than the other. *Fig.* 20. $\frac{P}{2}$. $-\frac{P}{2}$. ∞ P ∞. *Fig.* 21. $\frac{P}{2}$. $-\frac{P}{2}$. ∞ P. 2 P ∞. *Fig.* 22. Hemitrope; composition-face P. *Fig.* 23. Hemitrope of *Fig.* 18. *Fig.* 24. Twin crystal; composition-face P ∞. *Fig.* 25. Twin crystal of the form *Fig.* 17. *Pl.* XXXII. *Fig.* 1. Twin crystal, composition repeated four times. *Fig.* 2. Twin crystal; Cornwall. *Fig.* 3. Twin crystal; 0 P. $\frac{1}{3}$P. P. $\frac{3}{2}$P. $\frac{3}{2}$P ∞; composition-face P ∞.

Tennantite. 4 (Ɇu,Fe) S+As^2 S^3. Isometric.

Syn.—Arsenikalfahlerz, Kupferblende, Zincfahlerz.

Its dominant form is the rhombic dodecahedron, with small faces of the tetrahedron, *Pl.* XXXII. *Fig.* 8. It has traces of dodecahedral cleavage. Fracture, uneven. Lustre, metallic. Color, blackish lead-gray to iron-black, and in old fractures, dull black. Streak, dark reddish-gray. **H.**=3.5 to 4. **G.**=4.37 to 4.53. Composition, Cu 47.70, Fe 9.75, As 12.46, S 30.25.

Pyr. &c. B. P. In a closed tube, gives a sublimate of sulphide of arsenic. In an open tube, sulphurous fumes and arsenous acid. On Ch., fuses with intumescence to a magnetic globule. In the R. F. gives a brittle globule of copper.

There are two compounds of copper, having very nearly the same composition, in one of which arsenic predominates, and in the other antimony. The former is called Tennantite, and the latter Tetrahedrite. These two varieties show different crystalline types, although they both crystallize in the isometric system and both show the hemihedry with inclined faces.

Tetrahedrite. $4(\text{Ɵu}, Fe, Zn, Hg, Ag)S(Sb, As)^2S^3$. ISOMETRIC.

SYN.—Gray Copper Ore, Fahlerz, Cuivre gris.

It crystallizes in tetrahedra, which are usually modified by the hemi-tetragonal trisoctahedron. They are rarely simple, and show traces of a tetrahedral cleavage. Fracture, subconchoidal or uneven. Its lustre is bright and metallic, both on its fracture and on the natural faces of the crystal, and does not become black on exposure, and is never resinous like that of Bournonite. Its color is a blackish-gray, which is more or less dark. Its streak is gray, and much lighter than that of Tennantite. **H.**=3–4.5. **G.**=4.5–5.11. Composition, Cu 19–25, Fe 2–7, Zn 1–7, Ag 0–31, As 0–11, Sb 11–28, S 19–26.

Pyr. &c. B. P. In the O. F., it is roasted, giving a slight odor of arsenic, and fumes of antimony, and in the R. F., gives a brittle globule of copper. Decomposed by $\dot{\ddot{\text{N}}}$.

It is found in masses with or without gangue. They are easily distinguished by their color and lustre, and their granular appearance. They often have blue or green stains like other minerals of copper. It sometimes resembles compact Galenite, but its color is a much lighter gray, and it has no tinge of blue. It resembles Chalcocite, but this is sectile and less bright, and it does not have the same density. Bournonite has a conchoidal fracture, a resinous lustre, and is not so hard. It is distinguished by the color of its streak, from Hematite, Magnetite and the oxides of Manganese. The blowpipe distinguishes it from other minerals. It is worked for copper and often for silver.

It has been found in Arkansas, California, Arizona and Nevada.

FORMULÆ OF THE CRYSTALS.

Pl. XXXII.

Fig. 4. $\frac{O}{2}$. $-\frac{O}{2}$. *Fig.* 5. $\frac{O}{2}$. ∞O. *Fig.* 6. $\frac{O}{2}$. $\infty O \infty$. *Fig.* 7. $\frac{O}{2}$. $\frac{2O2}{2}$. *Fig.* 8. ∞O. $\infty O \infty$. $\frac{O}{2}$; with the rhombic dodecahedron predominating. *Fig.* 9. $\frac{2O2}{2}$. ∞O. *Fig.* 10. The preceding, with $\frac{3}{2}O$. *Fig.* 11. The form *Fig.* 7, with $-\frac{O}{2}$. and ∞O. *Fig.* 12. $\frac{2O2}{2}$. $\infty O \infty$. ∞O. $-\frac{2O2}{2}$. $\infty O3$. *Fig.* 13. The preceding, with $\frac{3}{2}O$. $\frac{O}{2}$ *Fig.* 14. Interpenetrating tetrahedra; composition-face O. *Fig.* 15. Interpenetrating tetrahedra. *Fig.* 16. Twin crystal, formed by repeated twinning similar to the preceding figure.

Chalcanthite. $\dot{C}u\,\dddot{S}+5\dot{H}$. TRICLINIC.

SYN.—Cyanosite, Blue Vitriol, Kupfervitriol, Cuivre sulfaté.

It is rarely found crystallized, but is generally amorphous, sometimes concretionary and mamelonated on the surface of gangues. Lustre, vitreous, Transparent, translucent. Color, different shades of blue. Streak, colorless. Brittle. Taste, metallic and astringent. **H.**=2.5. **G.**=2.13. Composition, $\dot{C}u$ 31.8, $\dddot{S}$ 32.1, $\dot{H}$ 36.1.

Pyr. &c. B. P. In a closed tube, yields water and then $\dddot{S}$. On Ch., in the R. F., gives metallic copper. Soluble in water.

It is frequently found in abandoned mines of copper as an accidental product resulting from the decomposition of Chalcopyrite, produced by hot damp air. It has all the characters of the sulphate of copper of the laboratories.

It has been found in the copper mines of Tennessee and Georgia.

FORMULÆ OF THE CRYSTALS.

Pl. XXXII.

Fig. 17. P'. $\infty P'$. $\infty' P$. $\infty \bar{P} \infty$. $\infty \breve{P} \infty$. $\infty' \bar{P} 2$. $2 \breve{P}' 2$.

Brochantite. $2\,\dot{C}u^3\,\dddot{S} + \dot{C}u\,\dot{H} + 4\,\dot{H}$. ORTHORHOMBIC.

SYN.—Krisuvigit, Warringtonite.

It crystallizes as a right rhombic prism of 104° 30′. It is generally found in groups of acicular crystals. It has an easy cleavage parallel to the brachypinacoid. Lustre, vitreous, pearly on the cleavage faces. Transparent, translucent. Color, emerald-green. Streak, paler than color. **H.**=3.5–4. **G.**=3.78–3.87. Composition, $\dot{C}u$ 69, $\dddot{S}$ 19.9, $\dot{H}$ 11.1.

Pyr. &c. B. P. In a closed tube yields water, and afterwards $\dddot{S}$. On Ch., in the R. F., gives metallic copper.

It resembles Malachite in its exterior characters. It is found in little crystals which can rarely be measured, as they are very generally capillary. It is also found in little mamelonated and fibrous masses. It is distinguished from Malachite, by not effervescing. They are both soluble in ammonia, but Brochantite gives a precipitate with chloride of barium. It resembles Atacamite, but has a lighter color. The blue color so easily shown before the blowpipe with Atacamite, is less easily produced with Brochantite. It has been found in Cornwall, Pa.

FORMULÆ OF THE CRYSTALS.

Pl. XXXII.

Fig. 18. $\infty \breve{P} \infty$. ∞P. $\breve{P} \infty$. $\bar{P} \infty$.

Atacamite. $Cu\,Cl\,\dot{H} + 3\,\dot{C}u\,\dot{H}$. ORTHORHOMBIC.

SYN.—Salzkupfererz, Cuivre chloruré.

It crystallizes as a right rhombic prism of 112° 20′. Cleavage perfect, parallel to the brachypinacoid, imperfect, parallel to the macropinacoid. The crystals are quite small. The fracture may be unequal, conchoidal or fibrous. Lustre, adamantine and vitreous. Color, dark emerald-green. It is much nearer olive-green than Malachite. Streak, apple-green. **H.**=3–3.5. **G.**=4–4.3. Composition, $\dot{C}u$ 53.6, CuCl 30.2, $\dot{H}$ 16.2.

Pyr. &c. B. P. In a closed tube gives off water. On Ch., fuses and gives an intensely blue color to the flame, before it is reduced. All the salts of copper when heated with HCl show this character, but as the chloride is much more volatile than the others, it shows it in a higher degree. It is soluble in acids, without effervescence and without residue. Ammonia dissolves it, and colors the solution blue.

It is found sometimes in little crystals, sometimes in little fibrous tufts covering a gangue, which is generally one of the oxides of iron. This coating is made up of crystals joined together, showing bands of different shades of green like Malachite. These fibers are closer together than in

Malachite, so that the lustre is not silky. It is also found as very small crystals, lining cavities in certain lavas. It is also found as sand, which can be distinguished by its color. It results from the action of water on large masses. It is found in this state, at Atacama, in Chili. When it is found abundantly, as in Chili, it is a valuable ore of copper.

FORMULÆ OF THE CRYSTALS.

Pl. XXXII.
Fig. 19. ∞P. $\breve{P}\infty$. $\infty\breve{P}\infty$.

Libethenite. $\dot{C}u^4\overset{\cdot\cdot\cdot\cdot\cdot}{P}+\dot{H}$. ORTHORHOMBIC.

SYN.—Aphérèse.

It crystallizes as a right rhombic prism of 92° 20′. Its usual forms are a combination of the prism, rhombic octahedron and brachydome, giving the crystals a roof-shaped appearance; the faces are very often curved. It has a very indistinct cleavage parallel to the macro- and brachypinacoids. Fracture, conchoidal or uneven. Lustre, resinous. Translucent, opaque. Color and streak, dark olive-green. **H.**=4. **G.**=3.6–3.8. Composition, $\dot{C}u$ 66.5, $\overset{\cdot\cdot\cdot\cdot\cdot}{P}$ 29.7, $\dot{H}$ 3.8.

Pyr. &c. B. P. In a closed tube, yields water and blackens. Fuses and colors the flame green. On Ch., gives metallic copper. Soluble in $\overset{\cdot\cdot\cdot\cdot\cdot}{N}$.

It is often found globular, or mamelonated, associated with Quartz and Chalcopyrite.

FORMULÆ OF THE CRYSTALS.

Pl. XXXII.
Fig. 20. ∞P. $\breve{P}\infty$. P.

Olivenite. $\dot{C}u^4(\overset{\cdot\cdot\cdot\cdot\cdot}{As}, \overset{\cdot\cdot\cdot\cdot\cdot}{P})+\dot{H}$. ORTHORHOMBIC.

SYN.—Olive Copper Ore, Olivenerz.

It crystallizes as a right rhombic prism of 92° 30′. Its crystals usually show a combination of the prism and the macro- and brachypinacoids and domes. It has an easy cleavage parallel to the prism. Its fracture is conchoidal or uneven. Translucent, opaque. Color, different shades of green, brown, yellow or grayish-white. Streak, olive-green or brown. Brittle. **H.**=3. **G.**=4.1–4.4. Composition, $\dot{C}u$ 57.4, $\overset{\cdot\cdot\cdot\cdot\cdot}{As}$ 35.7, $\overset{\cdot\cdot\cdot\cdot\cdot}{P}$ 3.7, $\dot{H}$ 3.2.

Pyr. &c. B. P. In a closed tube, gives off water. Fuses at 2, coloring the flame bluish-green and giving a crystalline mass on cooling. On Ch., deflagrates and gives arsenical fumes. With soda in the R. F., gives a button of copper. Soluble in $\overset{\cdot\cdot\cdot\cdot\cdot}{N}$.

It is often found in fibers of a very light green color, which are often very much bent; it is sometimes found lamellar or granular and rarely quite earthy, of a grayish color. It is usually found coating, or associated with Quartz. Some varieties resemble Libethenite, but they can generally be distinguished by the greater quantity of arsenic present and, when crystallized, by the general absence of octahedra. Some of the fibrous varieties resemble Atacamite, but are distinguished by the blowpipe.

Pl. XXXII. FORMULÆ OF THE CRYSTALS.
Fig. 21. ∞P. $\infty\bar{P}\infty$. $\infty\breve{P}\infty$. $\bar{P}\infty$. $\breve{P}\infty$.

Liroconite. $\dot{C}u^3(\overset{\cdot\cdot\cdot\cdot\cdot}{As}, \overset{\cdot\cdot\cdot\cdot\cdot}{P})+(\frac{1}{3}\dot{C}u^3+\frac{2}{3}\ddot{Al})\dot{H}^3+9\dot{H}$. MONOCLINIC.

SYN—Linsenerz, Chalcophacit.

It crystallizes as an inclined rhombic prism of 74° 21′. It generally shows the prism and clinodome, giving the crystals a roof-shaped appearance. The crystals are generally quite small. Its fracture is conchoidal or uneven. Lustre, vitreous. Translucent. Color and streak, blue or green. **H.**=2–2.5. **G.**=2.882–2.985. Composition, $\dot{C}u$ 36.38, $\ddot{A}s$ 23.05, $\ddot{P}$ 3.73. $\ddot{A}l$ 10.85, $\dot{H}$ 25.01.

Pyr. &c. B. P. In a closed tube gives off water, and turns olive-green. Cracks, but does not decrepitate, and fuses to a dark gray slag. On Ch., cracks, deflagrates and gives the reaction for arsenic and copper. Soluble in $\ddot{N}$.

It is usually associated with other ores of Copper. The blue varieties resemble Chalcanthite, but can be distinguished by their insolubility.

FORMULÆ OF THE CRYSTALS.

Pl. XXXII.

Fig. 22. ∞P. P̀∞.

Malachite. $\dot{C}u^2\ddot{C}+\dot{H}$. MONOCLINIC.

SYN.—Green Carbonate of Copper, Cuivre carbonaté vert.

It crystallizes as an inclined rhombic prism of 104° 28′. Crystals are generally very small and rare. They are not often sufficiently distinct to admit of being measured. It has as easy basal cleavage. It is found crystallized in silky tufts, made up of crystals which are more or less distinct; these crystals are sometimes translucent, and are always intensely green. These silky tufts are often scattered over a gangue. It is often found as large crystals, which are pseudomorphs after Cuprite or Azurite. Its fracture is unequal, conchoidal, or fibrous in the concretionary varieties. Lustre of the crystals adamantine, of the fibrous varieties, silky; sometimes dull or earthy. Translucent, opaque. Its color is green, and may be of different degrees of intensity. Streak, paler than color. **H.**=3.5–4. **G.**=3.7–4.01. Composition, $\dot{C}u$ 71.9, $\ddot{C}$ 19.9, $\dot{H}$ 8.2.

Pyr. &c. B. P. In a closed tube blackens and gives off water. It melts at 2, coloring the flame green, and gives a scoriaceous mass. In the R. F., a globule of metallic copper is easily produced. Soluble in acid with effervescence.

It is usually found in concretionary masses, which have a fibrous fracture, rarely conchoidal. Their lustre is silky and velvety. These masses present a variety of shades, which are owing to different causes, a change in the direction of the fibers, a difference in the state of hydration of the mineral, or an admixture of other substances. This banded, silky and velvety appearance, causes Malachite to be much sought for, as an object of ornament. It is often found mixed with other minerals, which cause a change in the tint of the green. Some varieties resemble Chrysocolla, but are distinguished by effervescing with acids. It is a valuable ore of copper, when found in large quantities. It is much used for articles of ornament, and is often employed for veneering large articles, such as tables, doors, &c. A mass weighing 40 tons, was found in Siberia. It has been found in small quantities, in most of the copper mines of the U. S. In Cornwall, Pa., it is sometimes found in quite thick pieces of a light green color.

FORMULÆ OF THE CRYSTALS.

Pl. XXXII.

Fig. 23. ∞P. ∞P̀∞. 0P; twin crystal, composition-face ∞P̀∞.

Azurite. 2 $\dot{C}u\ \ddot{C}+\dot{C}u\ \dot{H}$. MONOCLINIC.

SYN.—Blue Malachite, Chessy Copper, Kupferlasur, Cuivre carbonaté bleu.

It crystallizes as an inclined rhombic prism of 99° 32′. It has traces of cleavage parallel to the clinodome. The primitive form of the crystal is very rare, but complex crystals are often found. Detached crystals are not common; they are usually scattered over a gangue or lining cavities. Its fracture is conchoidal. Lustre, vitreous, nearly adamantine. It is transparent, translucent or opaque. Its color is azure-blue, which is more or less dark. Streak, lighter than color. Brittle. **H.**=3.5–4.25. **G.**= 3.5–3.831. Composition, $\dot{C}u$ 69.2, $\ddot{C}$ 25.6, $\dot{H}$ 5.2.

Pyr. &c. B. P. Like Malachite. In acids it is soluble with effervescence, which is not immediate but requires heat.

It is sometimes found as concretionary masses in mamelons, which are sometimes so close together as to become joined. The little crystals which protrude from the surface, then give it a velvety lustre. It has been found in little balls which are evidently the product of decomposition, since they contain some oxide in their interior. The massive pieces do not always have a uniform color, but are found as concentric zones, made up of different shades of blue. They are also very frequently associated with other substances such as sandstone, &c. It is also found earthy, when it has no other character than its blue color, which is lighter than ordinary, as it is frequently mixed with clay, marl and sand. The crystalline varieties have a color, which prevents their being mistaken for anything else. The earthy varieties often resemble Lapis-Lazuli or Vivianite. Before the blowpipe it can be immediately distinguished from the former. The lustre of the earthy varieties of Vivianite is quite different, but it can easily be distinguished by the blowpipe or by acids. It has been reproduced artificially, with the same crystalline types, by submitting Cu Cl, and an alkaline carbonate to a high temperature, in a closed vessel for a long time. When found abundantly it is a valuable ore of copper. It has been found in many of the copper mines of the U. S.

FORMULÆ OF THE CRYSTALS.

Pl. XXXII.

Fig. 24. ∞P. 0P. —P. $\frac{1}{3}\grave{P}\infty$. *Fig.* 25. 0P. ∞P. $\infty\bar{P}\infty$. —P. $\frac{1}{2}$P. $\frac{1}{2}\bar{P}\infty$. *Pl.* XXXIII. *Fig* 1. 0P. $\infty\bar{P}\infty$. ∞P. $\frac{1}{2}\bar{P}\infty$; tabular crystal. *Fig.* 2. 0P. $\frac{1}{2}\bar{P}\infty$. ∞P. —P. $\frac{1}{3}\grave{P}\infty$.

MERCURY.

Mercury. Hg. ISOMETRIC.

SYN.—Quicksilver, Gediegen Queeksilber, Mercure natif.

When solidified it crystallizes in octahedra. Lustre, metallic. Opaque. Color, tin-white. **G.**=13.568. Composition, pure mercury, sometimes with a little silver.

Pyr. &c. B. P. On Ch., completely volatilized. Soluble in $\dot{\ddot{N}}$.

Mercury is found in some rocks, having all the properties of the ordinary metal. It is usually found in globules in cavities associated with Cinnabar. These globules are sometimes strung together and are not very fluid, as the mercury is not pure, but frequently contains silver. It sometimes occurs not associated with Cinnabar. In California, geodes oı Quartz containing several pounds have been found. It is a very rare mineral, and has been found in states of different geological ages. It is also found in pockets.

Cinnabar. Hg S. HEXAGONAL.

SYN.—Cinnabarite, Zinnober, Cinabre, Mercure sulfuré.

It crystallizes in a rhombohedron of 92° 36′, with cleavages parallel to the hexagonal prism. It is frequently found in crystals lining the sides of cracks. Besides the rhombohedron R, it frequently has a large number of faces belonging to other rhombohedra on the edges of the hexagonal prism. The flat crystals are usually transparent. It is also found in lamellar and fibrous masses. Generally it is in large granular masses, sometimes fibrous or compact. Its fracture is conchoidal or uneven, both in the crystals and crystalline masses. Lustre, adamantine on the faces of crystals, dull on the compact varieties. The crystals are transparent, translucent, or opaque. The color is cochineal-red, inclining to violet. Streak, characteristic vermilion-red. When it is impure the color is often black, but the streak is always red. It absorbs light easily, which often makes it opaque. It is the most refringent of all known bodies. **H.**=2–2.5. **G.**=8.998. Composition, Hg 86.2, S 13.8.

Pyr. &c. B. P. On Ch., it volatilizes without residue. In a tube, gives a red sublimate. It is not attacked by acids and is the only sulphide which is not acted on by aqua regia.

When it is pure it is easily recognized by its color and its high density, It is often impure, being scattered through bituminous limestones, when it is very black. It might be mistaken for Proustite, Rutile, Realgar or Crocoite, but it can always be recognized by its density and streak. It is very rare that the black varieties do not show some filaments, which have the natural color of the mineral.

Cinnabar is the ore of mercury, from which most of that used in commerce is obtained. It is sometimes ground and used as a pigment, called vermilion. The principal supply in the U. S. comes from California.

FORMULÆ OF THE CRYSTALS.

Pl. XXXIII.

Fig. 3. R. 0 R. ∞ R. *Fig.* 4. ∞ R. $-\frac{1}{2}$ R. –4 R. –2 R. $-\frac{2}{3}$ R. R. $\frac{2}{3}$ R. 2 P 2. 0 R. *Fig.* 5. R. 0 R. ∞ R. $\frac{2}{5}$ R.

Calomel. Hg^2 Cl. TETRAGONAL.

SYN.—Horn Mercury, Quecksilberhornerz, Chlormercur, Mercure chloruré.

The usual form is the prism of the second order, with the octahedron of the first order. Fracture, conchoidal. Lustre, adamantine. Translucent. Cleavage, parallel to the prism of the second order. Color, white, gray or brown. Streak, yellowish-white. It has the consistency of horn and is easily cut into thin, transparent shavings. **H.**=1–2. **G.**=6.482. Composition, Hg 84.9, Cl 15.1.

Pyr. &c. B. P. In a closed tube volatilizes without fusion. On Ch., volatile without residue. Soluble in aqua regia; decomposed by alkalies.

The mineralogical assay of all the minerals of Hg, is very simple. By heating them in a tube with a little soda, almost to fusion of the glass, globules of metallic mercury condense. It is sometimes found in the cavities of Cinnabar, or on the rocks which contain it.

FORMULÆ OF THE CRYSTALS.

Pl. XXXIII.

Fig. 6. ∞ P ∞. P.

SILVER.

Silver. Ag. Isometric.

Syn.—Gediegen Silber, Argent natif.

It is usually found in cubes and octahedra, alone or together, or sometimes associated with the rhombic dodecahedron, and very often in tetrahexahedra. Twins are quite common. The crystals are usually distorted. Arborescent forms are also found, which are made up of distorted octahedra, branching from a given direction. Lustre, metallic. Color and streak, silver-white, but liable to tarnish. Ductile, sectile. **H.**=2.5–3. **G.**=10.1–11.1. Composition when pure, silver, but it is usually associated with gold or other metals.

Pyr. &c. B. P. Native Silver has all the characteristics of Ag. On Ch., fuses easily to a metallic globule. In the O. F., gives a brown coating. Soluble in N̈; the addition of a drop of HCl, shows a precipitate of chloride of silver.

The density of Native Silver is not generally equal to that of melted silver, but is somewhat less and sometimes it has no more ring than pasteboard. These same characters are shown by silver precipitated by the galvanic battery, which has given rise to the opinion that this may have been the way in which it was formed. It is often found in filaments which are more or less white, and bent in every direction. These filaments, as well as the crystals, are ordinarily in a gangue of Calcite. Those which are seen disengaged in collections, have been separated by acids. The silver which is used in the arts is never pure. For coins in this country, it is alloyed with 10 % of Cu. Pure silver in the arts, is said to be 12 pennyweights fine. It is said to be 11, 10, 8 or 9 pennyweights fine, according to the quantity of alloy it contains. A mass of Native Silver from one of the Mexican mines, weighed 400 lbs., and at Freiberg, a mass weighing 1400 lbs. was found. It is found associated with a large number of minerals. In Lake Superior it occurs with Native Copper. It occurs in Idaho in large masses, associated with Cerargyrite. It is usually the result of the decomposition of other silver ores.

FORMULÆ OF THE CRYSTALS.

Pl. XXXIII.

Fig. 7. ∞ O ∞. *Fig.* 8. 3 O 3. *Fig.* 9. Hemitrope, composition-face O. *Fig.* 10. Hemitrope of the cube. *Fig.* 11. The same, but upon the other face of the octahedron. *Fig.* 12. The hemitrope, *Fig.* 10, flattened in the direction of the diagonal.

Amalgam. Ag Hg^2. Isometric.

Syn.—Silberamalgam, Amalgam natif.

It is always crystallized as a rhombic dodecahedron, very highly modified, but the faces are rarely distinct, as a separation of silver and mercury takes place after a certain time, so that the crystals become distorted and the edges rounded; sometimes it spreads out, and is imbibed by the rock. It shows traces of cleavage, parallel to the faces of the rhombic dodecahedron. Its fracture is conchoidal. Lustre, metallic. Opaque. Color and streak, silver-white. Brittle; makes a creaking noise when cut with a knife. **H.**=3–3.5. **G.**=10.5–14. Composition, Ag 34.8, Hg 65.2. There is no invariable formula for this mineral, but the Ag and Hg oscillate around the one given. These two metals become alloyed in different proportions, without losing the property of crystallizing.

Pyr. &c. B. P. On Ch., the Hg is driven off, and a globule of Ag remains. In a closed tube the Hg sublimes and leaves the Ag. A strip of copper is whitened and an amalgam of Cu and Hg is formed. Soluble in $\dot{N}$.

It is found associated with Cerargyrite and Calomel. It resembles Native Silver, and is almost always crystallized. It does not lose its color or become black as easily as the former. Still the blowpipe must be used to discover the presence of mercury, before it can be distinguished with certainty.

FORMULÆ OF THE CRYSTALS.

Pl. XXXIII.

Fig. 13. ∞O; characteristic form. *Fig.* 14. ∞O. O. *Fig* 15. ∞O∞. ∞O. *Fig.* 16. 2O2. ∞O. *Fig.* 17. ∞O. ∞O∞. 2O2. ∞O3. 3O$\frac{3}{2}$. O.

Argentite. Ag S. Isometric.

Syn.—Vitreous Silver, Silver Glance, Glaserz, Silberglanz, Argent sulfuré.

It is often found in cubes, octahedra, tetragonal trisoctahedra and hexoctahedra alone, but usually combined one with the other. The crystals are sometimes very distinct and are often grouped. It is also found in wiry filaments. It has indications of a dodecahedral cleavage, which are not very distinct. Fracture, conchoidal or uneven. Lustre, metallic. Opaque. Color, deep iron-black with very little lustre on the natural faces. The lustre is however bright on the fracture. Streak, same as color and shining. It is ductile, receives and retains the mark of the hammer. It is flexible, but not elastic. When it is pure it can be cut with a knife, giving curved shavings. **H.**=2–2.5. **G.**=7.196–7.365. Composition, Ag 87.1, S 12.9.

Pyr. &c. B. P. Melts in a flame without the aid of a blowpipe. In the O. F., it is roasted. In the R. F., gives a metallic globule. Soluble in $\dot{N}$.

It is also found as amorphous masses disseminated in gangues, which are usually limestone; these masses have a variable, conchoidal fracture which is rarely lamellar. It often shows in the fracture that it is both ductile and tenacious. Its softness and the look of the fracture make it resemble Bournonite, but this is more brittle, while Argentite can be cut with a knife. The blowpipe gives with this a globule of silver, with that a globule of copper and lead fumes. The specific gravity is much higher than any of the copper ores which it resembles. Its a very valuable ore of silver. It is mined as an ore in Nevada.

FORMULÆ OF THE CRYSTALS.

Pl. XXXIII.

Fig. 18. ∞O∞. O; a very frequent form. *Fig.* 13. ∞O. *Fig.* 19. ∞O∞. 2O2. *Fig.* 20. 2O2. *Fig.* 21. 2O2. ∞O∞. O *Fig.* 22. 2O2; lengthened vertically.

Proustite. $3\,Ag\,S+As^2\,S^3$. HEXAGONAL.

SYN.—Ruby Silver Ore, Arsensilberblende, Argent rouge arsenicale.

It crystallizes as a rhombohedron of 107° 48′; its usual forms are the hexagonal prism and rhombohedron, the scalenhedron or hexagonal prism, and hexagonal pyramid. The forms are almost as numerous as those of Calcite. The number of faces is so large, that the crystals usually have a rounded appearance. Fracture, conchoidal or uneven. Lustre, adamantine. Transparent, translucent; highly refringent. Color and streak, cochineal-red. **H.**=2–2.5. **G.**=5.422–5.56. Composition, Ag 65.4, S 19.4, As 15.2.

Pyr. &c. B. P. In a closed tube decrepitates and fuses easily, giving a slight sublimate of sulphide of arsenic. In an open tube, gives sulphurous fumes and Äs. On Ch., gives the smell of sulphur and arsenic. After long treatment in the O. F., it gives in the R. F. a globule of silver. Soluble with difficulty in the oxidizing acids.

It is sometimes found in capillary masses, and often amorphous or crystalline, associated with other ores of silver. There are two extreme types of Red Silver, having the same general formulæ, and the same crystalline form, except that in one the arsenic is replaced by antimony. There is every possible gradation between the two limits. The arsenical variety, or Proustite, is called Light Red Silver, and the antimonial, or Pyrargyrite, Dark Red Silver. It has been found in the U. S. in Nevada, Idaho and N. C.

Pyrargyrite. $3\,Ag\,S+Sb^2\,S^3$. HEXAGONAL.

SYN.—Ruby Silver Ore, Antimonsilberblende, Argent rouge antimoniale. Argent antimonié sulfuré.

Its primitive form is a rhombohedron of 108° 42′. It very frequently occurs as hexagonal prisms so highly modified as to appear rounded. Simple crystals are quite rare. It has an indistinct cleavage parallel to the rhombohedron. Fracture, conchoidal. Lustre, metallic, adamantine. Translucent. Color, black, or very dark red. Streak, cochineal-red. **H.**=2–2.5. **G.**=5.7–5.9. Composition, Ag 59.8, Sb 22.5, S 17.7.

Pyr. &c. B. P. In a closed tube, gives a red sublimate of sulphide of antimony. In an open tube, sulphurous fumes and a white sublimate of antimony. On Ch., fuses and coats the coal. Heated for some time in the O. F., in the R. F., a globule of silver is obtained. Decomposed by N̈.

It is sometimes found in capillary crystals or as amorphous masses associated with other ores of silver. It resembles Proustite. The color of the two minerals is red, which is more or less bright. In Pyrargyrite it is so intense, that the mineral appears black and opaque. But there are almost always some thin points, where the color can be seen. Proustite is much lighter and is often transparent. When compact or capillary, it can easily be distinguished from Cinnabar, Cuprite and Hematite, by the streak and blowpipe reactions.

It is a valuable ore of silver and has been found in considerable quantities in Nevada and Idaho.

FORMULÆ OF THE CRYSTALS.

Pl. XXXIII.

Fig. 23. $\infty P 2$. R. *Fig.* 24. $\infty P 2$. $-\frac{1}{2}R$. $\frac{\infty R}{2}$. *Fig.* 25. R 3. *Pl.* XXXIV. *Fig.* 1. $\infty P 2$. $\frac{\infty R}{2}$. $-\frac{1}{2}R$. $-2R$. R 3. $\frac{1}{4}R 3$. $\frac{5}{8}R 3$. $-\frac{1}{2}R 3$. *Fig.* 2. $\infty P 2$. R3. $\frac{1}{4}R 3$. $\frac{\infty R}{2}$; hemitrope, composition-face 0 R. *Fig.* 3. $-\frac{1}{2}R$. $\infty P 2$; twin consisting of four individuals, composition-face $-\frac{1}{2}R$.

Stephanite. $5 Ag S + Sb^2 S^3$. ORTHORHOMBIC.

SYN.—Brittle Silver Ore, Sprödglaserz, Melanglanz, Argent sulfuré fragile.

It crystallizes as a right rhombic prism of 115° 39′. Its forms are usually the prism and brachypinacoid. The prisms are always striated. It has an indistinct cleavage, parallel to the brachyprism and pinacoid. Fracture, conchoidal or uneven. It has a bright, metallic lustre. Translucent on the edges. Color and streak, black. **H.**=2–2.5. **G.**=6.269. Composition, Ag 68.5, S 16.2, Sb 15.3.

Pyr. &c. B. P. In a closed tube, decrepitates and fuses; in an open tube gives fumes of sulphur and antimony. On Ch., decrepitates and fuses, giving the rose-colored coating of silver and antimony. After long treatment in the O. F., in the R. F. a globule of silver is obtained.

By its form, color, streak and lustre, it can easily be recognized. It is a valuable ore of silver and is mined in Nevada.

FORMULÆ OF THE CRYSTALS.

Pl. XXXIV.

Fig 4. 0P. P. $2\breve{P}\infty$; resembles a truncated hexagonal pyramid *Fig.* 5. 0P. ∞P. $\infty\breve{P}\infty$. *Fig.* 6. ∞P. $\infty\breve{P}\infty$. P. $2\breve{P}\infty$. 0P. $\frac{1}{2}P$. *Fig.* 7. The preceding, with $\infty\breve{P}\infty$ and 2P.

Polybasite. $9 (Ag, \text{Ɇu}) S + (Sb, As)^2 S^3$. ORTHORHOMBIC.

SYN.—Eugenglanz.

It crystallizes as a right rhombic prism, the angle of which is very near 120°. The crystals are usually hexagonal in shape, and are frequently quite flat, showing but little of the prism. Fracture, uneven. Lustre, metallic. Translucent in thin splinters. Opaque. Color, black, but in thin splinters by transmitted light, cherry-red. Streak, black. **H.**=2–3. **G.**=6.214. Composition, Ag 64.7, Cu 9.8, S 14.8, Sb 9.7.

Pyr. &c. B. P. In an open tube, fuses and gives the sublimates of sulphur and antimony. On Ch., fuses, gives reactions for sulphur and antimony. With fluxes, gives reaction for copper. Decomposed by $\ddot{N}$.

Quite a number of minerals are frequently described under this name; they are all quite analogous to Red Silver in their external characters. They are more opaque however, and their streak also has a brown color It has been found in Nevada.

Cerargyrite. Ag Cl. ISOMETRIC.

SYN.—Horn Silver, Silberhornerz, Hornsilber, Chlorsilber, Argent corné, Argent chloruré.

It crystallizes in cubes, octahedra, rhombic dodecahedra, or combinations of two or more of them. Crystals are quite rare. The largest known came from the Poorman's lode, Idaho. Fracture, conchoidal. Lustre, resinous. Transparent, translucent. Opaque. Color, white, gray, grayish-green, or colorless when perfectly pure. Streak, colorless and shining. Sectile, having about the consistency of horn; when cut with a knife it gives curved shavings. **H.**=1–1.5. **G.**=5.31–5.43. Composition, Ag 75.3, Cl 24.7.

Pyr. &c. B. P. In a closed tube fuses without decomposition. Fuses in a flame of a candle. On Ch., gives a globule of silver. Insoluble in N̈, but soluble in Ammonia.

It is usually found as a coating, which is sometimes wax-like and sometimes fibrous. It is subject to alteration when exposed to the light; it then loses its color and becomes brown. It is mined as an ore of silver in South America. Extensive deposits have been found in Arizona and Nevada.

FORMULÆ OF THE CRYSTALS.

Pl. XXXIV.

Fig. 8. ∞O∞. *Fig.* 9. O. *Fig.* 10. ∞O. *Fig.* 11. ∞O∞. O.

Bromyrite. Ag Br. ISOMETRIC.

SYN.—Bromic Silver, Bromsilber, Bromit, Bromargyrite, Bromure d'argent.

It crystallizes in cubes, octahedra, or a combination of the cube and rhombic dodecahedron. Crystals are quite rare. Lustre, splendent. Color, when pure, yellowish-green, often a darker green when altered. **H.**=2–3. **G.**=5.8–6. Composition, Ag 57.4, Br 42.6.

Pyr. &c. B. P. In a closed tube acts like Cerargyrite. On Ch., gives vapors of bromine. Insoluble in N̈, soluble with difficulty in ammonia.

It is usually found as small concretions or coatings. It occurs in S. America, associated with other ores of Silver, but it is a very rare mineral.

FORMULÆ OF THE CRYSTALS.

Pl. XXXIV.

Fig. 8. ∞O∞. *Fig.* 9. O. *Fig.* 12. ∞O. ∞O∞.

Embolite. Ag (Cl, Br). ISOMETRIC.

SYN.—Chlorobromide of silver, Chlorbromsilber, Megabromit, Microbromit, Chlorobromure d'argent.

It is found in combinations of the cube and octahedron, but crystals are quite rare. Lustre, resinous. Translucent, opaque. Color, various shades of green, usually with a yellow tint, which becomes darker on exposure. **H.**=1–1.5. **G.**=5.31–5.81. Composition, Ag 69.28, Br 14.30, Cl 16.42; the relation of the chlorine to the bromine may vary indefinitely.

Pyr. &c. B. P. Acts like Cerargyrite, but gives the reaction for Bromine.

It is found associated with Cerargyrite and Bromyrite generally as a coating, which has sometimes a columnar structure.

FORMULÆ OF THE CRYSTALS.

Pl. XXXIV.

Fig. 8. ∞O∞. *Fig* 11. ∞O∞. O. *Fig.* 12. ∞O. ∞O∞.

Iodyrite. Ag I. HEXAGONAL.

SYN.—Iodic Silver, Iodsilber, Iodit, Iodargyrit, Iodure d'argent.

The crystals are quite small and rare. Cleavage perfect, parallel to the base. Lustre, adamantine, resinous. Translucent. Color, sulphur-yellow, canary-yellow, sometimes greenish or brownish. Streak, yellow. In thin plates it is flexible and sectile, but in masses it is brittle. **H.**=1–1.5. **G.**=5.5–5.71. Composition, Ag 46, I 54.

Pyr. &c. B. P. In a closed tube, fuses and is orange-yellow while hot, but resumes its color on cooling. Fuses in the flame of a candle. On Ch., gives fumes of iodine and a globule of silver.

It is usually compact, with a more or less lamellar structure. It has been found in Arizona.

GOLD.

Gold. Au. ISOMETRIC.

SYN.—Gediegen Gold, Or natif.

It is found crystallized in cubes, octahedra, tetrahexahedra and rhombic dodecahedra. Sometimes faces of the tetragonal trisoctahedron and hexoctahedron may be observed. The octahedron is sometimes hemitrope or hollow and the crystals are often rounded. Lustre, metallic. Color and streak, different shades of yellow. Ductile, malleable. **H.**=2.5–3. **G.**=15.16–19.34. Composition, when pure, Au, but generally it contains traces of other metals.

Pyr. &c. B. P. Fuses easily, but gives no reactions with fluxes. Soluble in aqua regia.

Native Gold has all the characters of ordinary gold; it is sometimes of a light tint, because it is alloyed with silver or lead. It is found in wiry masses, made up of chains of octahedra or cubes joined together, resembling those of Native Silver. It is also found in thin sheets or dendrites on the surface of some rocks. The ordinary gangue of Gold is Quartz, but it is also found in slates and metallic sulphides. It is not combined with them, but is in the native state in little pellets. The sulphides are the gangue of the Gold as the Quartz sometimes is. The rocks which contain Gold, have often become decomposed and broken up, and the products of this alteration have been washed out, thus forming the alluvial auriferous sands. The scales in these sands are flattened, the size of the scales depending on the condition of the Gold in the vein. When these sands are treated for Gold by washing, the scales are easily separated, on account of their high density. The large rounded fragments found in rich sands are called nuggets. Some of these nuggets are of considerable size. The following table gives the weights of a few of the principal ones.

	Weight. lbs. oz.		Weight. lbs.
Welcome Nugget, Ballarat, Australia, value, $41,822	184 8	Miask, Urals	27
		" "	16
		Paraguay	50
Blanch Barkley Nugget,	146	Cabarrus Co., N. C.,	37
Miask, Urals	96	California	27
" "	27	"	17

In the Eastern states Gold has been found principally in Virginia, N. and S. Carolina, and Georgia.

FORMULÆ OF THE CRYSTALS.

Pl. XXXIV.

Fig. 13. O, with cavernous faces. *Fig.* 14. ∞O2. *Fig.* 15. 3O3. O. *Fig.* 16. ∞O. O. 4O2. *Fig.* 17. O. ∞O; flattened parallel to a face of the octahedron.

PLATINUM.

Platinum. Pt. Isometric.

Syn.—Platina, Platin.

When crystallized it is found in cubes and octahedra. Fracture, hackly. Lustre, metallic. Opaque. Color and streak, steel-gray. Streak, shining. Ductile, occasionally showing polar magnetism. **H.**=4–4.5. **G.**=16–19. Composition, Pt. combined with iron, iridium, osmium, and other metals.

Pyr. &c. B. P. Infusible and gives no reactions with the fluxes. Soluble in aqua regia.

Its density is very great, but is less than that of Platinum which has been worked. It is between 16 and 19, while that of worked Platinum is 22. In nature, Platinum is very frequently associated with other metals, which are rarely, if ever, found except associated with it, such as palladium, rhodium, iridium and osmium. The ores of Platinum have sometimes been found in place in Serpentines and Syenites. It is disseminated in these rocks either as little grains, or masses, which are more or less large. The largest found weighed 21 lbs. In the collection of minerals at the School of Mines, Columbia College, there is a mass weighing 22.5 ozs. These masses have a metallic lustre, and the grayish-white color, peculiar to Platinum. Sometimes this mass is quite brilliant, but it is always more or less cavernous or filled with holes, which sometimes go completely through it. Sometimes it is quite black. The color varies with the place where it is found. In Russia it is found black and white. In California it is white and brilliant, while that from Oregon is darker. Besides the ores in place, which are quite rare, it is frequently found in sands, produced by the wash of the altered rocks which contained it. These sands contain not only Platinum and the metals which accompany it, but Magnetite, Hematite, Menaccanite and Chromite, and are more or less auriferous, because Pyrite containing gold is often found in it. The minerals found with Platinum vary with the locality. They can be very nearly separated from it by mechanical preparation. They are known by the following characters. Platinum is found in small irregular grains, which are more or less flattened and tin-white. Palladium is in small grains more or less rounded, with a radiated fracture, which can be easily seen. These grains look like silver. These two are soluble in aqua regia, while the others are not. The residue from the acid is made up of grains which are bent, and contain rhodium and iridium. In these grains some crystals can be distinguished, which have two parallel faces, with a regular hexagonal contour and a grayish color. These are Iridosmine. Ruthenium was found in these grains of Iridosmine, in the proportion of about 2 %. The ores of Platinum usually contain Pt. 90 %, insoluble residue 10 %, in which ½ % is Iridosmine, and 2 % Ruthenium. All of these substances are infusible, except in the oxyhydrogen flame.

IRIDIUM.

Iridosmine. Ir, Os. HEXAGONAL.

SYN.—Osmiridium, Newjanskit, Iridosmium, Sisserskit, Osmiure d'iridium.

It is usually found in hexagonal tables, rarely as hexagonal prisms, with the basal edges modified. Lustre, metallic. Opaque. Color tin-white or steel-gray. Malleable with difficulty. **H.**=6–7. **G.**=19.3–21.12. Composition, Ir 43–70, Rd 0.63–12.3, Pt 0–2.8, Ru 0–8.49, Os 17.2–48.85.

Pyr. &c. B. P. At a very high temperature, gives off fumes of osmium. With nitre gives the reactions for osmium.

It is found generally in irregular, flattened grains. In this country it is found principally in California.

FORMULÆ OF THE CRYSTALS.

Pl. XXXIV.

Fig. 18. ∞P. 0P. *Fig.* 19. ∞P. P. 0P.

INDEX.

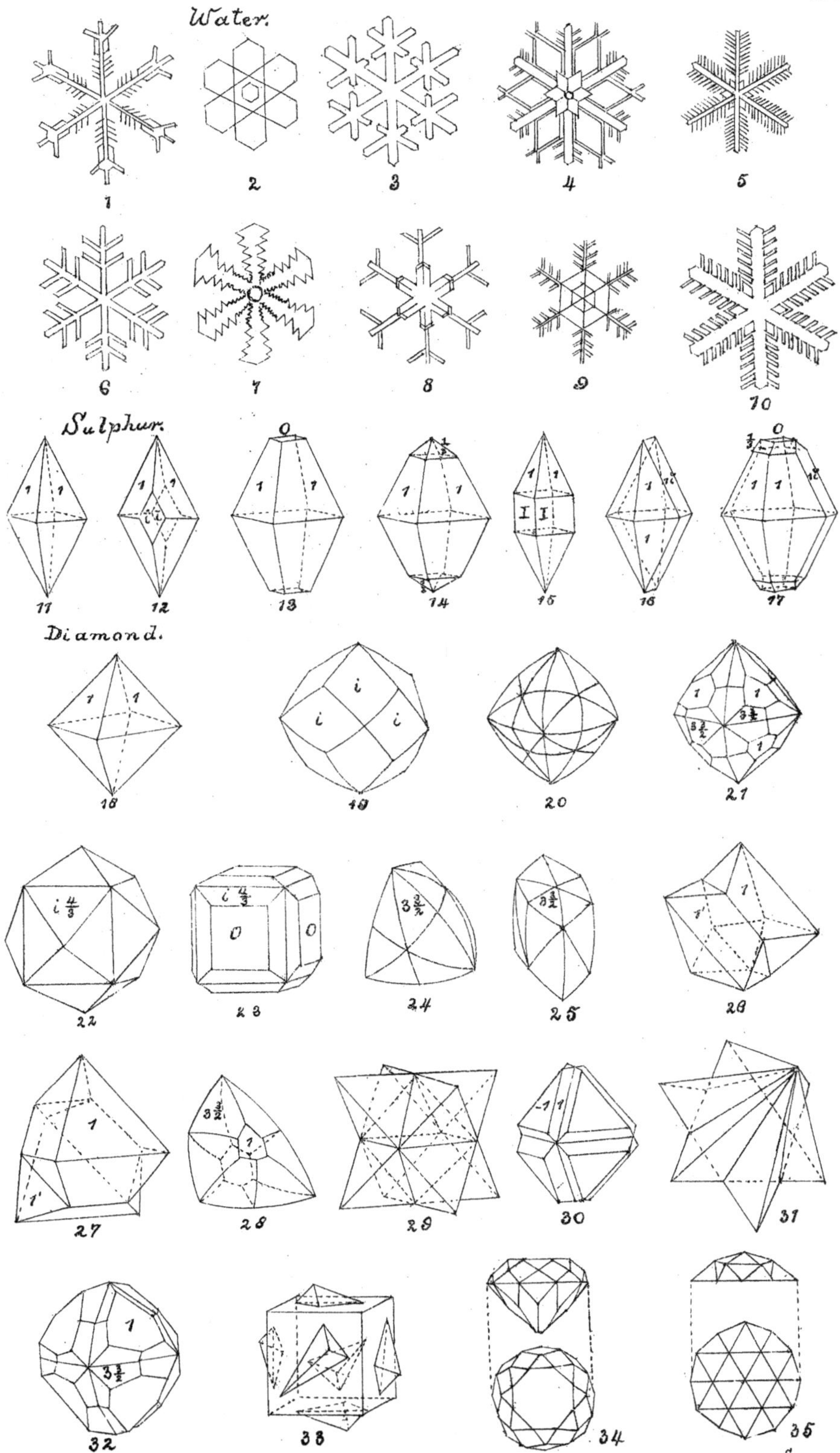
Water.
1
2
3
4
5
6
7
8
9
10
Sulphur.
11
12
13
14
15
16
17
Diamond.
18
19
20
21
22
23
24
25
26
27
28
29
30
31
32
33
34
35

Pl. 2.

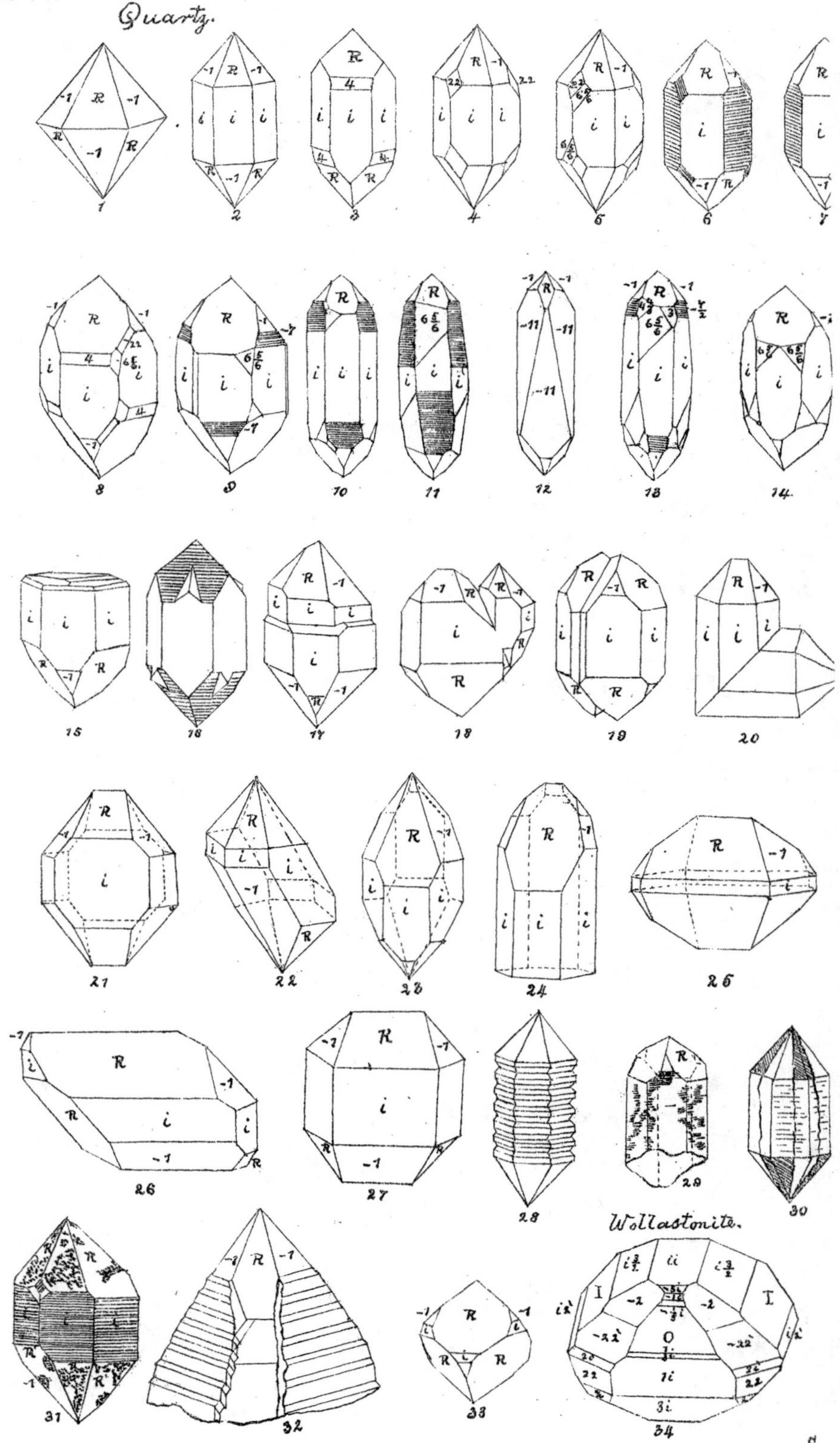

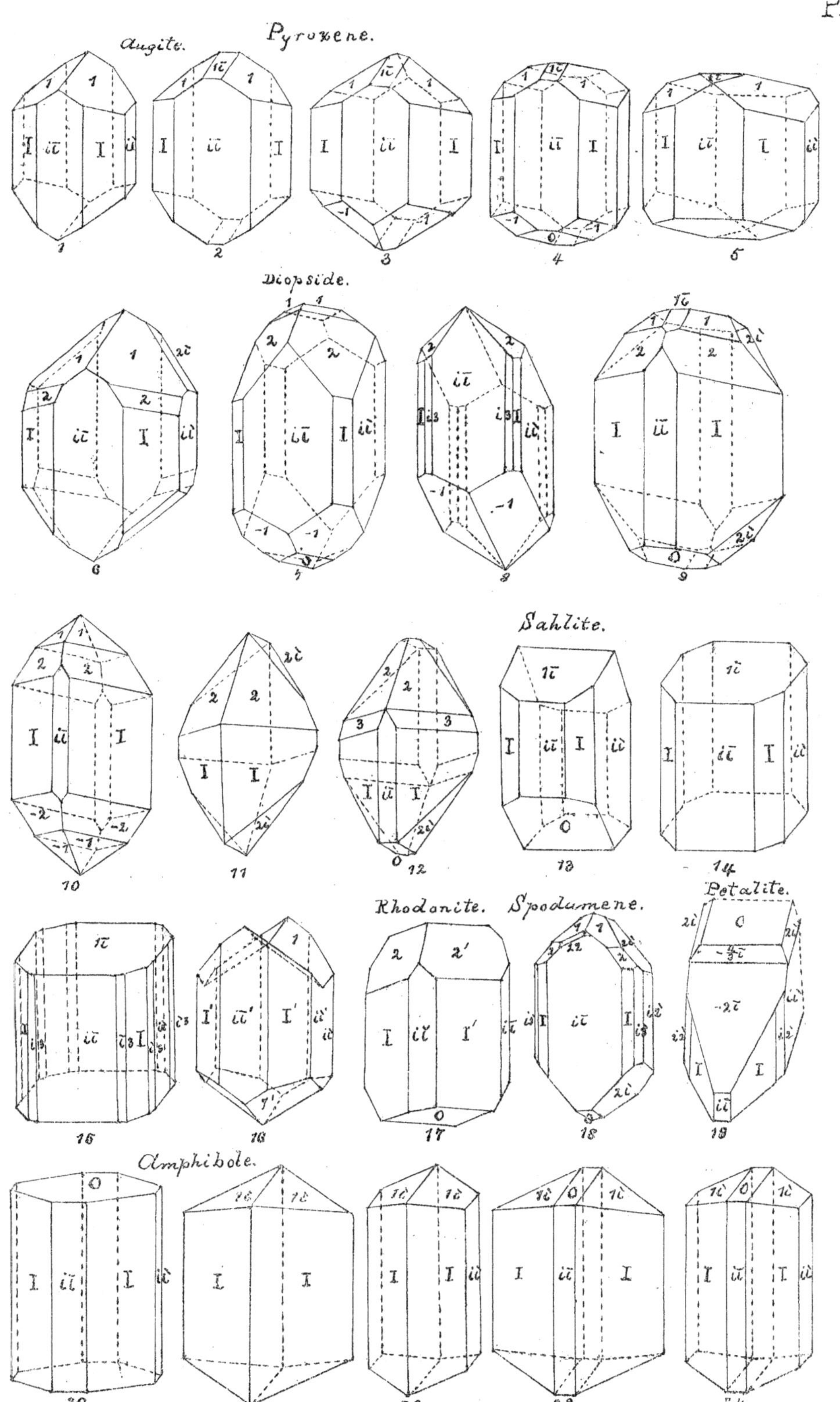
Pyroxene.
Augite.
1
2
3
4
5
Diopside.
6
7
8
9
Sahlite.
10
11
12
13
14
Rhodonite.
Spodumene.
Petalite.
15
16
17
18
19
Amphibole.
20
21
22
23
24

Pl. 4.

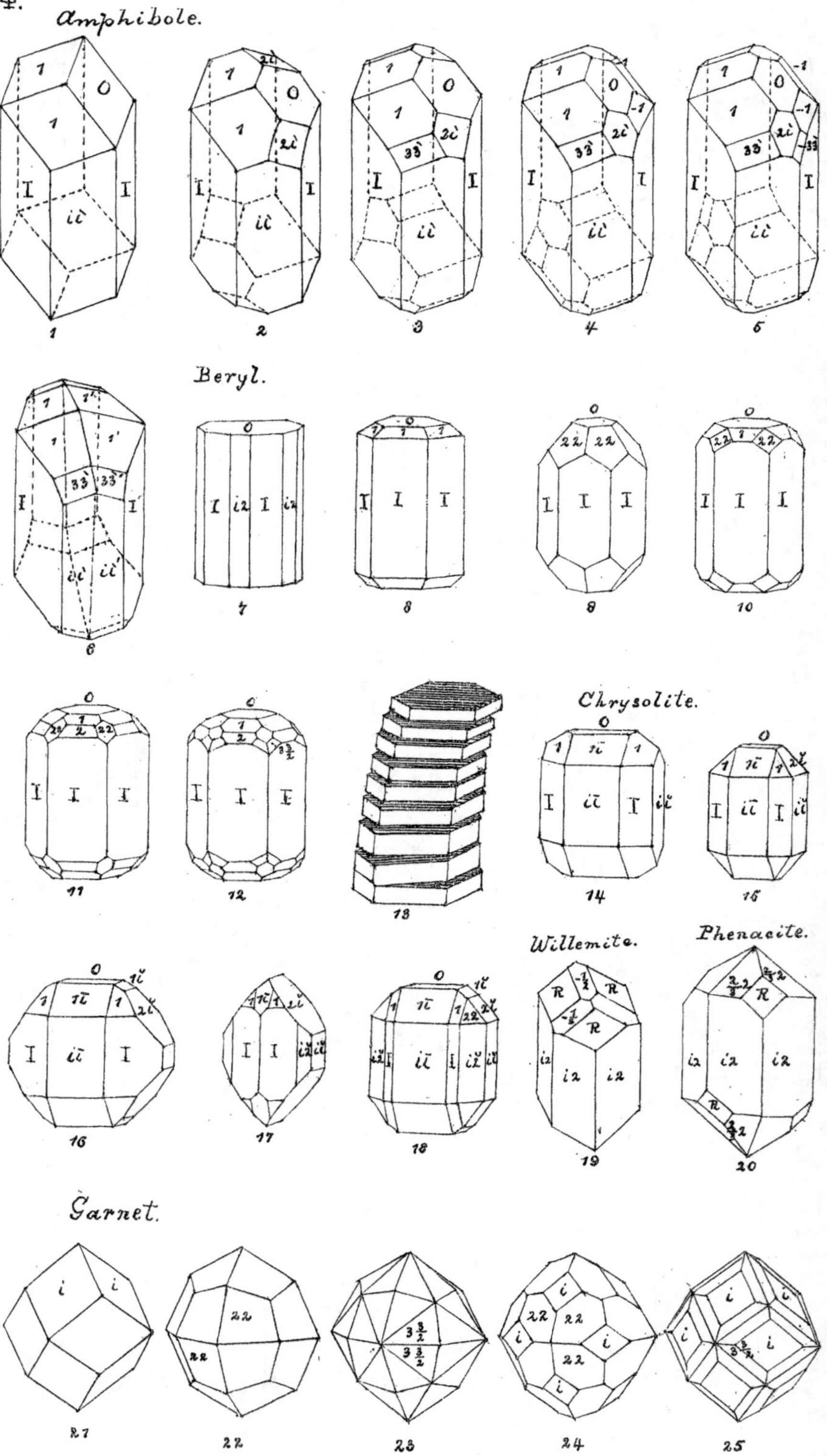

C.

Pl. 5.

Garnet.
1
2
3
4
Zircon.
5
6
7
8
9
10
11
12
13
14
15
Vesuvianite.
16
17
18
19
c.

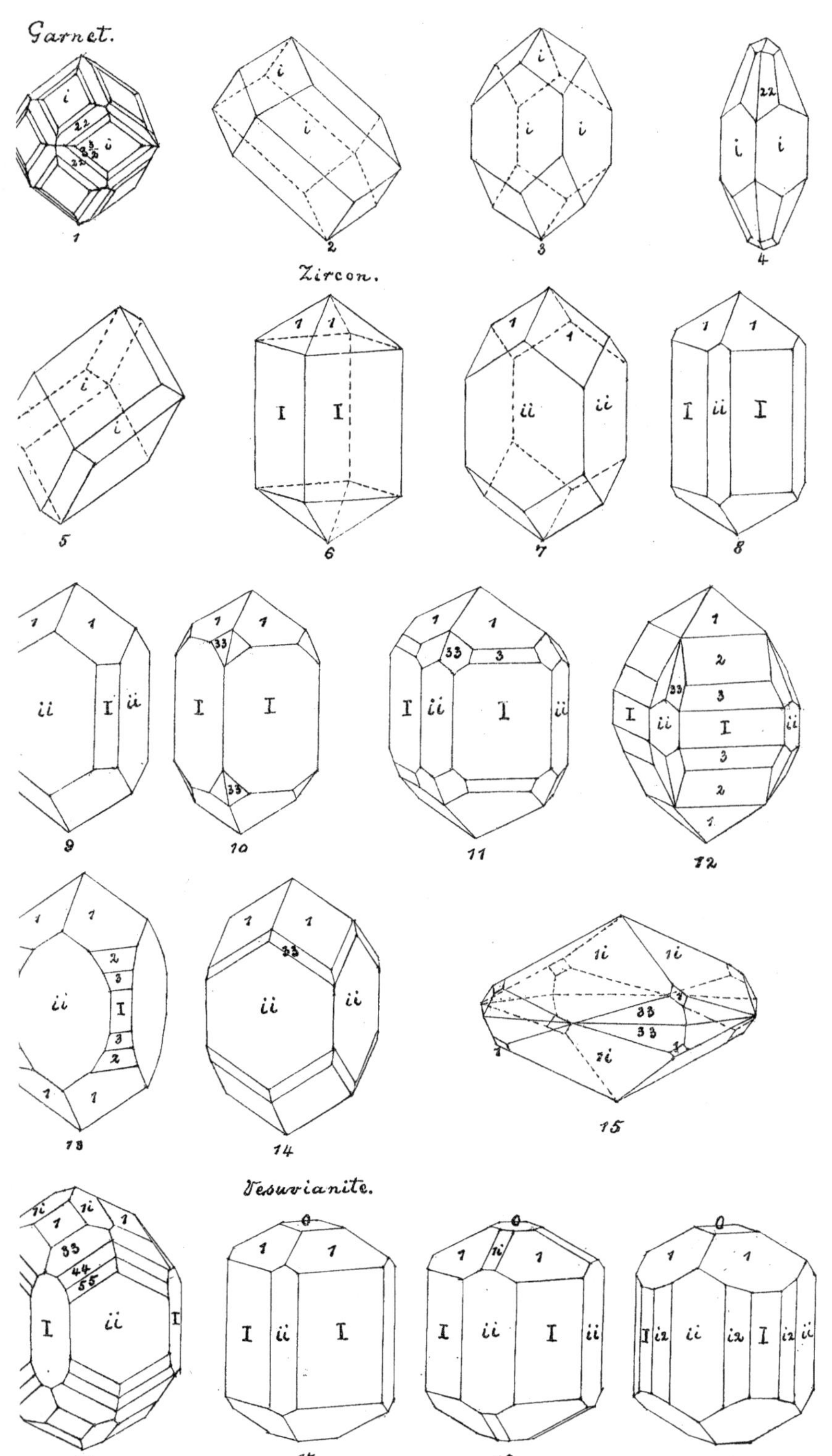

Pl. 6.

Vesuvianite.
1
2
3
4
5
6
Epidote.
7
8
9
10
11
Iolite.
12
13
14
Biotite.
15
Muscovite.
16
17
18
19
20
21

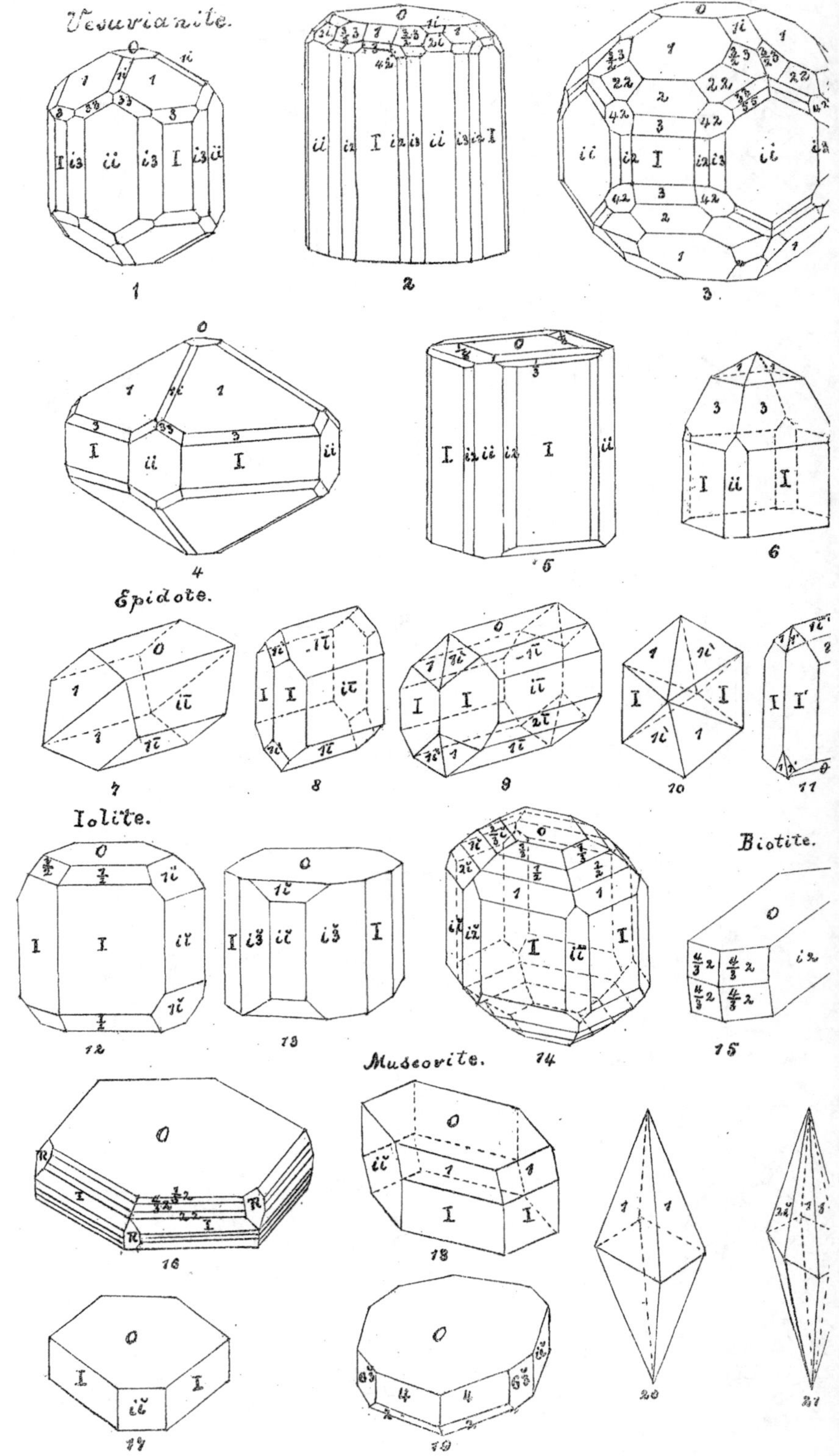

Pl. 7.

Wernerite.
1
2
3
4
Nephelite.
5
6
Haüynite.
7
8
9
Leucite.
10
11
Anorthite.
12
13
Oligoclase.
14
Albite.
15
16
17
18
19
20
21
Pericline.
22
23

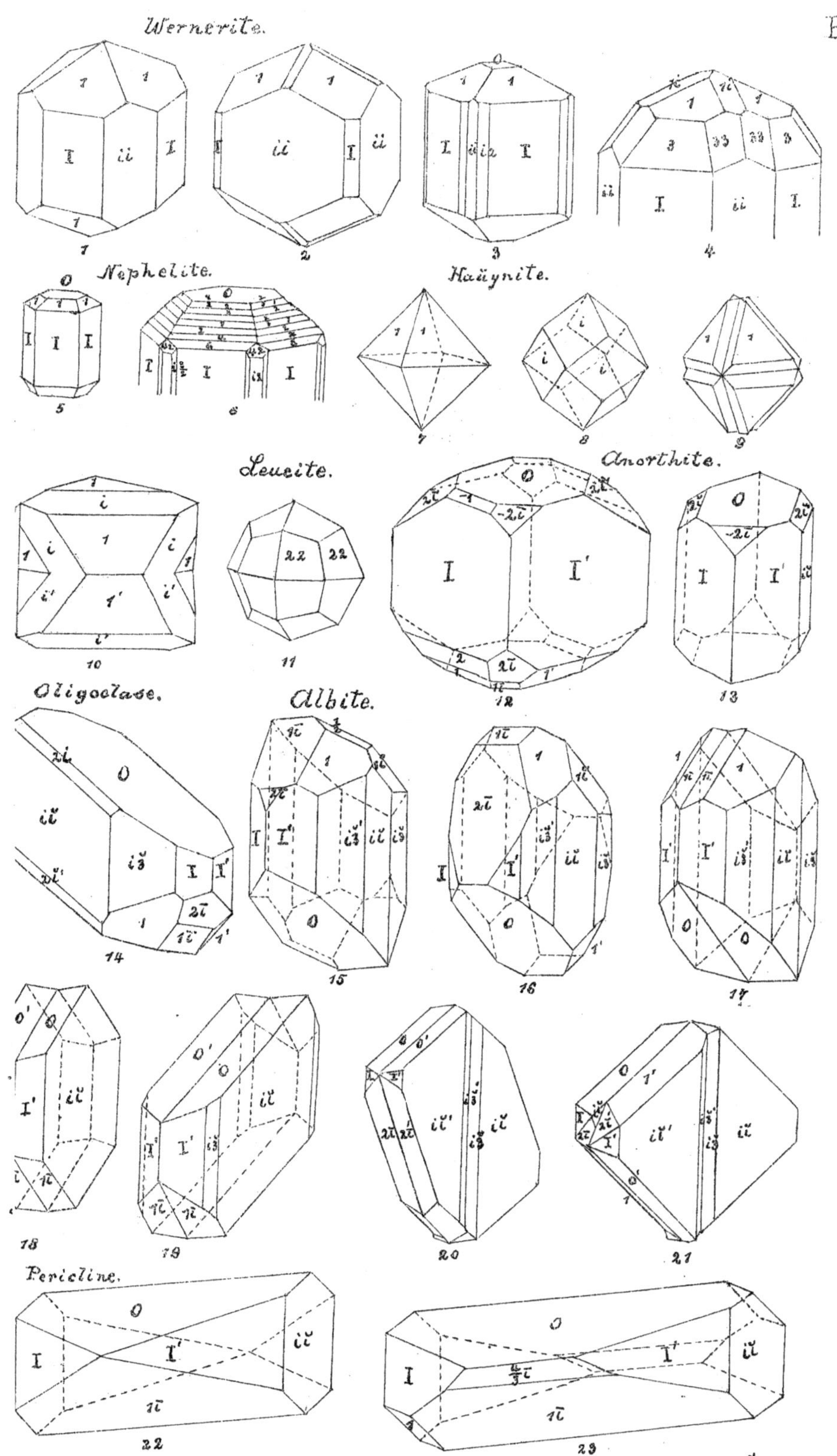

Pl. 8.

Pericline.
1
2
orthoclase.
3
4
5
6
7
8
9
10
11
12
13
14
15
16
17
18
19
20
21
22
Chondrodite.
23
Tourmaline.
24
25
26
27

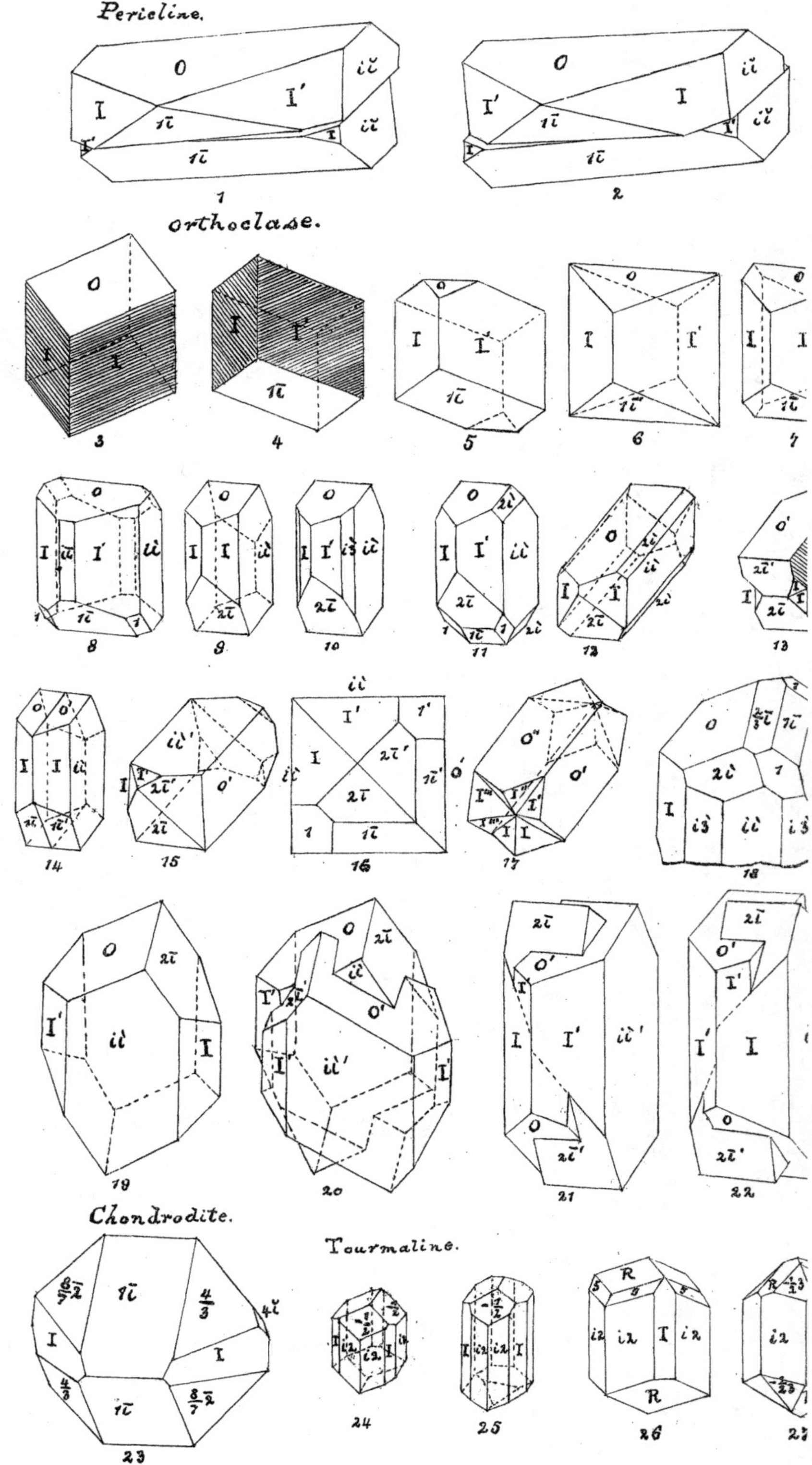

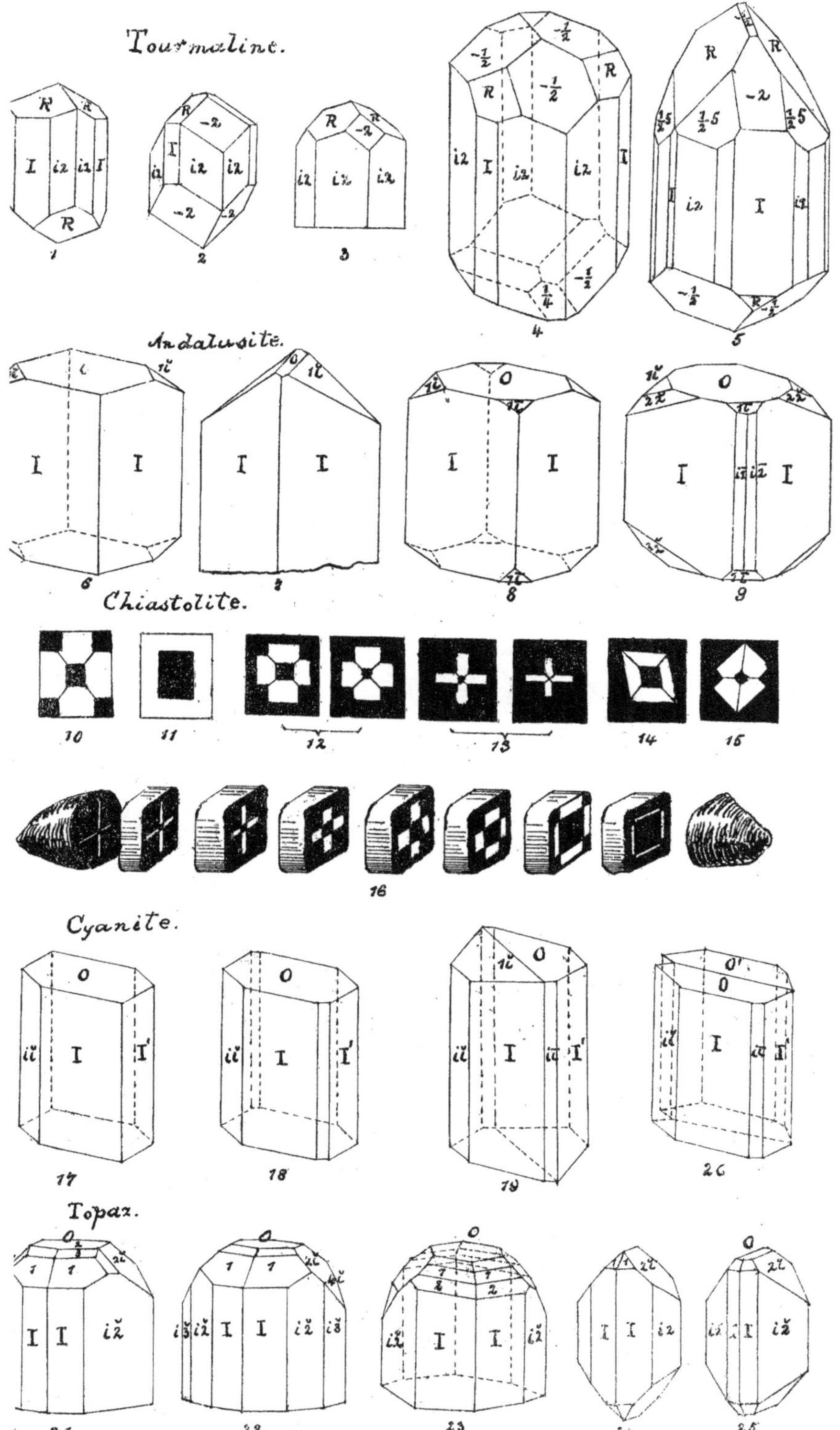
Tourmaline.
Andalusite.
Chiastolite.
Cyanite.
Topaz.

Pl. 10.

Topaz.
Euclase.
Datolite.
Titanite.
Spinthere.
Semeline.
Greenovite.
Lederite.
Staurolite.
1
2
3
4
5
6
7
8
9
10
11
12
13
14
15
16
17
18
19
20
C.

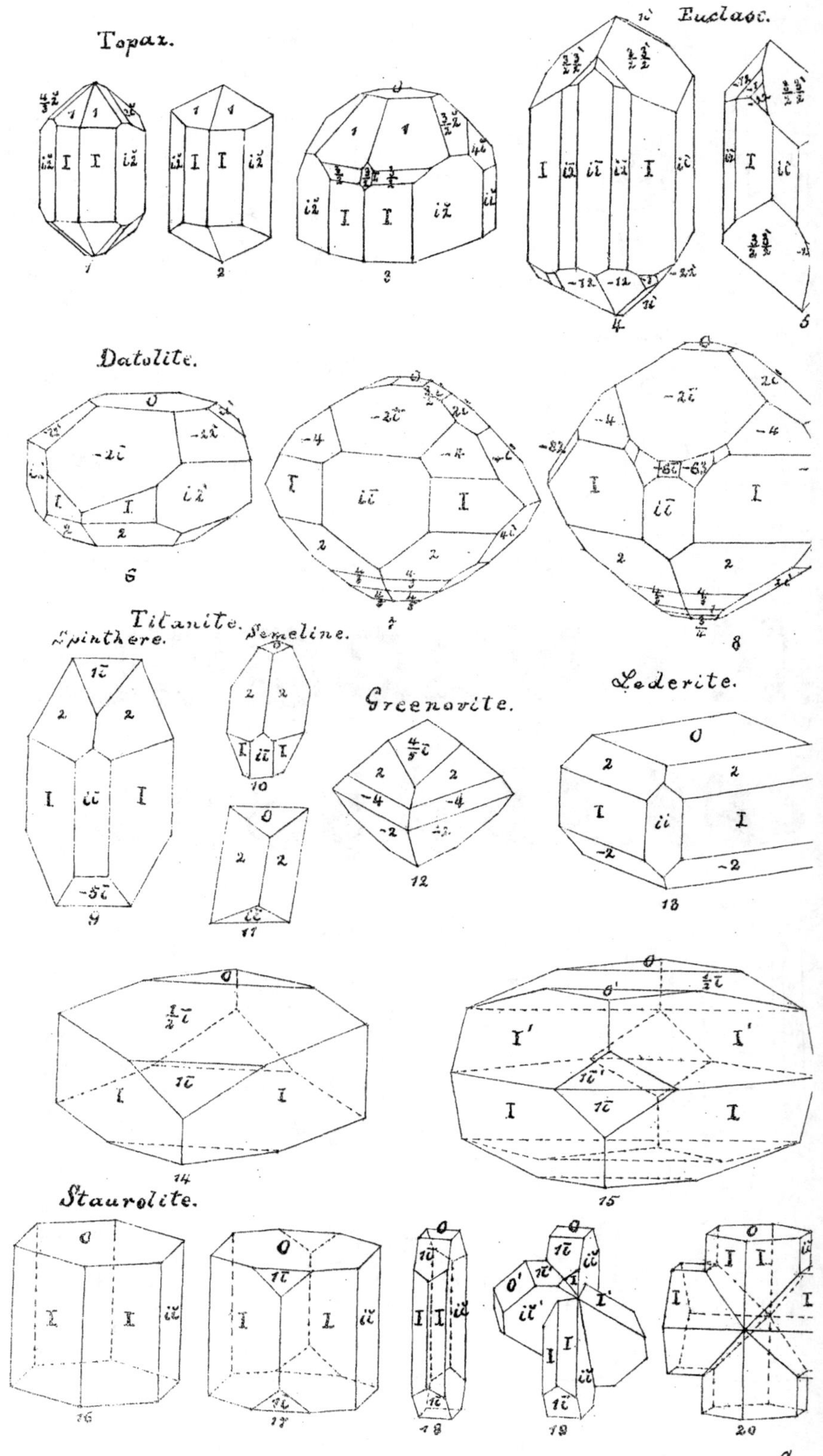

Staurolite.
Laumontite.
Dioptase.
Calamine.
rehnite.
Apophyllite.

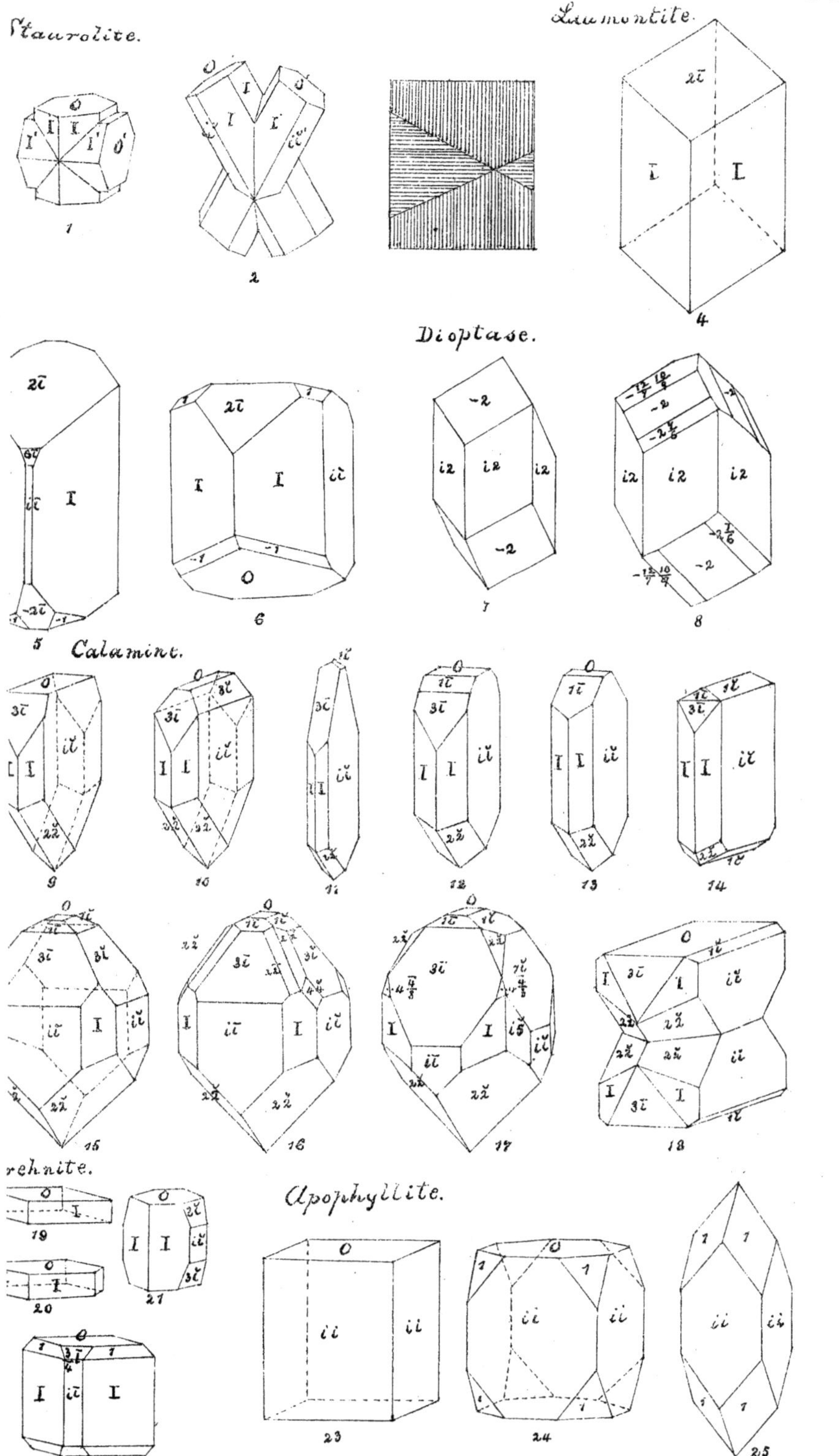

c.

Pl. 12.

Apophyllite.
Natrolite.
Analcite.
Chabazite.
Harmotome.
Stilbite.
Heulandite.
C.

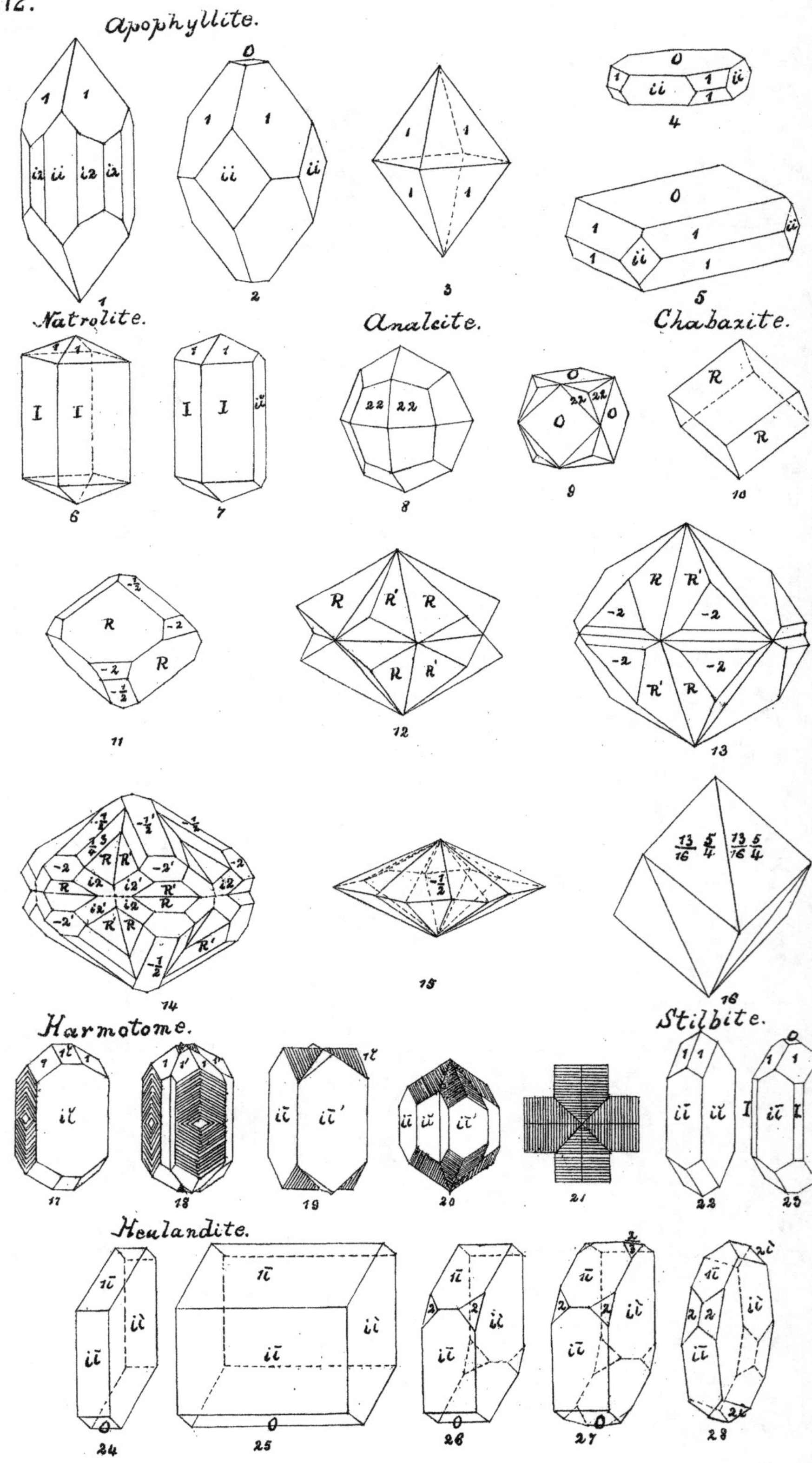

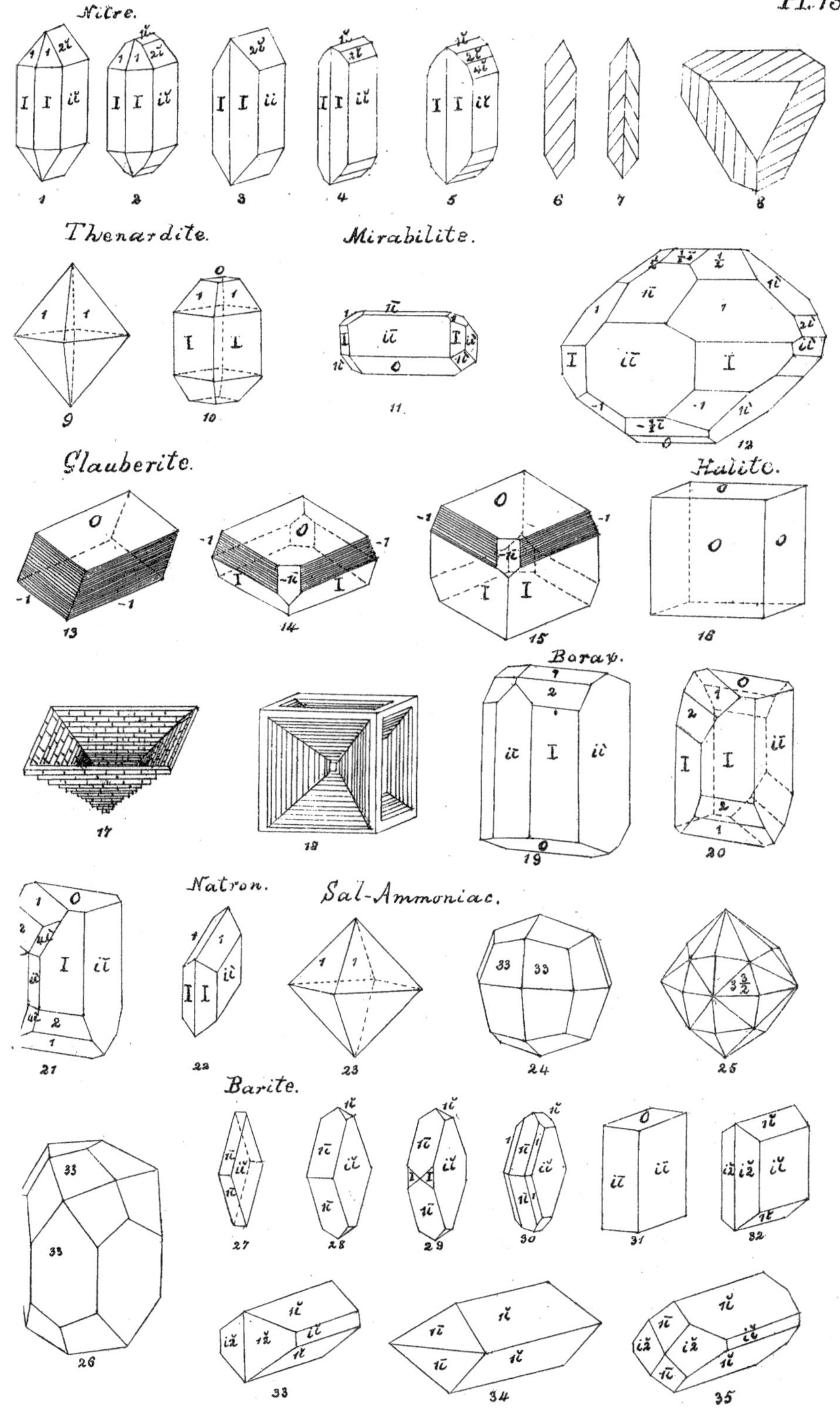
Nitre.
Thenardite.
Mirabilite.
Glauberite.
Halite.
Borax.
Natron.
Sal-Ammoniac.
Barite.

Pl. 14.

Barite.
Witherite.
Barytocalcite.
Celestite.
Strontianite.
Anhydrite.
Gypsum.

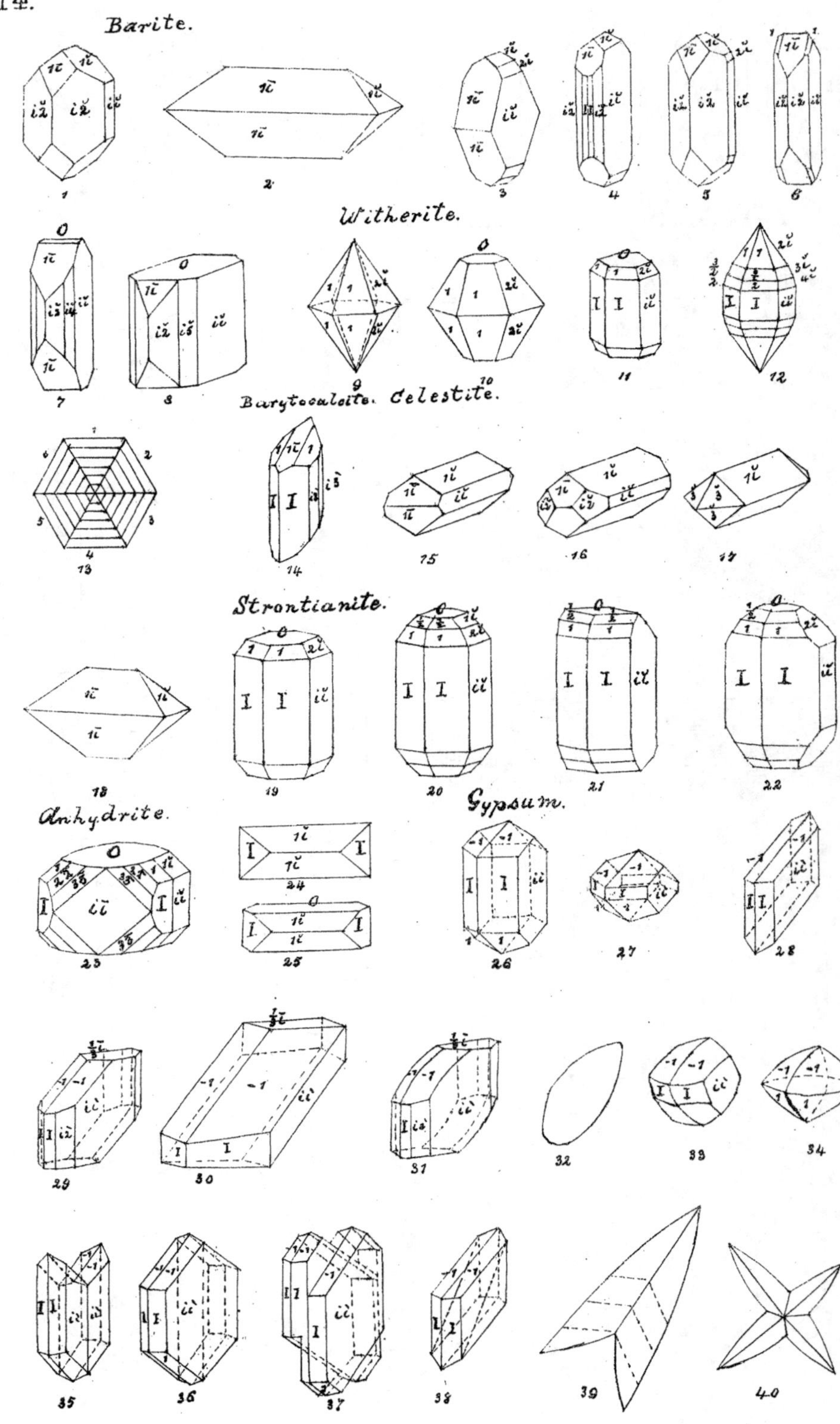

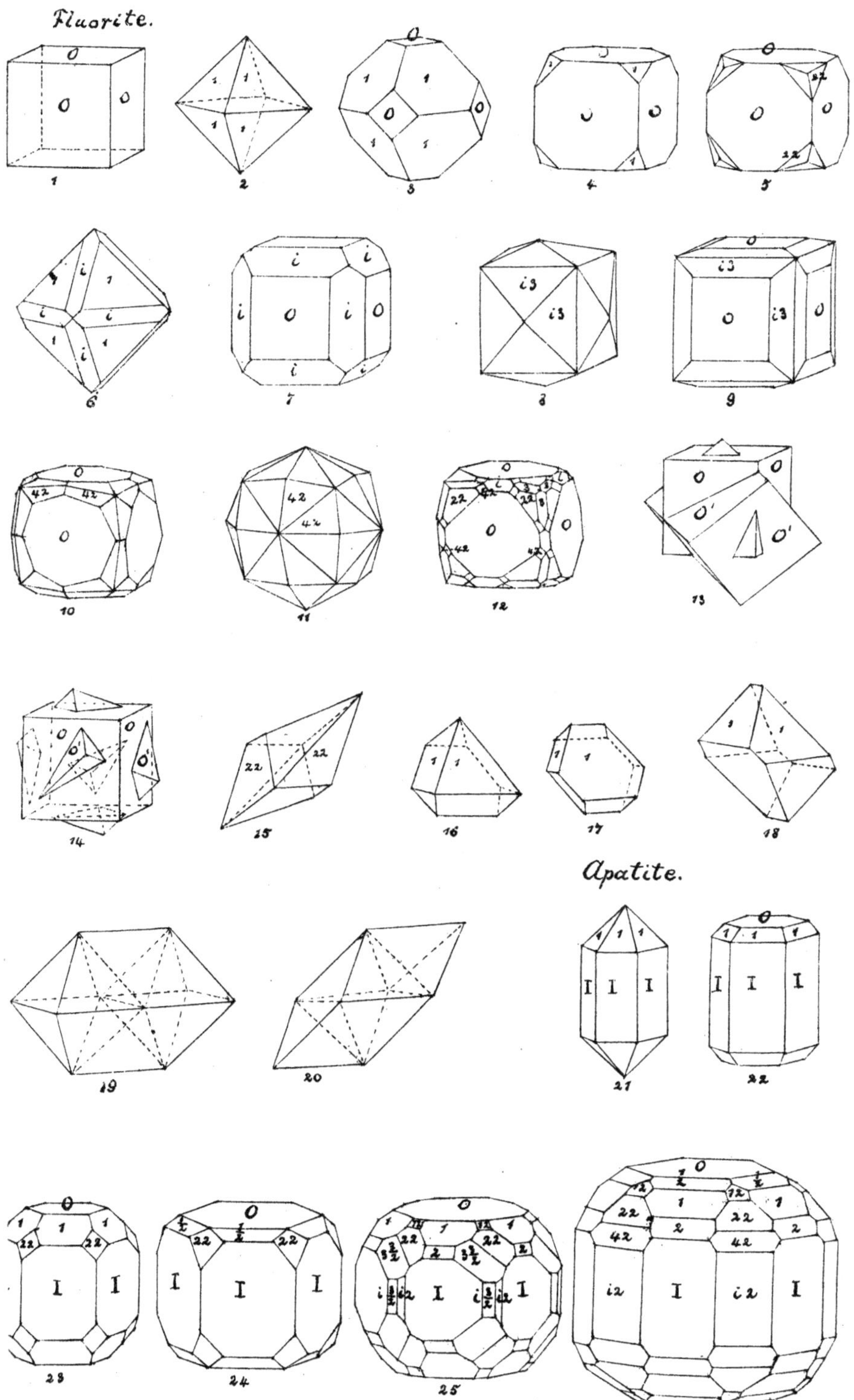
Fluorite.
1
2
3
4
5
6
7
8
9
10
11
12
13
14
15
16
17
18
19
20
Apatite.
21
22
23
24
25
26

Pl. 16.

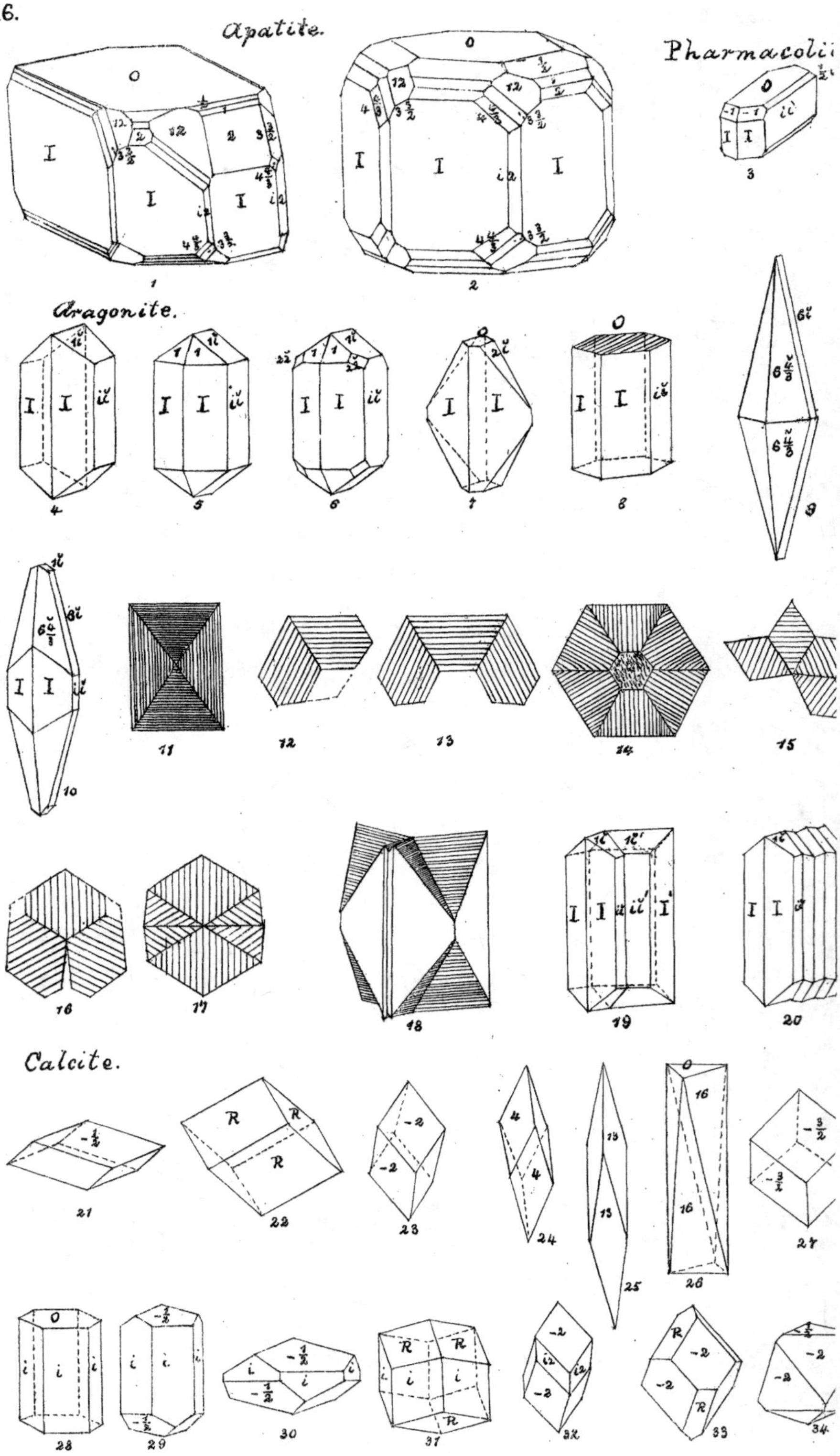

Apatite.

1 2

Pharmacoli

3

Aragonite.

4 5 6 7 8 9

10 11 12 13 14 15

16 17 18 19 20

Calcite.

21 22 23 24 25 26 27

28 29 30 31 32 33 34

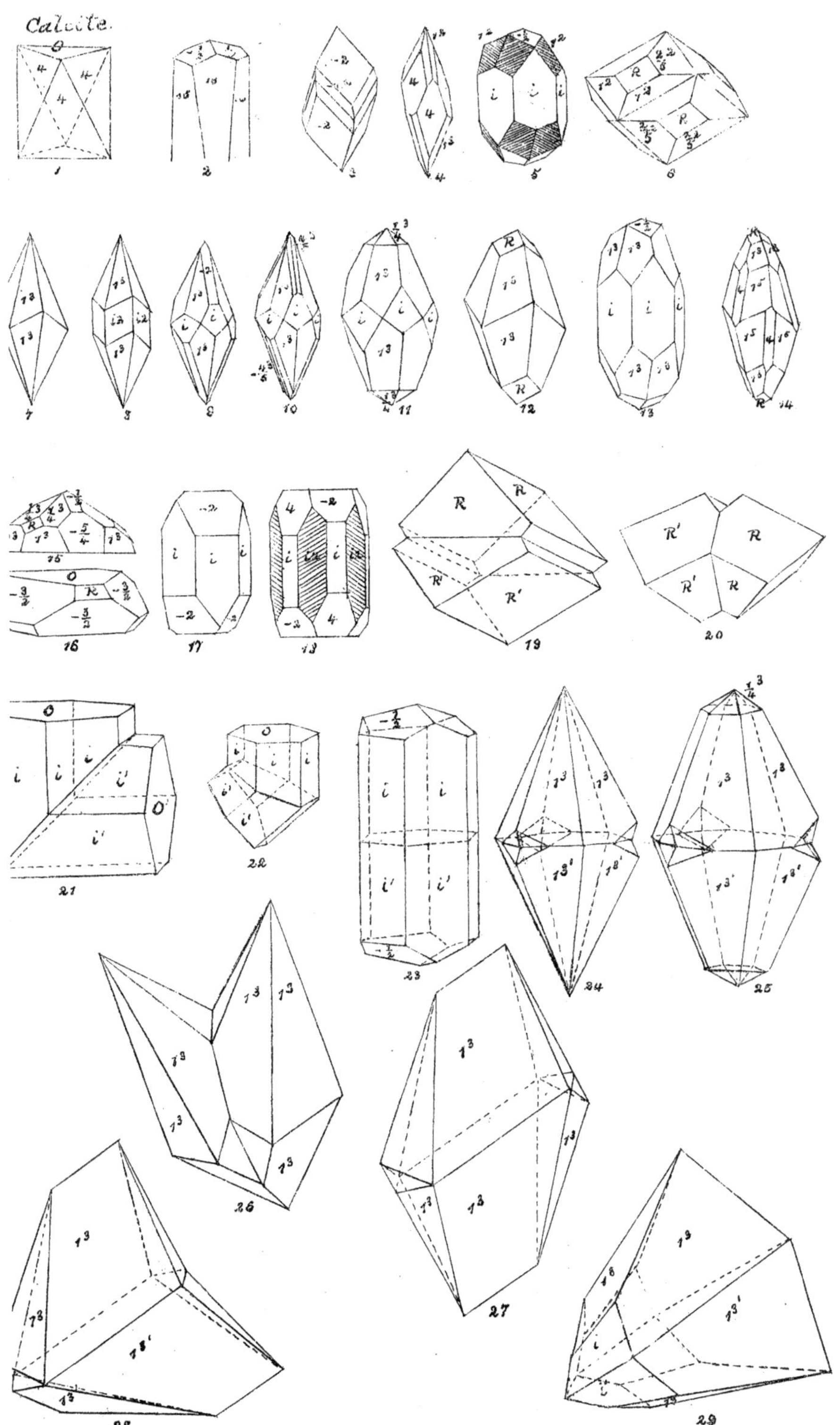

c.

Pl. 18.

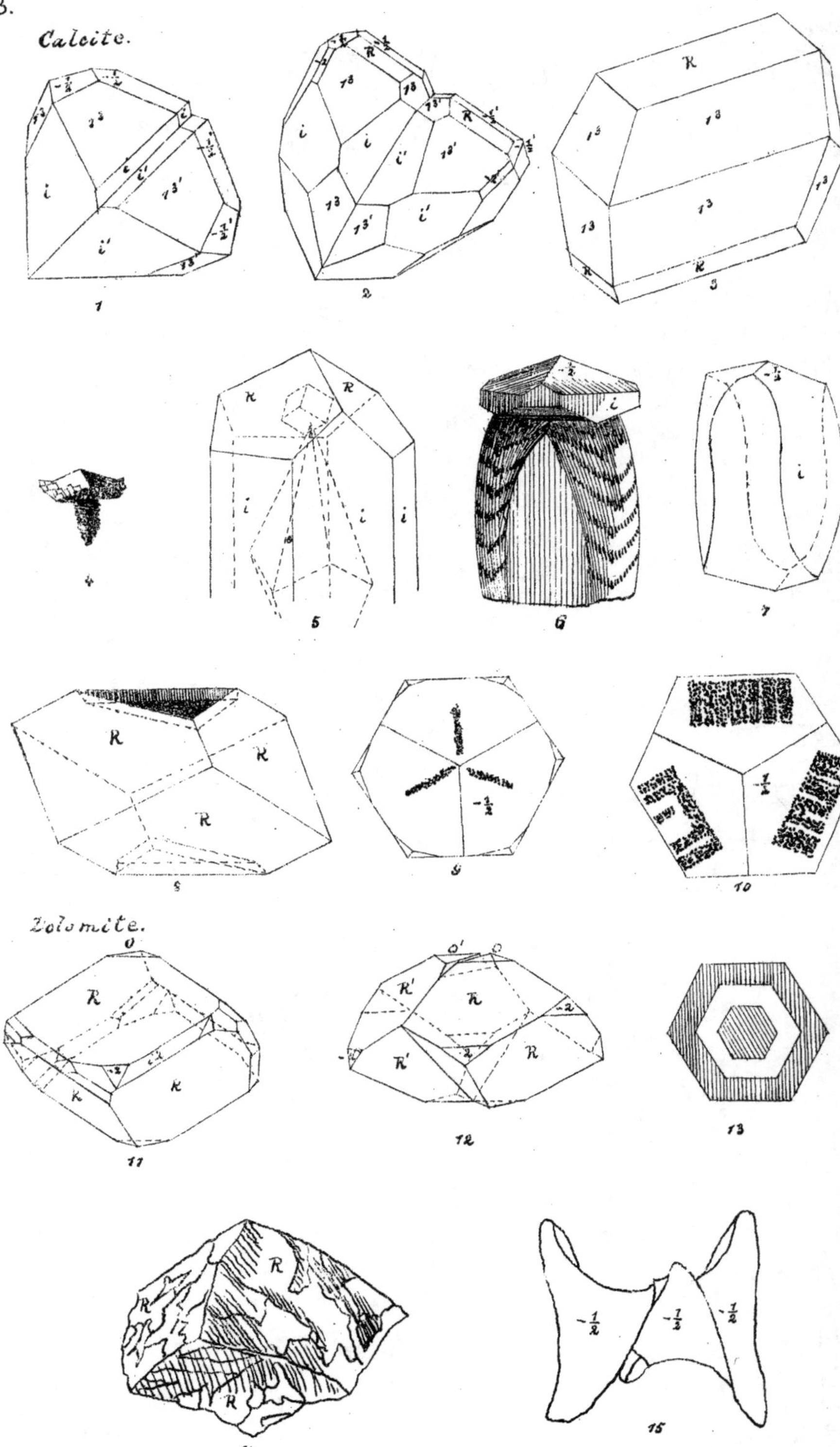

C.

Scheelite.
1
3
4
5
6
7
8
9
10
Brucite.
11
12
Epsomite.
13
14
Boracite.
15
16
17
18
19
20
21
Spinel.
22
23
24
25
26

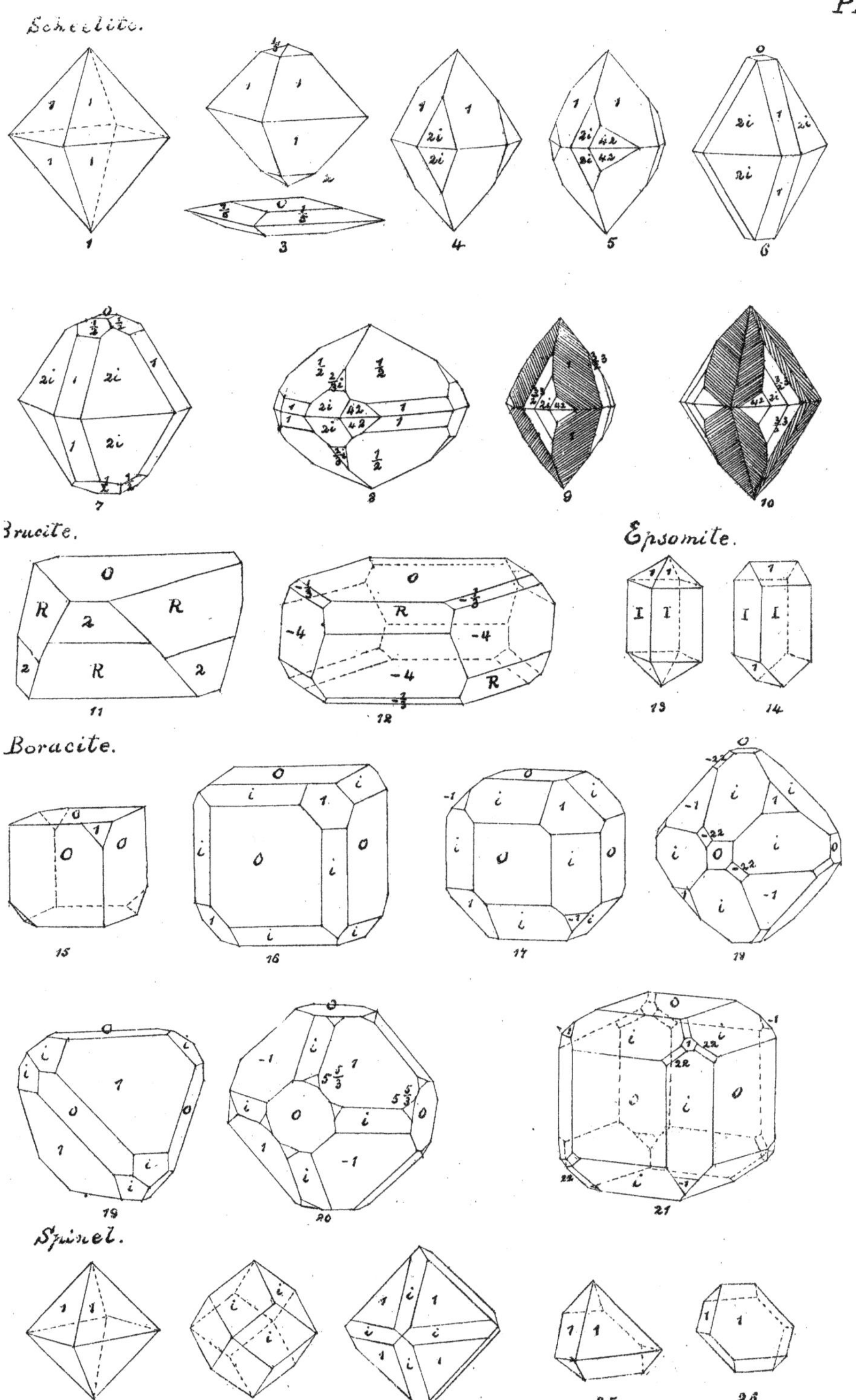

c.

Pl. 20.

Spinel.
1
2
3
4
5
Corundum.
6
7
8
9
10
11
12
13
14
Diaspore.
15
16
Alunite.
17
18
19
Cryolite.
20
21
Wavellite.
22

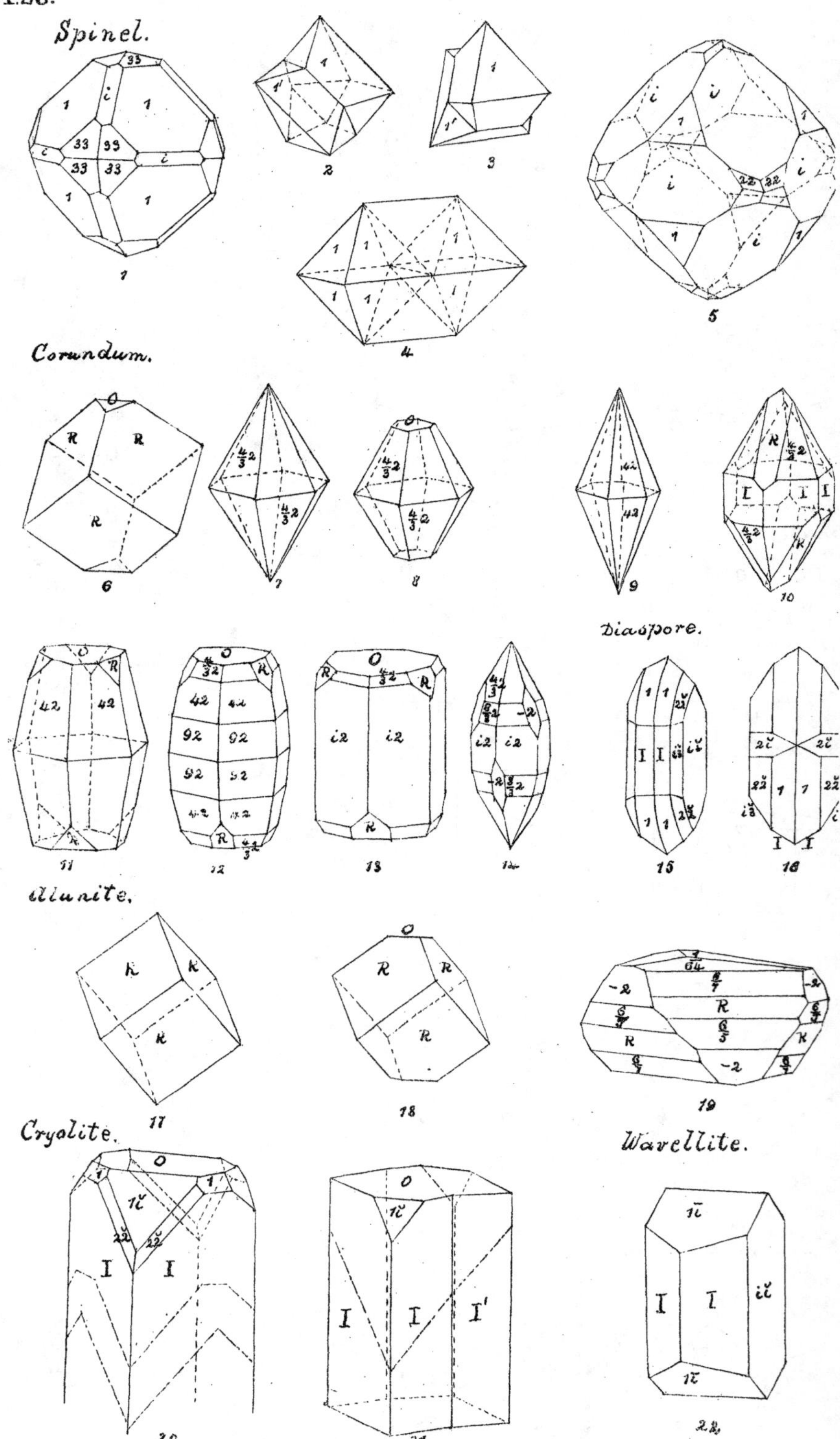

c.

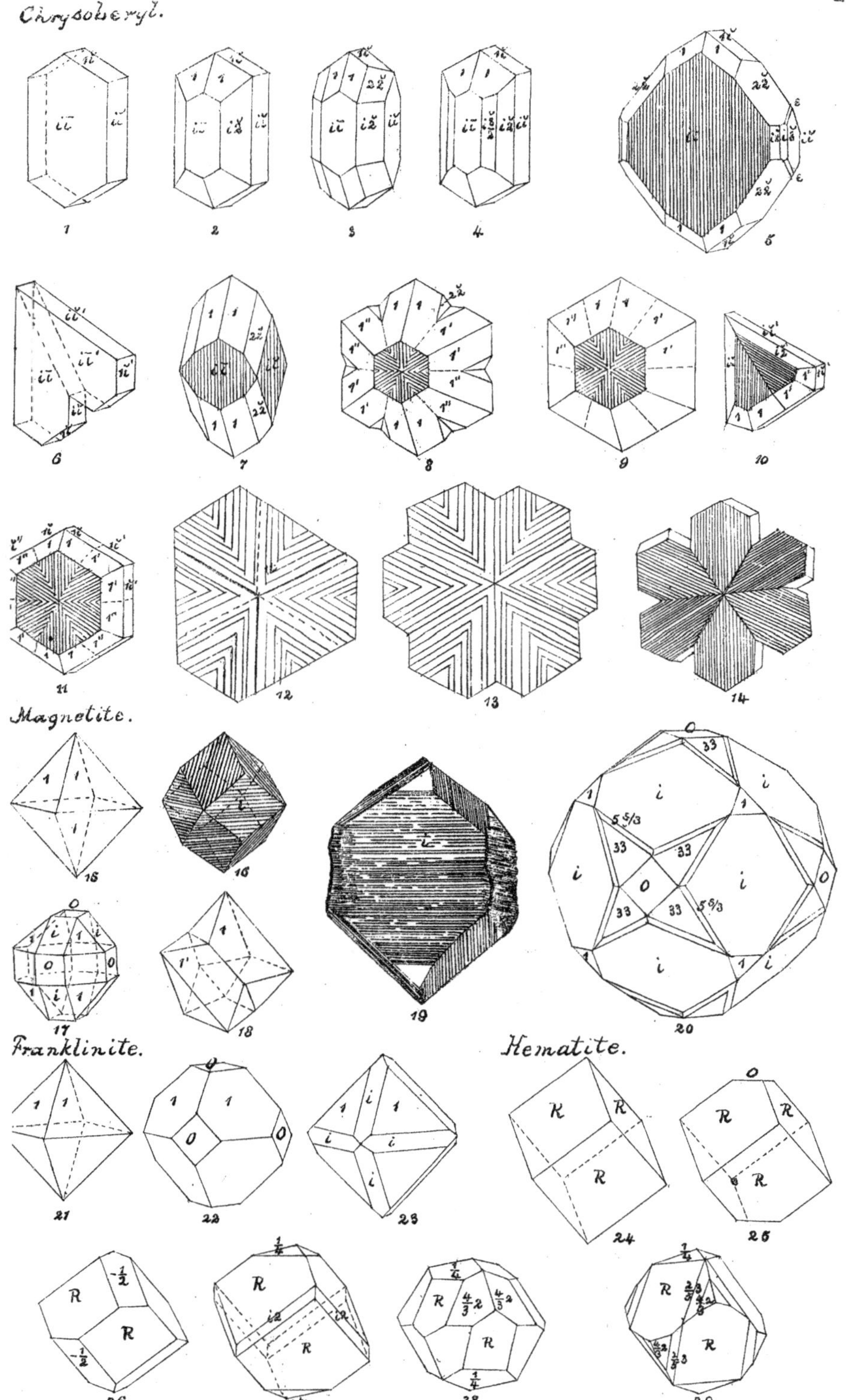
Chrysoberyl.
Magnetite.
Franklinite.
Hematite.

Pl.22.

Hematite.
1
2
3
4
5
6
7
8
Goethite.
9
10
11
Pyrrhotite.
12
Pyrite.
13
14
15
16
17
18
19
20
21
22
23
24
25

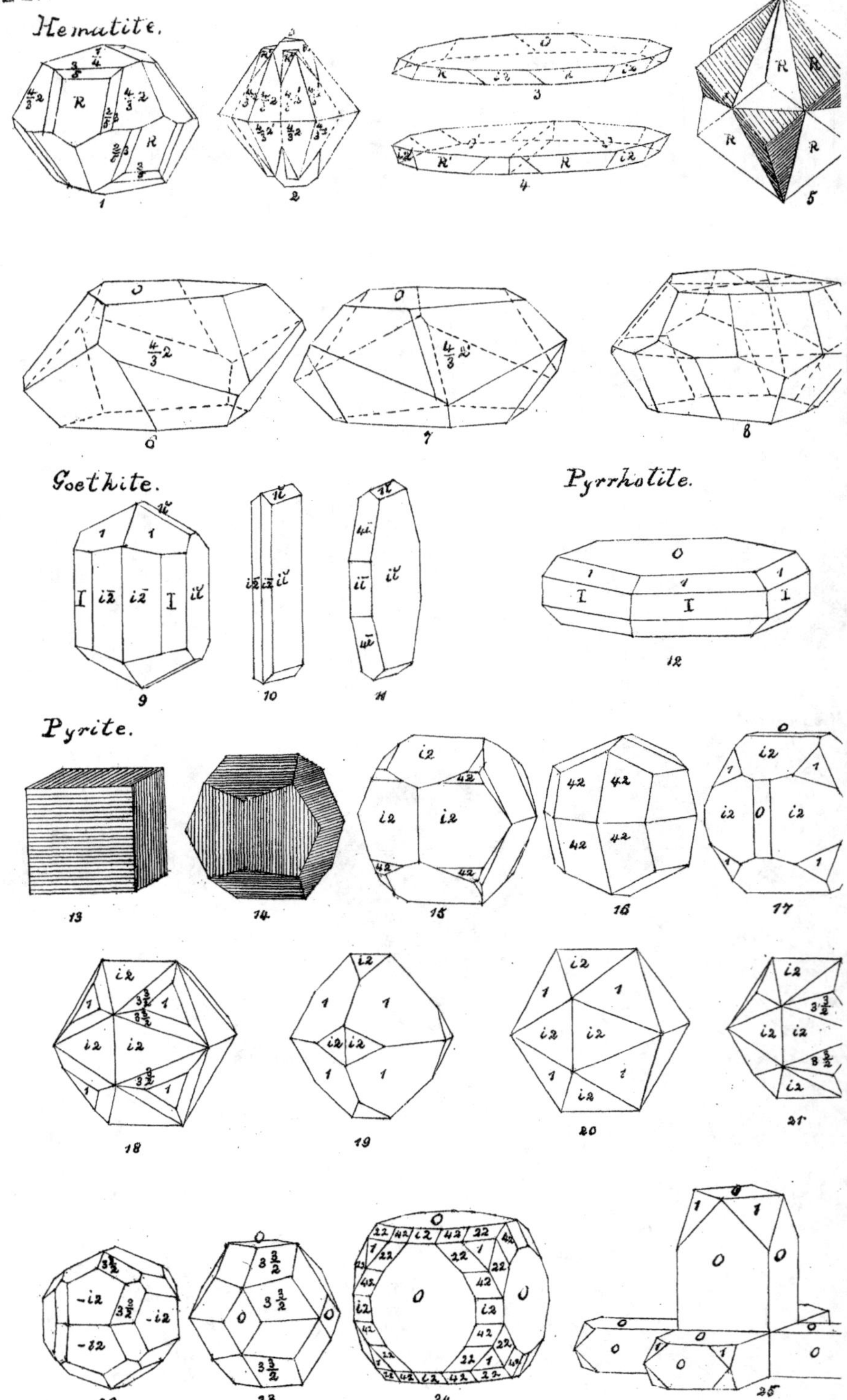

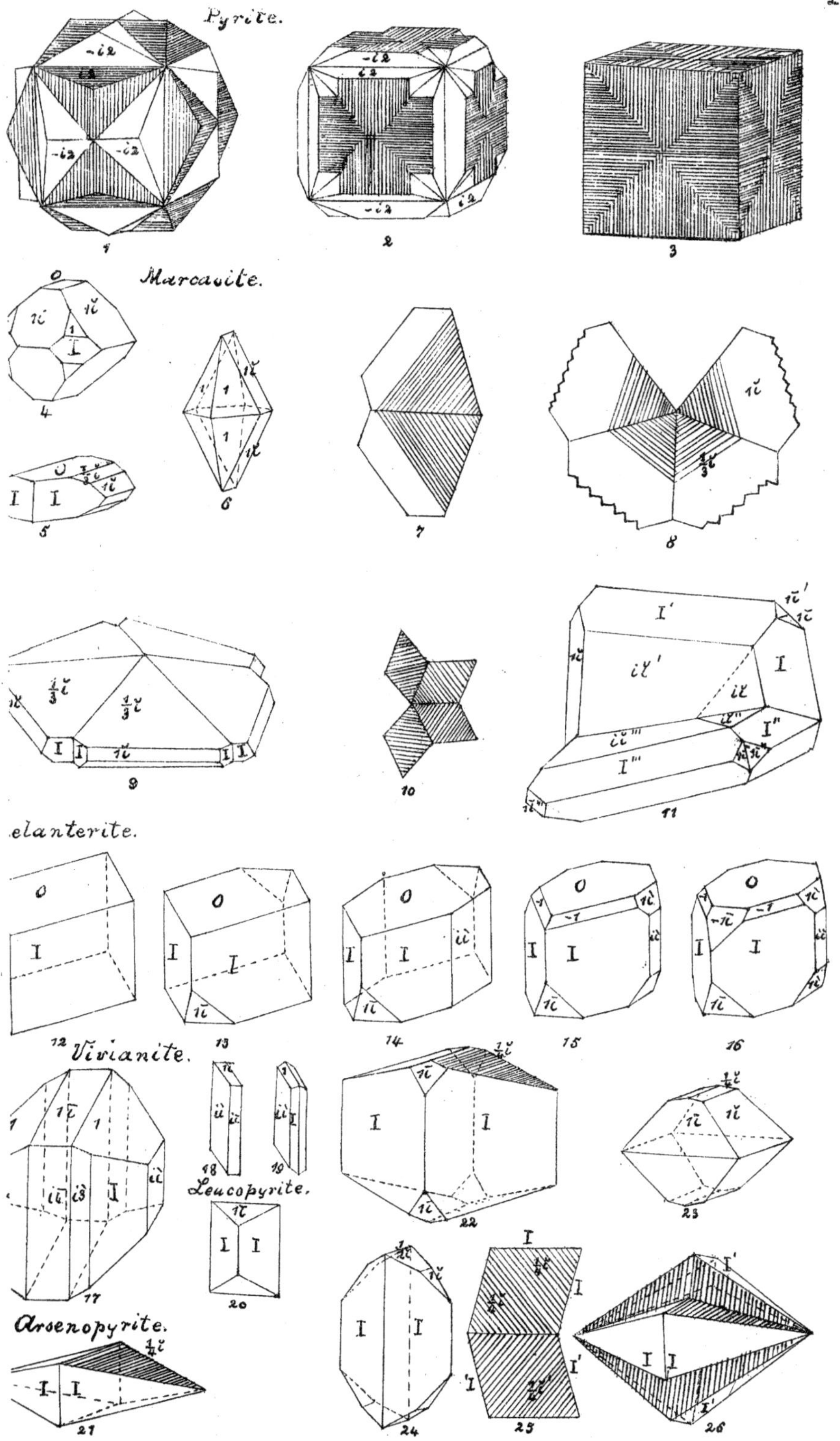
Pyrite.
Marcasite.
elanterite.
Vivianite.
Leucopyrite.
Arsenopyrite.

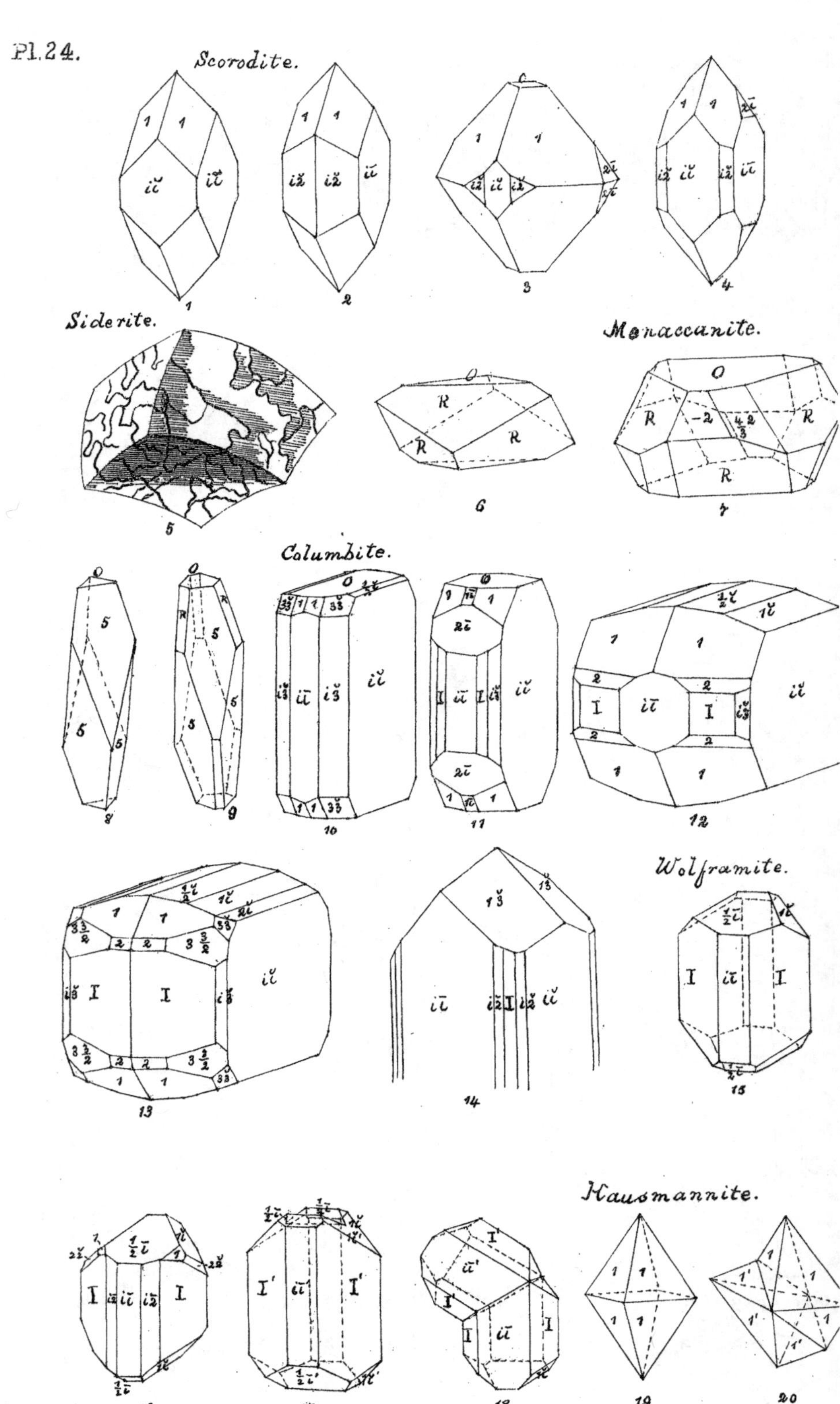
Pl. 24.
Scorodite.
Siderite.
Menaccanite.
Columbite.
Wolframite.
Hausmannite.

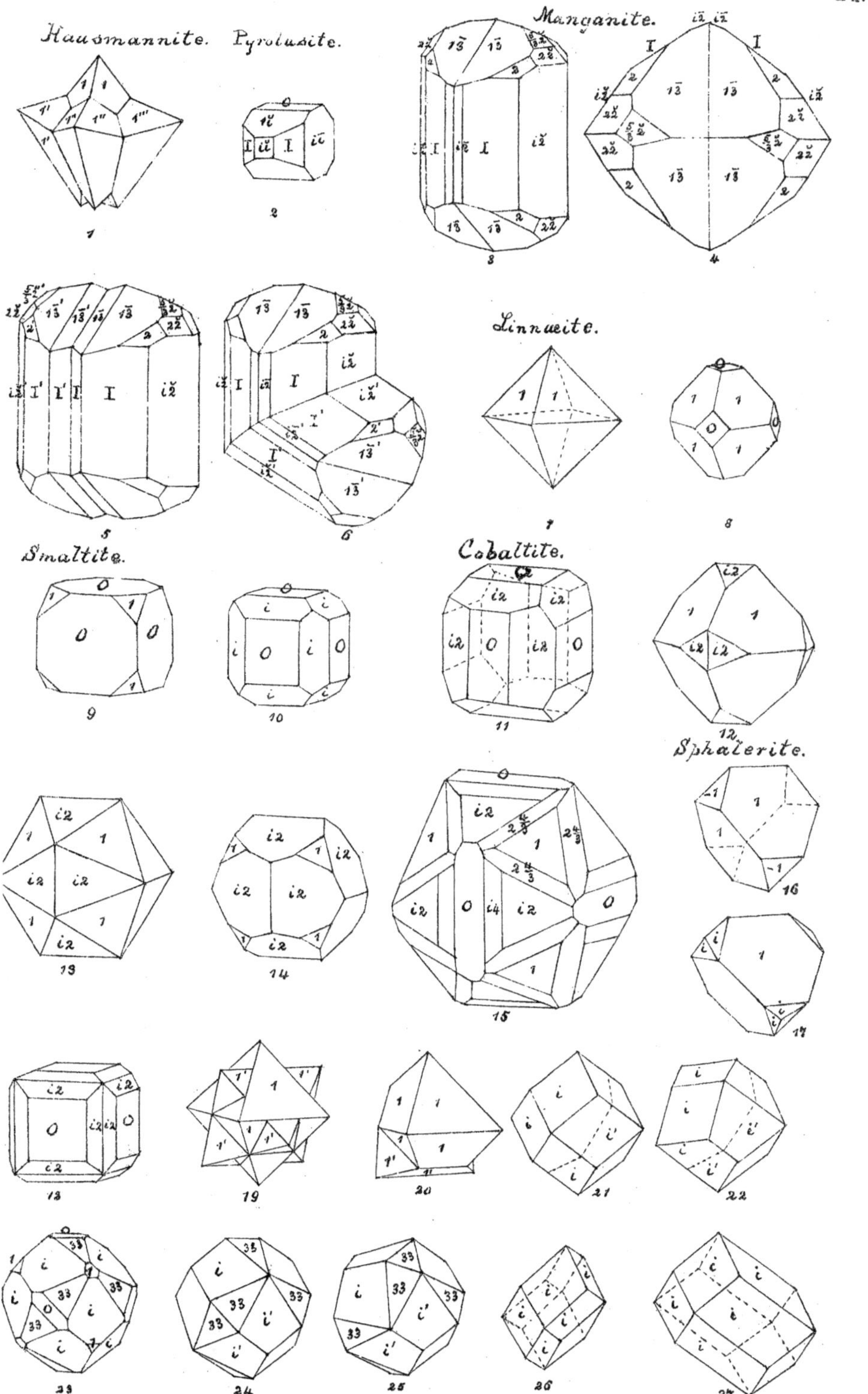
Hausmannite.
Pyrolusite.
Manganite.
1
2
3
4
5
6
Linnæite.
7
8
Smaltite.
Cobaltite.
9
10
11
12
Sphalerite.
13
14
15
16
17
18
19
20
21
22
23
24
25
26
27

Cassiterite.

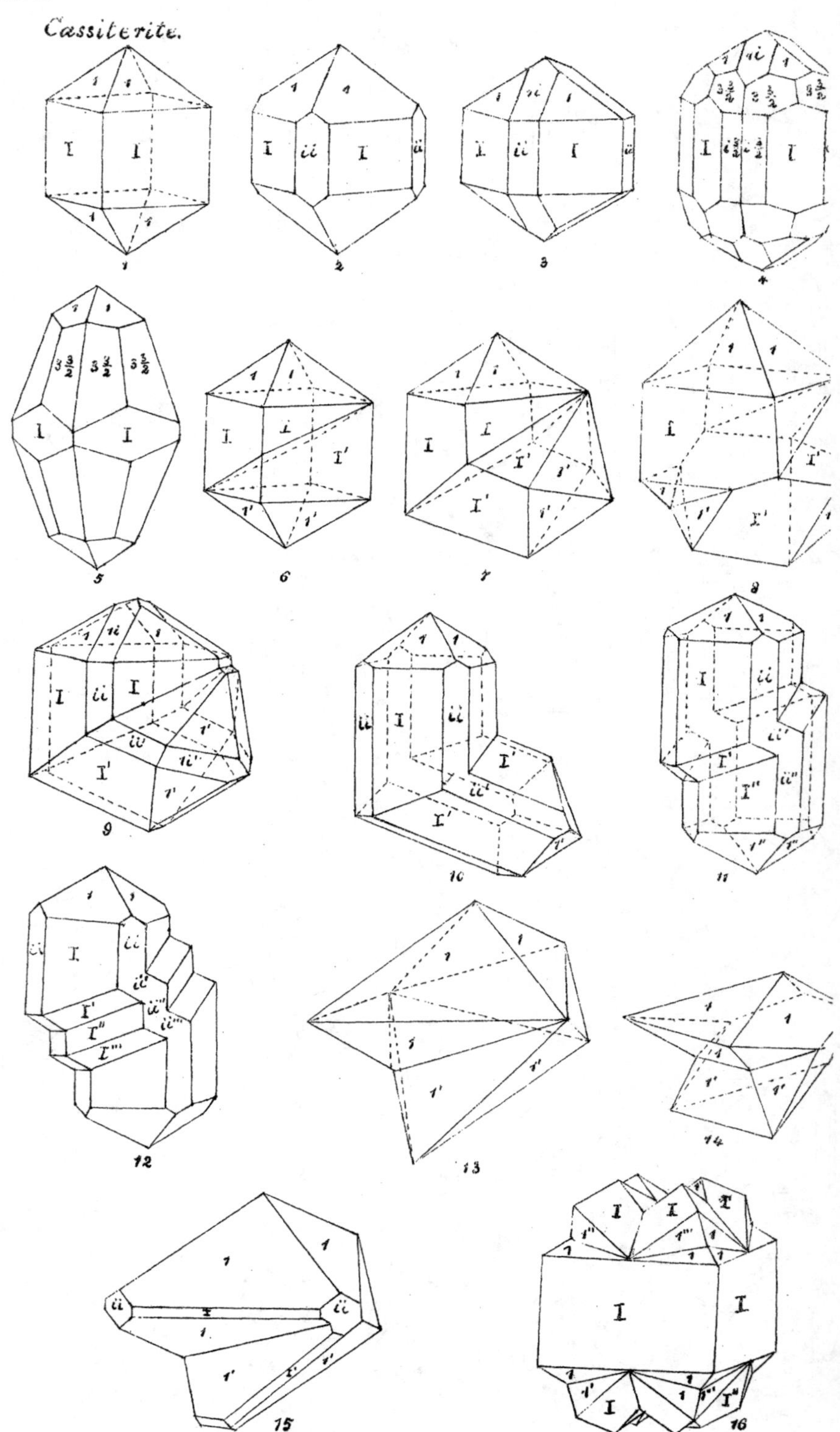

C.

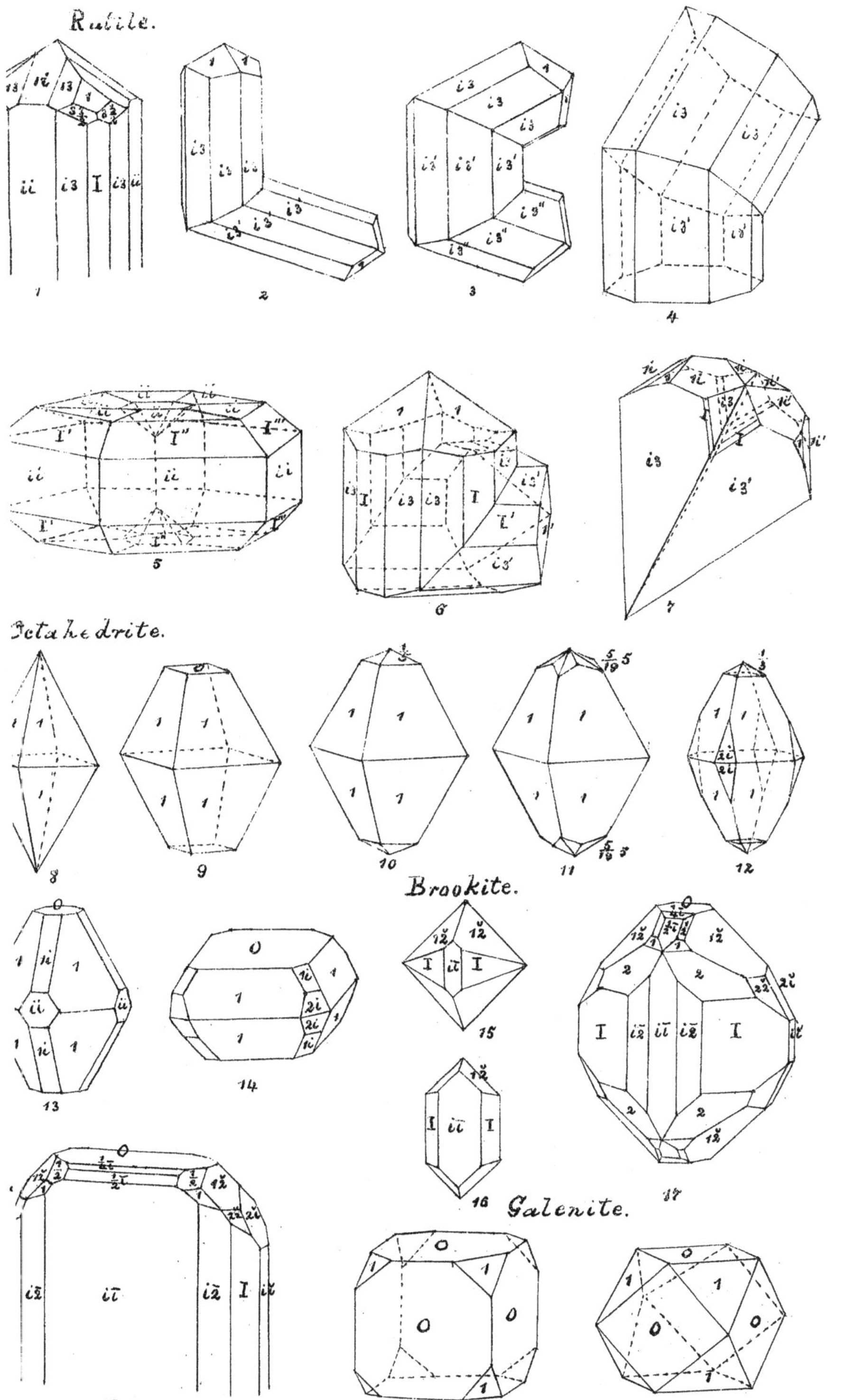
Rutile.
1
2
3
4
5
6
7
Octahedrite.
8
9
10
11
12
Brookite.
13
14
15
16
17
18
Galenite.
19
20

Pl. 38.

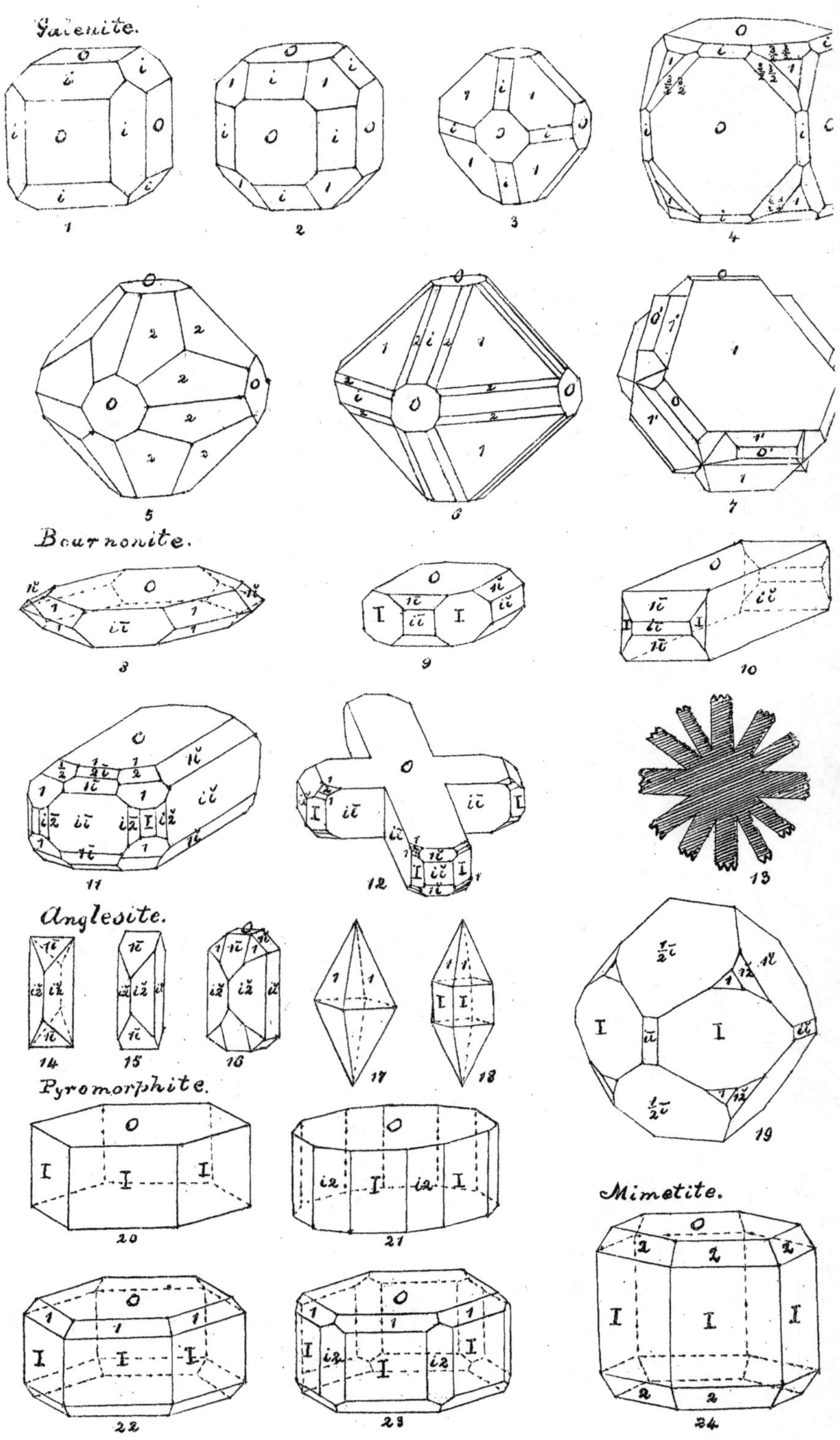

c.

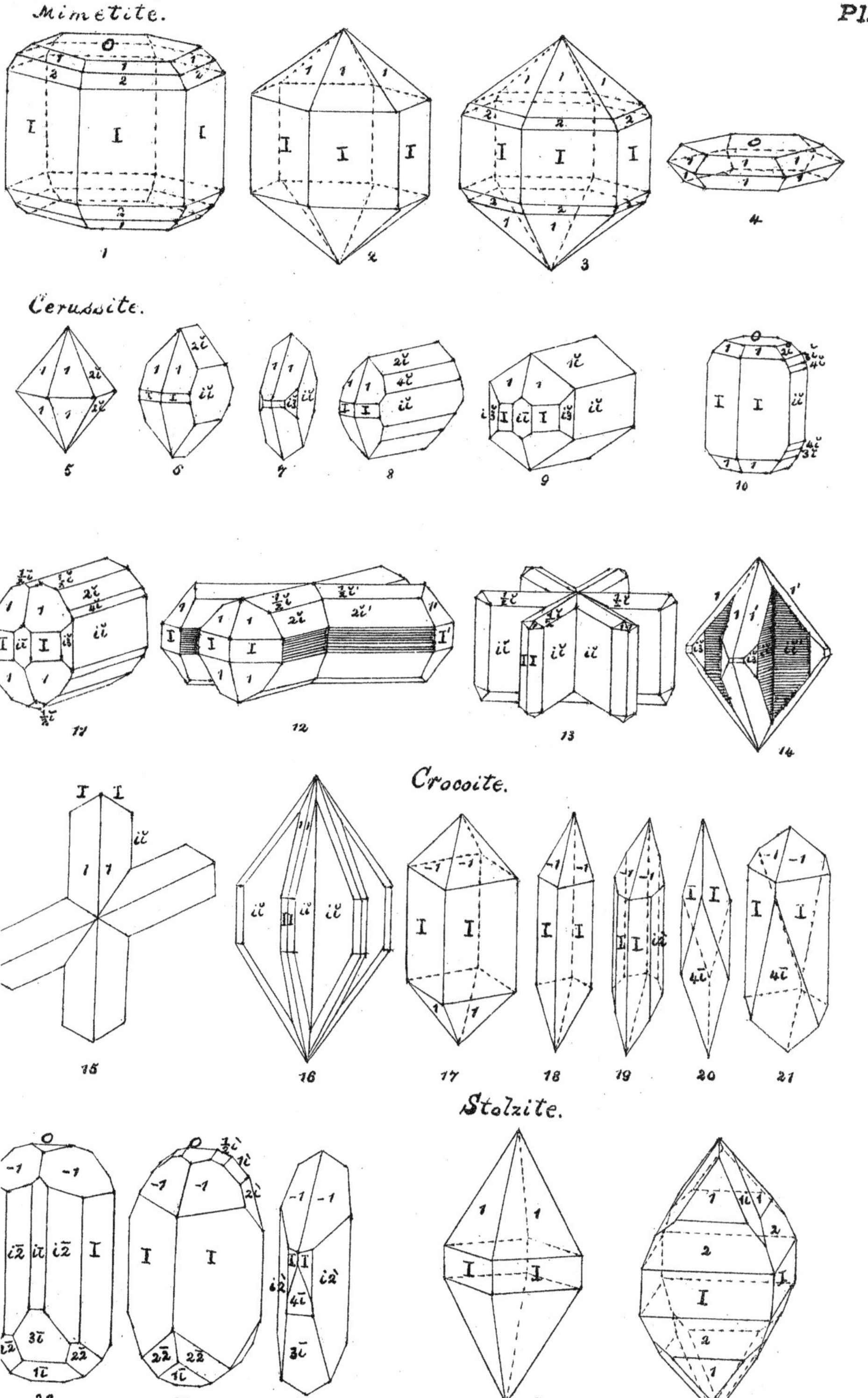
Mimetite.
Cerussite.
Crocoite.
Stolzite.

Pl. 30.

Wulfenite.
Tetradymite.
Realgar.
Orpiment.
Senarmontite.
Valentinite.
Stibnite.
Torbernite.

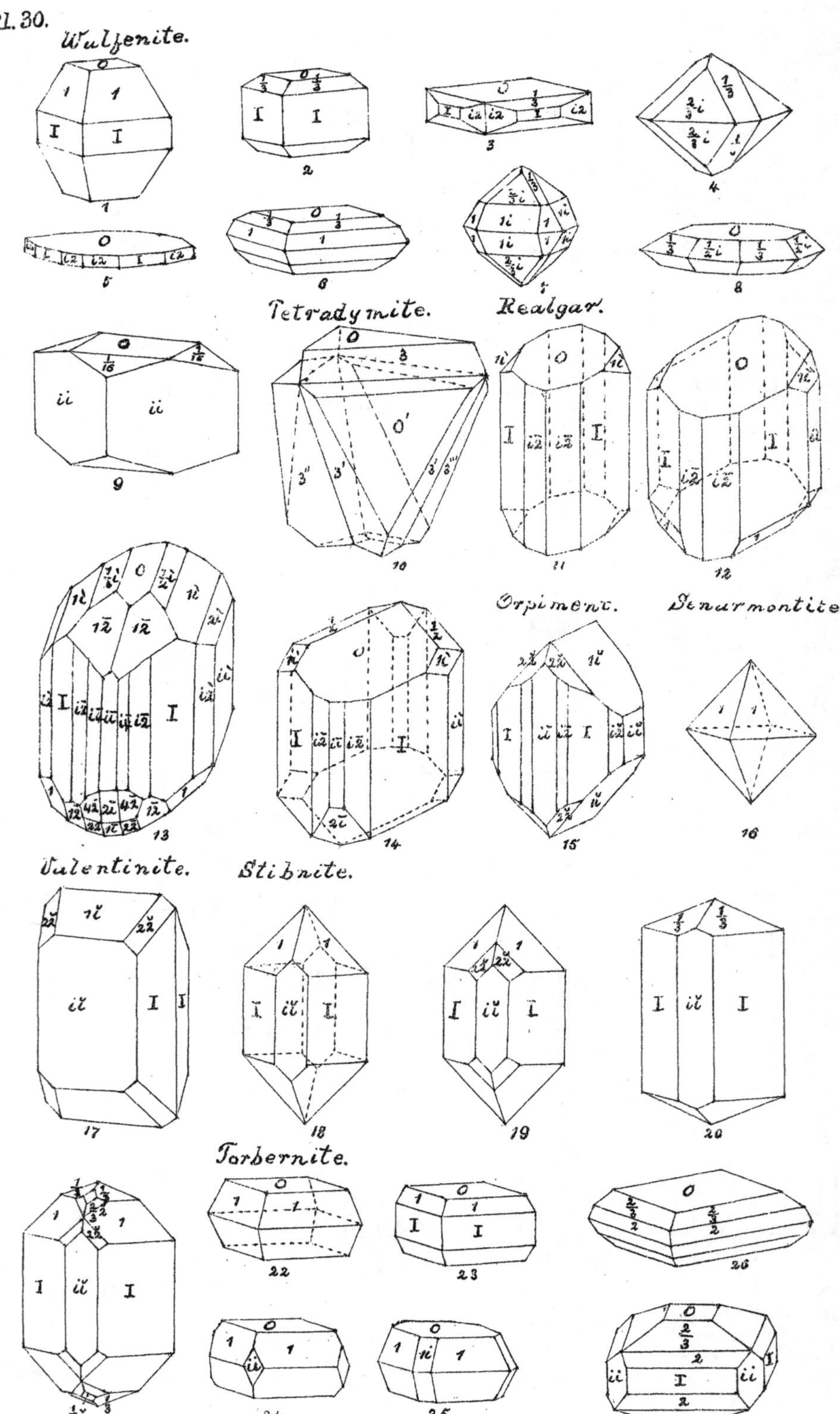

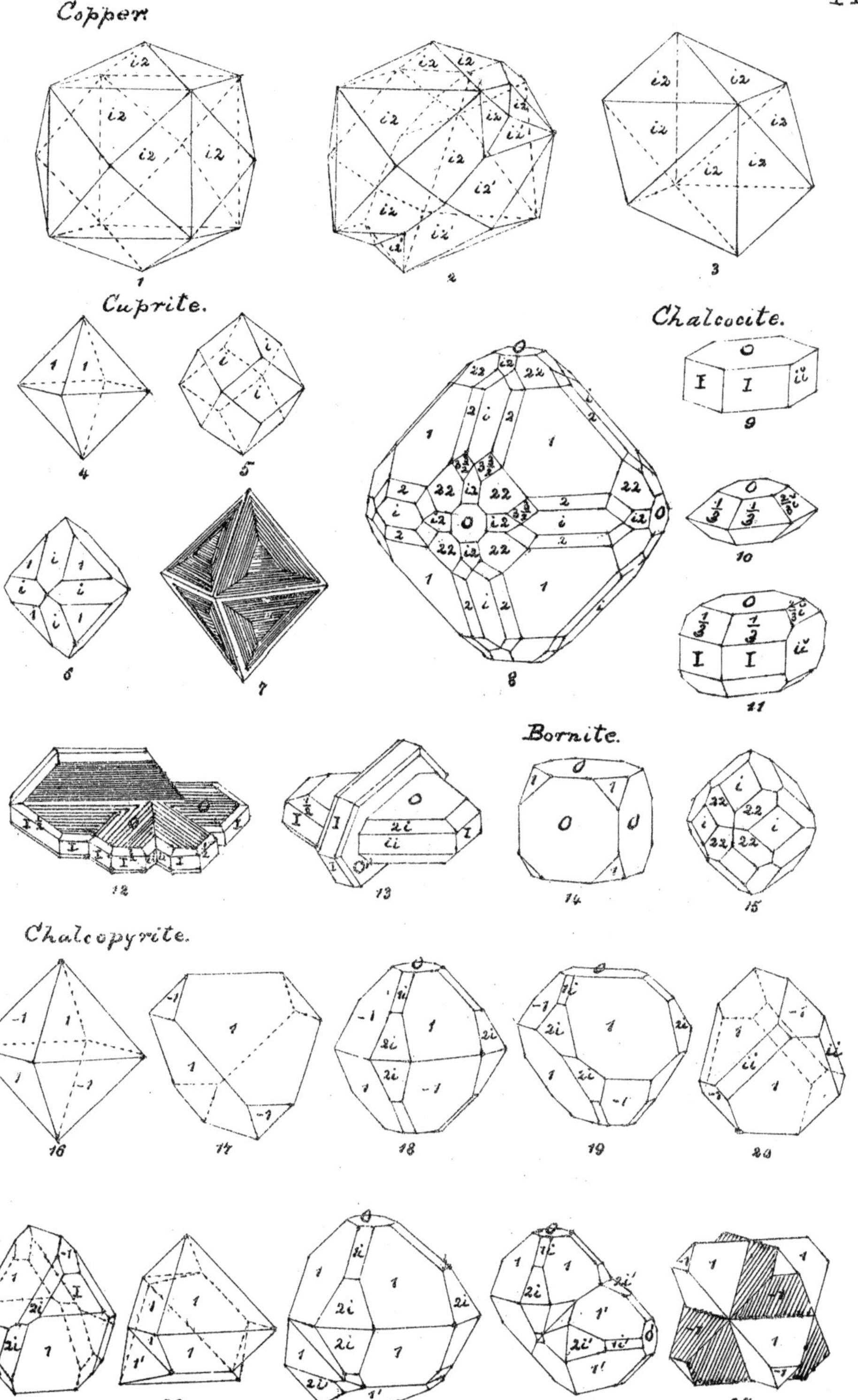
Copper.
Cuprite.
Chalcocite.
Bornite.
Chalcopyrite.

Pl. 32.
Chalcopyrite.
Tetrahedrite.
Chalcanthite.
Brochantite.
Atacamite.
Libethenite.
Olivenite.
Malachite.
Liroconite.
Azurite.

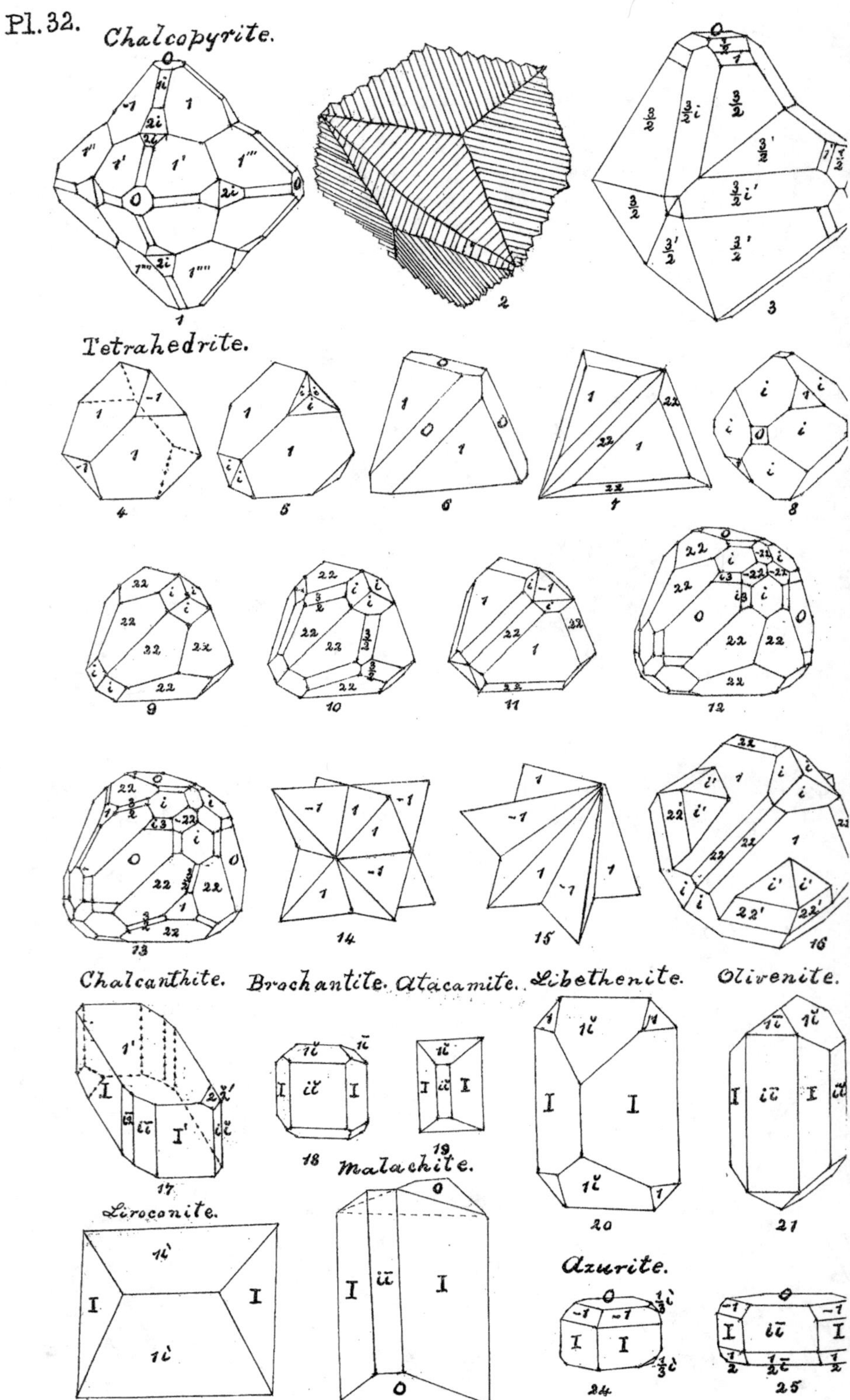

c.

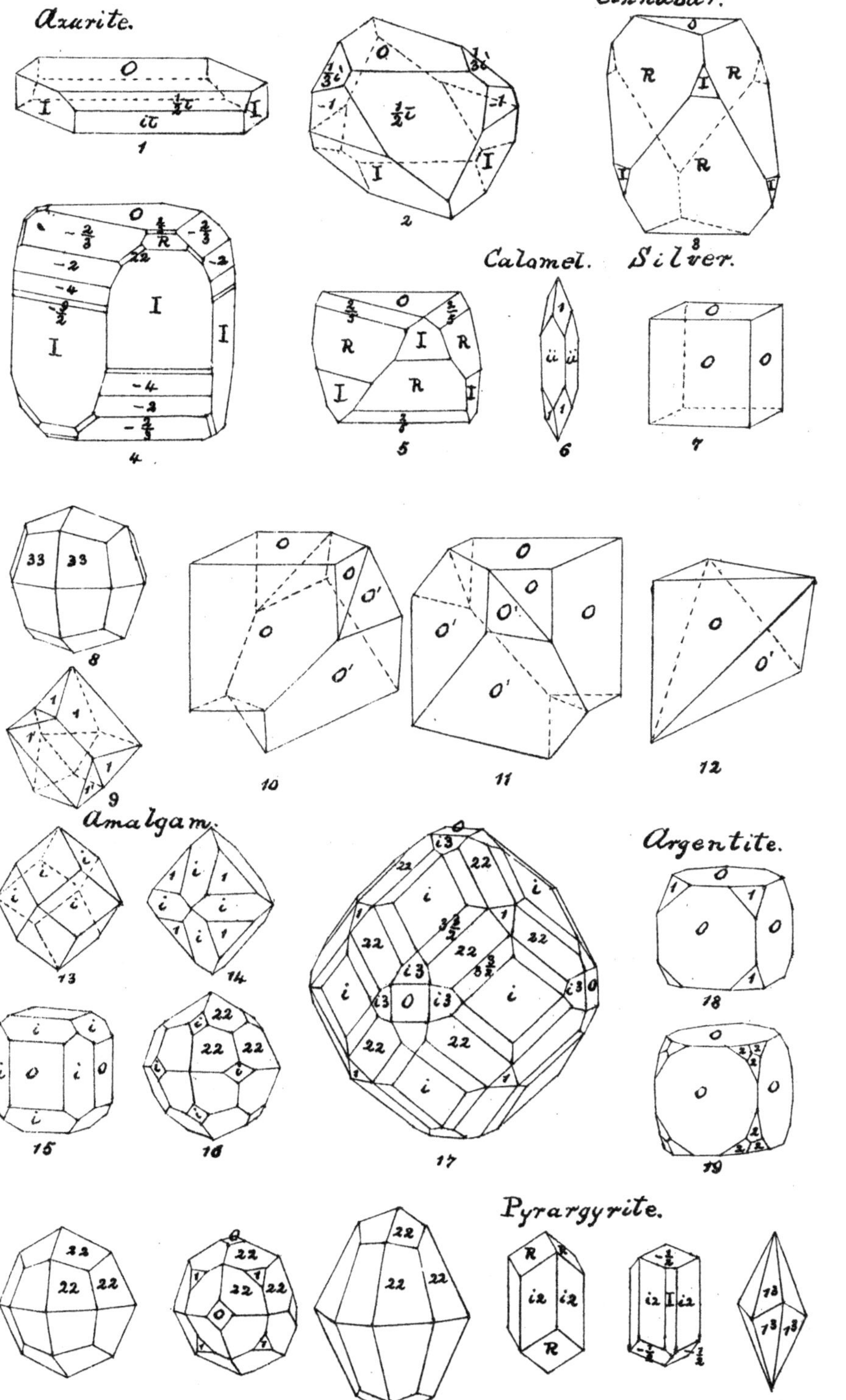
Azurite.
1
2
Cinnabar.
3
4
5
Calomel.
6
Silver.
7
8
9
10
11
12
Amalgam.
13
14
15
16
17
Argentite.
18
19
Pyrargyrite.
22
25

Pl.34.

Pyrargyrite.
Stephanite.
Cerargyrite.
Bromyrite.
Embolite.
Gold.
Iridosmine.
J.H.C.
1870.

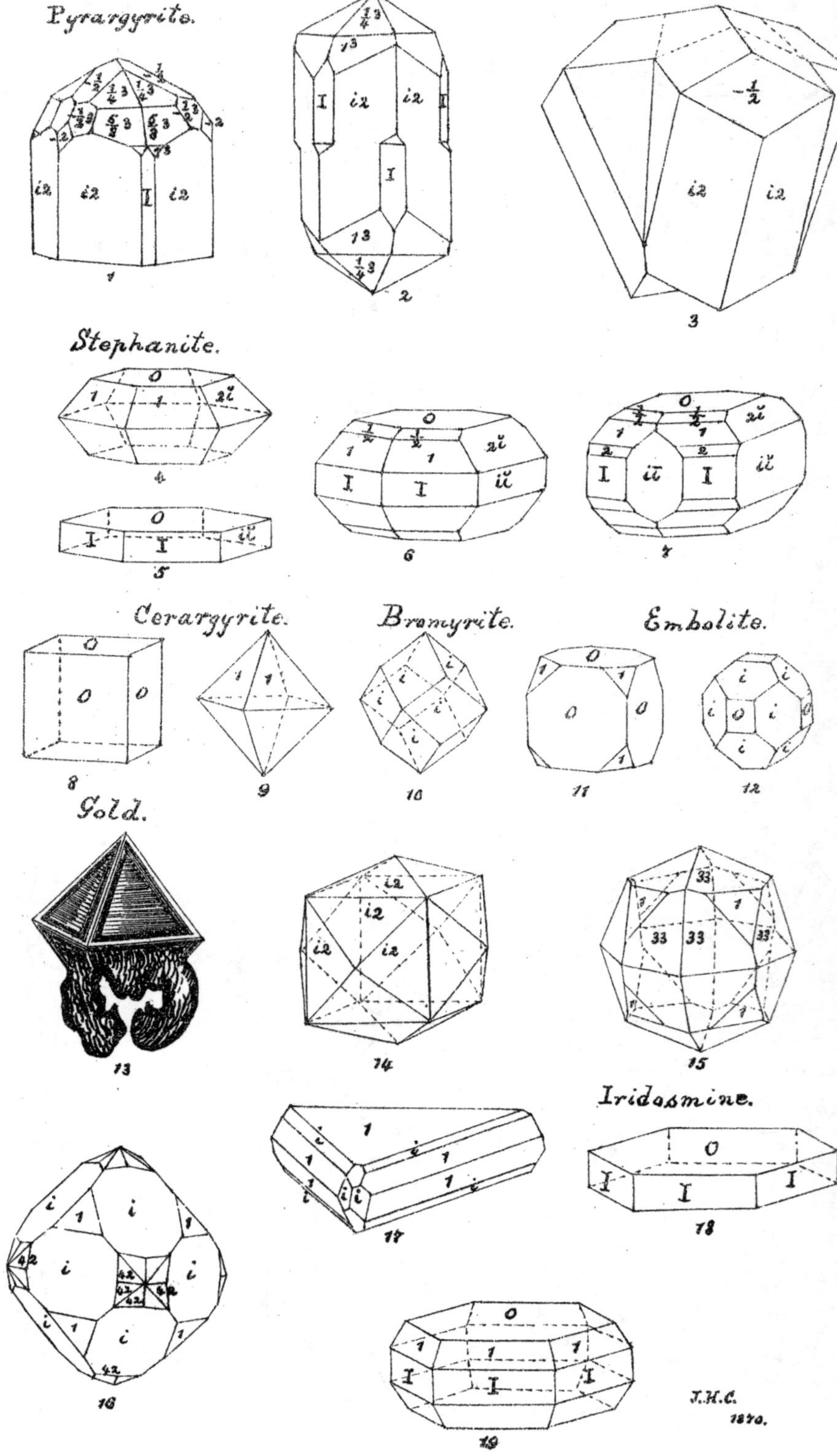

INDEX TO PLATES.

SCIENTIFIC BOOKS,

PUBLISHED BY

D. VAN NOSTRAND,

23 Murray and 27 Warren St., New York.

FRANCIS' (J. B.) Hydraulic Experiments. Lowell Hydraulic Experiments—being a Selection from Experiments on Hydraulic Motors, on the Flow of Water over Weirs, and in Open Canals of Uniform Rectangular Section, made at Lowell, Mass. By J. B. FRANCIS, Civil Engineer. Second edition, revised and enlarged, including many New Experiments on Gauging Water in Open Canals, and on the Flow through Submerged Orifices and Diverging Tubes. With 23 copperplates, beautifully engraved, and about 100 new pages of text. 1 vol., 4to. Cloth. $15.

Most of the practical rules given in the books on hydraulics have been determined from experiments made in other countries, with insufficient apparatus, and on such a minute scale, that in applying them to the large operations arising in practice in this country, the engineer cannot but doubt their reliable applicability. The parties controlling the great water-power furnished by the Merrimack River at Lowell, Massachusetts, felt this so keenly, that they have deemed it necessary, at great expense, to determine anew some of the most important rules for gauging the flow of large streams of water, and for this purpose have caused to be made, with great care, several series of experiments on a large scale, a selection from which are minutely detailed in this volume.

The work is divided into two parts—PART I., on hydraulic motors, includes ninety-two experiments on an improved Fourneyron Turbine Water-Wheel, of about two hundred horse-power, with rules and tables for the construction of similar motors:—Thirteen experiments on a model of a centre-vent water-wheel of the most simple design, and thirty-nine experiments on a centre vent water-wheel of about two hundred and thirty horse-power.

PART II. includes seventy-four experiments made for the purpose of determining the form of the formula for computing the flow of water over weirs; nine experiments on the effect of back water on the flow over weirs; eighty-eight experiments made for the purpose of determining the formula for computing the flow over weirs of regular or standard forms, with several tables of comparisons of the new formula with the results obtained by former experimenters; five experiments on the flow over a dam in which the crest was of the same form as that built by the Essex Company across the Merrimack River at Lawrence, Massachusetts; twenty-one experiments on the effect of observing the depths of water on a weir at different distances from the weir; an extensive series of experiments made for the purpose of determining rules for gauging streams of water in open canals, with tables for facilitating the same; and one hundred and one experiments on the discharge of water through submerged orifices and diverging tubes, the whole being fully illustrated by twenty-three double plates engraved on copper.

In 1855 the proprietors of the Locks and Canals on Merrimack River, at whose expense most of the experiments were made, being willing that the public should share the benefits of the scientific operations promoted by them, consented to the publication of the first edition of this work, which contained a selection of the most important hydraulic experiments made at Lowell up to that time. In this second edition the principal hydraulic experiments made there, subsequent to 1855, have been added, including the important series above mentioned, for determining rules for the gauging the flow of water in open canals, and the interesting series on the flow through a submerged Venturi's tube, in which a larger flow was obtained than any we find recorded.

FRANCIS (J. B.) On the Strength of Cast-Iron Pillars, with Tables for the use of Engineers, Architects, and Builders. By JAMES B. FRANCIS, Civil Engineer. 1 vol., 8vo. Cloth. $2.

PEIRCE'S SYSTEM OF ANALYTIC MECHANICS. Physical and Celestial Mechanics, by BENJAMIN PEIRCE, Perkins Professor of Astronomy and Mathematics in Harvard University, and Consulting Astronomer of the American Ephemeris and Nautical Almanac. Developed in four systems of Analytic Mechanics, Celestial Mechanics, Potential Physics, and Analytic Morphology. 1 vol., 4to. Cloth. $10.

GILLMORE. Practical Treatise on Limes, Hydraulic Cements, and Mortars. Papers on Practical Engineering, U. S. Engineer Department, No. 9, containing Reports of numerous experiments conducted in New York City, during the years 1858 to 1861, inclusive. By Q. A. GILLMORE, Brig.-General U. S. Volunteers, and Major U. S. Corps of Engineers. With numerous illustrations. One volume, octavo. Cloth. $4.

ROGERS (H. D.) Geology of Pennsylvania. A complete Scientific Treatise on the Coal Formations. By HENRY D. ROGERS, Geologist. 3 vols., 4to., plates and maps. Boards. $30.00.

BURGH (N. P.) Modern Marine Engineering, applied to Paddle and Screw Propulsion. Consisting of 36 colored plates, 259 Practical Woodcut Illustrations, and 403 pages of Descriptive Matter, the whole being an exposition of the present practice of the following firms: Messrs. J. Penn & Sons; Messrs. Maudslay, Sons, & Field; Messrs. James Watt & Co.; Messrs. J. & G. Rennie; Messrs. R. Napier & Sons; Messrs. J. & W. Dudgeon; Messrs. Ravenhill & Hodgson; Messrs. Humphreys & Tenant; Mr. J. T. Spencer, and Messrs. Forrester & Co. By N. P. BURGH, Engineer. In one thick vol., 4to. Cloth. $25.00. Half morocco. $30.00.

KING. Lessons and Practical Notes on Steam, the Steam-Engine, Propellers, &c., &c., for Young Marine Engineers, Students, and others. By the late W. R. KING, U. S. N. Revised by Chief-Engineer J. W. KING, U. S. Navy. Twelfth edition, enlarged. 8vo. Cloth. $2.

WARD. Steam for the Million. A Popular Treatise on Steam and its Application to the Useful Arts, especially to Navigation. By J. H. WARD, Commander U. S. Navy. New and revised edition. 1 vol., 8vo. Cloth. $1.

WALKER. Screw Propulsion. Notes on Screw Propulsion, its Rise and History. By Capt. W. H. WALKER, U. S. Navy. 1 vol., 8vo. Cloth. 75 cents.

THE STEAM-ENGINE INDICATOR, and the Improved Manometer Steam and Vacuum Gauges: Their Utility and Application. By PAUL STILLMAN. New edition. 1 vol., 12mo. Flexible cloth. $1.

ISHERWOOD. Engineering Precedents for Steam Machinery. Arranged in the most practical and useful manner for Engineers. By B. F. ISHERWOOD, Civil Engineer U. S. Navy. With illustrations. Two volumes in one. 8vo. Cloth. $2.50.

HOLLEY'S RAILWAY PRACTICE. American and European Railway Practice, in the Economical Generation of Steam, including the materials and construction of Coal-burning Boilers, Combustion, the Variable Blast, Vaporization, Circulation, Superheating, Supplying and Heating Feed-water, &c., and the adaptation of Wood and Coke-burning Engines to Coal-burning; and in Permanent Way, including Road-bed, Sleepers, Rails, Joint Fastenings, Street Railways, &c., &c. By ALEXANDER L. HOLLEY, B. P. With 77 lithographed plates. 1 vol., folio. Cloth. $12.

"This is an elaborate treatise by one of our ablest civil engineers, on the construction and use of locomotives, with a few chapters on the building of Railroads. * * * All these subjects are treated by the author, who is a first-class railroad engineer, in both an intelligent and intelligible manner. The facts and ideas are well arranged, and presented in a clear and simple style, accompanied by beautiful engravings, and we presume the work will be regarded as indispensable by all who are interested in a knowledge of the construction of railroads and rolling stock, or the working of locomotives."—*Scientific American.*

HENRICI (OLAUS). Skeleton Structures, especially in their Application to the Building of Steel and Iron Bridges. By OLAUS HENRICI. With folding plates and diagrams. 1 vol., 8vo. Cloth. $3.

WHILDEN (J. K.) On the Strength of Materials used in Engineering Construction. By J. K. WHILDEN. 1 vol., 12mo. Cloth. $2.

"We find in this work tables of the tensile strength of timber, metals, stones, wire, rope, hempen cable, strength of thin cylinders of cast-iron; modulus of elasticity, strength of thick cylinders, as cannon, &c., effects of reheating, &c., resistance of timber, metals, and stone to crushing; experiments on brick-work; strength of pillars; collapse of tube; experiments on punching and shearing; the transverse strength of materials; beams of uniform strength; table of coefficients of timber, stone, and iron; relative strength of weight in cast-iron, transverse strength of alloys; experiments on wrought and cast-iron beams: lattice girders, trussed cast-iron girders; deflection of beams; torsional strength and torsional elasticity."—*American Artisan.*

CAMPIN (F.) On the Construction of Iron Roofs. A Theoretical and Practical Treatise. By FRANCIS CAMPIN. With wood-cuts and plates of Roofs lately executed. Large 8vo. Cloth. $3.

BROOKLYN WATER-WORKS AND SEWERS. Containing a Descriptive Account of the Construction of the Works, and also Reports on the Brooklyn, Hartford, Belleville, and Cambridge Pumping Engines. Prepared and printed by order of the Board of Water Commissioners. With illustrations. 1 vol., folio. Cloth. $15.

ROEBLING (J. A.) Long and Short Span Railway Bridges. By JOHN A. ROEBLING, C. E. Illustrated with large copperplate engravings of plans and views. Imperial folio, cloth. $25.

CLARKE (T. C.) Description of the Iron Railway Bridge across the Mississippi River at Quincy, Illinois. By THOMAS CURTIS CLARKE, Chief Engineer. Illustrated with numerous lithographed plans. 1 vol., 4to. Cloth. $7.50.

WILLIAMSON (R. S.) On the Use of the Barometer on Surveys and Reconnaissances. Part I. Meteorology in its Connection with Hypsometry. Part II. Barometric Hypsometry. By R. S. WILLIAMSON, Bvt. Lieut.-Col. U. S. A., Major Corps of Engineers. With Illustrative Tables and Engravings. Paper No. 15, Professional Papers, Corps of Engineers. 1 vol., 4to. Cloth. $15.

"SAN FRANCISCO, CAL., *Feb.* 27, 1867.

"Gen. A. A. HUMPHREYS, Chief of Engineers, U. S. Army:

"GENERAL—I have the honor to submit to you, in the following pages, the results of my investigations in meteorology and hypsometry, made with the view of ascertaining how far the barometer can be used as a reliable instrument for determining altitudes on extended lines of survey and reconnaissances. These investigations have occupied the leisure permitted me from my professional duties during the last ten years, and I hope the results will be deemed of sufficient value to have a place assigned them among the printed professional papers of the United States Corps of Engineers. Very respectfully, your obedient servant,

"R. S. WILLIAMSON,

"Bvt. Lt.-Col. U. S. A., Major Corps of U. S. Engineers."

TUNNER (P.) A Treatise on Roll-Turning for the Manufacture of Iron. By PETER TUNNER. Translated and adapted. By JOHN B. PEARSE, of the Pennsylvania Steel Works. With numerous engravings and wood-cuts. 1 vol., 8vo., with 1 vol. folio of plates. Cloth. $10

SHAFFNER (T. P.) Telegraph Manual. A Complete History and Description of the Semaphoric, Electric, and Magnetic Telegraphs of Europe, Asia, and Africa, with 625 illustrations. By TAL. P. SHAFFNER, of Kentucky. New edition. 1 vol., 8vo. Cloth. 850 pp. $6.50.

MINIFIE (WM.) Mechanical Drawing. A Text-Book of Geometrical Drawing for the use of Mechanics and Schools, in which the Definitions and Rules of Geometry are familiarly explained; the Practical Problems are arranged, from the most simple to the more complex, and in their description technicalities are avoided as much as possible. With illustrations for Drawing Plans, Sections, and Elevations of Buildings and Machinery; an Introduction to Isometrical Drawing, and an Essay on Linear Perspective and Shadows. Illustrated with over 200 diagrams engraved on steel. By WM. MINIFIE, Architect. Seventh edition. With an Appendix on the Theory and Application of Colors. 1 vol., 8vo. Cloth. $4.

"It is the best work on Drawing that we have ever seen, and is especially a text-book of Geometrical Drawing for the use of Mechanics and Schools. No young Mechanic, such as a Machinist, Engineer, Cabinet-Maker, Millwright, or Carpenter should be without it."—*Scientific American.*

"One of the most comprehensive works of the kind ever published, and cannot but possess great value to builders. The style is at once elegant and substantial."—*Pennsylvania Inquirer.*

"Whatever is said is rendered perfectly intelligible by remarkably well-executed diagrams on steel, leaving nothing for mere vague supposition; and the addition of an introduction to isometrical drawing, linear perspective, and the projection of shadows, winding up with a useful index to technical terms."—*Glasgow Mechanics' Journal.*

☞ The British Government has authorized the use of this book in their schools of art at Somerset House, London, and throughout the kingdom.

MINIFIE (WM.) Geometrical Drawing. Abridged from the octavo edition, for the use of Schools. Illustrated with 48 steel plates. New edition, enlarged. 1 vol., 12mo, cloth. $2.

"It is well adapted as a text-book of drawing to be used in our High Schools and Academies where this useful branch of the fine arts has been hitherto too much neglected."—*Boston Journal.*

Scientific Books.

POOK'S METHOD OF COMPARING THE LINES AND DRAUGHTING VESSELS PROPELLED BY SAIL OR STEAM, including a Chapter on Laying off on the Mould-Loft Floor. By SAMUEL M. POOK, Naval Constructor. 1 vol., 8vo. With illustrations. Cloth. $5.

SWEET (S. H.) Special Report on Coal; showing its Distribution, Classification and Cost delivered over different routes to various points in the State of New York, and the principal cities on the Atlantic Coast. By S. H. SWEET. With maps. 1 vol., 8vo. Cloth. $3.

ALEXANDER (J. H.) Universal Dictionary of Weights and Measures, Ancient and Modern, reduced to the standards of the United States of America. By J. H. ALEXANDER. New edition. 1 vol., 8vo. Cloth. $3.50.

"As a standard work of reference this book should be in every library; it is one which we have long wanted, and it will save us much trouble and research."—*Scientific American.*

CRAIG (B. F.) Weights and Measures. An Account of the Decimal System, with Tables of Conversion for Commercial and Scientific Uses. By B. F. CRAIG, M. D. 1 vol., square 32mo. Limp cloth. 50 cents.

"The most lucid, accurate, and useful of all the hand-books on this subject that we have yet seen. It gives forty-seven tables of comparison between the English and French denominations of length, area, capacity, weight, and the centigrade and Fahrenheit thermometers, with clear instructions how to use them; and to this practical portion, which helps to make the transition as easy as possible, is prefixed a scientific explanation of the errors in the metric system, and how they may be corrected in the laboratory."—*Nation.*

BAUERMAN. Treatise on the Metallurgy of Iron, containing outlines of the History of Iron manufacture, methods of Assay, and analysis of Iron Ores, processes of manufacture of Iron and Steel, etc., etc. By H. BAUERMAN. First American edition. Revised and enlarged, with an appendix on the Martin Process for making Steel, from the report of Abram S. Hewitt. Illustrated with numerous wood engravings. 12mo. Cloth. $2.00.

"This is an important addition to the stock of technical works published in this country. It embodies the latest facts, discoveries, and processes connected with the manufacture of iron and steel, and should be in the hands of every person interested in the subject, as well as in all technical and scientific libraries."—*Scientific American.*

HARRISON. Mechanic's Tool Book, with practical rules and suggestions, for the use of Machinists, Iron Workers, and others. By W. B. HARRISON, associate editor of the "American Artisan." Illustrated with 44 engravings. 12mo. Cloth. $1.50.

"This work is specially adapted to meet the wants of Machinists and workers in iron generally. It is made up of the work-day experience of an intelligent and ingenious mechanic, who had the faculty of adapting tools to various purposes. The practicability of his plans and suggestions are made apparent even to the unpractised eye by a series of well-executed wood engravings."—*Philadelphia Inquirer.*

PLYMPTON. The Blow-Pipe: A System of Instruction in its practical use, being a graduated course of Analysis for the use of students, and all those engaged in the Examination of Metallic Combinations. Second edition, with an appendix and a copious index. By GEORGE W. PLYMPTON, of the Polytechnic Institute, Brooklyn. 12mo. Cloth. $2.

"This manual probably has no superior in the English language as a text-book for beginners, or as a guide to the student working without a teacher. To the latter many illustrations of the utensils and apparatus required in using the blow-pipe, as well as the fully illustrated description of the blow-pipe flame, will be especially serviceable."—*New York Teacher.*

NUGENT. Treatise on Optics: or, Light and Sight, theoretically and practically treated; with the application to Fine Art and Industrial Pursuits. By E. NUGENT. With one hundred and three illustrations. 12mo. Cloth. $2.

"This book is of a practical rather than a theoretical kind, and is designed to afford accurate and complete information to all interested in applications of the science."—*Round Table.*

SILVERSMITH (Julius). A Practical Hand-Book for Miners, Metallurgists, and Assayers, comprising the most recent improvements in the disintegration, amalgamation, smelting, and parting of the Precious Ores, with a Comprehensive Digest of the Mining Laws. Greatly augmented, revised, and corrected. By JULIUS SILVERSMITH. Fourth edition. Profusely illustrated. 1 vol., 12mo. Cloth. $3.

LARRABEE'S CIPHER AND SECRET LETTER AND TELEGRAPHIC CODE. By C. S. LARRABEE. 18mo. Cloth. $1.

BRÜNNOW. Spherical Astronomy. By F. BRÜNNOW, Ph. Dr. Translated by the Author from the Second German edition. 1 vol., 8vo. Cloth. $6.50.

CHAUVENET (Prof. Wm.) New method of Correcting Lunar Distances, and Improved Method of Finding the Error and Rate of a Chronometer, by equal altitudes. By WM. CHAUVENET, LL.D. 1 vol., 8vo. Cloth. $2.

POPE. Modern Practice of the Electric Telegraph. A Handbook for Electricians and Operators. By FRANK L. POPE. Seventh edition. Revised and enlarged, and fully illustrated. 8vo. Cloth. $2.

GAS WORKS OF LONDON. By ZERAH COLBURN. 12mo. Boards. 60 cents.

HEWSON. Principles and Practice of Embanking Lands from River Floods, as applied to the Levees of the Mississippi. By WILLIAM HEWSON, Civil Engineer. 1 vol., 8vo. Cloth. $2.

"This is a valuable treatise on the principles and practice of embanking lands from river floods, as applied to Levees of the Mississippi, by a highly intelligent and experienced engineer. The author says it is a first attempt to reduce to order and to rule the design, execution, and measurement of the Levees of the Mississippi. It is a most useful and needed contribution to scientific literature."—*Philadelphia Evening Journal.*

Scientific Books.

THE MECHANIC'S AND STUDENT'S GUIDE in the Designing and Construction of General Machine Gearing, as Eccentrics, Screws, Toothed Wheels, etc., and the Drawing of Rectilineal and Curved Surfaces; with Practical Rules and Details. Edited by FRANCIS HERBERT JOYNSON. Illustrated with 18 folded plates. 8vo. Cloth. $2.00.

"The aim of this work is to be a guide to mechanics in the designing and construction of general machine-gearing. This design it well fulfils, being plainly and sensibly written, and profusely illustrated."—*Sunday Times.*

FREE-HAND DRAWING: a Guide to Ornamental, Figure, and Landscape Drawing. By an Art Student. 18mo. Cloth. 75 cents.

THE EARTH'S CRUST: a Handy Outline of Geology. By DAVID PAGE. Fourth edition. 18mo. Cloth. 75 cents.

"Such a work as this was much wanted—a work giving in clear and intelligible outline the leading facts of the science, without amplification or irksome details. It is admirable in arrangement, and clear and easy, and, at the same time, forcible in style. It will lead, we hope, to the introduction of Geology into many schools that have neither time nor room for the study of large treatises."—*The Museum.*

HISTORY AND PROGRESS OF THE ELECTRIC TELEGRAPH, with Descriptions of some of the Apparatus. By ROBERT SABINE, C. E. Second edition, with additions. 12mo. Cloth. $1.25.

IRON TRUSS BRIDGES FOR RAILROADS. The Method of Calculating Strains in Trusses, with a careful comparison of the most prominent Trusses, in reference to economy in combination, etc., etc. By Brevet Colonel WILLIAM E. MERRILL, U. S. A., Major Corps of Engineers. With illustrations. 4to. Cloth. $5.00.

USEFUL INFORMATION FOR RAILWAY MEN. Compiled by W. G. HAMILTON, Engineer. Fourth edition, revised and enlarged. 570 pages. Pocket form. Morocco, gilt. $2.00.

REPORT ON MACHINERY AND PROCESSES OF THE INDUSTRIAL ARTS AND APPARATUS, OF THE EXACT SCIENCES. By F. A. P. BARNARD, LL. D.—Paris Universal Exposition, 1867. 1 vol., 8vo. Cloth. $5.00.

THE METALS USED IN CONSTRUCTION: Iron, Steel, Bessemer Metal, etc., etc. By FRANCIS HERBERT JOYNSON. Illustrated. 12mo. Cloth. 75 cents.

"In the interests of practical science, we are bound to notice this work; and to those who wish further information, we should say, buy it; and the outlay, we honestly believe, will be considered well spent."—*Scientific Review.*

DICTIONARY OF MANUFACTURES, MINING, MACHINERY, AND THE INDUSTRIAL ARTS. By GEORGE DODD 12mo Cloth. $2.00.

AUCHINCLOSS. Application of the Slide Valve and Link Motion to Stationary, Portable, Locomotive, and Marine Engines, with new and simple methods for proportioning the parts. By WILLIAM S. AUCHINCLOSS, Civil and Mechanical Engineer. Designed as a hand-book for Mechanical Engineers, Master Mechanics, Draughtsmen, and Students of Steam Engineering. All dimensions of the valve are found with the greatest ease by means of a PRINTED SCALE, and proportions of the link determined *without* the assistance of a model. Illustrated by 37 woodcuts and 21 lithographic plates, together with a copperplate engraving of the Travel Scale. 1 vol. 8vo. Cloth. $3.

HUMBER'S STRAINS IN GIRDERS. A Handy Book for the Calculation of Strains in Girders and Similar Structures, and their Strength, consisting of Formulæ and Corresponding Diagrams, with numerous details for practical application. By WILLIAM HUMBER. 1 vol. 18mo. Fully illustrated. Cloth. $2.50.

GLYNN ON THE POWER OF WATER, as applied to drive Flour Mills, and to give motion to Turbines and other Hydrostatic Engines. By JOSEPH GLYNN, F. R. S. Third edition, revised and enlarged, with numerous illustrations. 12mo. Cloth. $1.25.

THE KANSAS CITY BRIDGE, with an Account of the Regimen of the Missouri River, and a description of the Methods used for Founding in that River. By O. CHANUTE, Chief Engineer, and GEORGE MORISON, Assistant Engineer. Illustrated with five lithographic views and 12 plates of plans. 4to. Cloth. $6.

TREATISE ON ORE DEPOSITS. By BERNHARD VON COTTA, Professor of Geology in the Royal School of Mines, Freidberg, Saxony. Translated from the second German edition, by FREDERICK PRIME, Jr., Mining Engineer, and revised by the author, with numerous illustrations. 1 vol. 8vo. Cloth, $4.

A TREATISE ON THE RICHARDS STEAM-ENGINE INDICATOR, with directions for its use. By CHARLES T. PORTER. Revised, with notes and large additions as developed by American Practice, with an Appendix containing useful formulæ and rules for Engineers. By F. W. BACON, M. E., member of the American Society of Civil Engineers. 12mo. Illustrated. Cloth. $1

THE ART OF GRAINING. How Acquired and How Produced. By Charles Pickert and Abraham Metcalf. 4to. Beautifully Illustrated. Tinted paper. $10.00.

INVESTIGATIONS OF FORMULAS, for the Strength of the Iron Parts of Steam Machinery. By J. D. VAN BUREN, Jr., C. E. 1 vol. 8vo. Illustrated. Cloth. $2.

www.ingramcontent.com/pod-product-compliance
Lightning Source LLC
LaVergne TN
LVHW010238110826
845151LV00004B/1326

* 9 7 8 1 4 2 5 5 2 3 9 3 0 *